NITROGEN CYCLING IN THE AMERICAS: NATURAL AND ANTHROPOGENIC INFLUENCES AND CONTROLS

NITROGEN CYCLING IN THE AMERICAS: NATURAL AND ANTHROPOGENIC INFLUENCES AND CONTROLS

Edited by

LUIZ A. MARTINELLI
Cena-University of São Paulo, Brazil

&

ROBERT W. HOWARTH
Cornell University, U.S.A.

Reprinted from *Biogeochemistry* Volume 79, Nos. 1–2, 2006

Springer

A C.I.P. catalogue record for this book is available from the Library of Congress.

ISBN 1-4020-4717-7

Published by Kluwer Academic Publishers,
P.O. Box 17, 3300 AA Dordrecht, The Netherlands.

Sold and distributed in the U.S.A. and Canada
by Kluwer Academic Publishers
101 Philip Drive, Norwell, MA 02061, U.S.A.

In all other countries, sold and distributed
by Kluwer Academic Publishers,
P.O. Box 322, 3300 AH Dordrecht, The Netherlands.

Printed on acid-free paper.

Printed in the Netherlands.

Table of Contents

Biogeochemistry (2006) 79: 1–2
DOI 10.1007/s10533-006-9012-z

Preface

This special issue presents the products of three workshops designed to compare differences in nitrogen cycling between tropical and temperate ecosystems across the Americas, as well as differences in how human societies shave transformed the nitrogen cycle according to economic, social and cultural characteristics of the diverse countries of the Americas. The first workshop on "N fluxes and processes in tropical and temperate systems" was held in the city of Ubatuba, State of São Paulo, Brazil, in March 2003. From this workshop, a strong consensus developed to form an ongoing "Inter American Nitrogen Network." Towards this end, two more workshops were held, one in San Juan, Puerto Rico in May 2004, and one in May of 2005 in the city of Brasília, Brazil. This effort to build the Inter American Nitrogen Network is a joint effort of the Latin American Nitrogen Center and the North American Nitrogen Center, two of the five continental-scale regional centers of the International Nitrogen Initiative (INI) created in 2003 by SCOPE and the IGBP. The INI's main objective is "to optimize nitrogen's beneficial role in sustainable food production and minimize nitrogen's negative effects on human health and the environment resulting from food and energy production." We hope the readers will agree with us that this issue of *Biogeochemistry* is an excellent start for the Inter American Nitrogen Network and the efforts of the INI in the Americas.

The first paper presented in this issue by Martinelli and other participants of the San Juan workshop is an overview of the nitrogen sources across the Americas and how they vary from one region to another. This paper is an introductory paper for a series of six following papers that detailed the nitrogen cycle in different countries or in important sub-regions of these countries. Among these papers, Schindler et al. discuss N-deposition related issues in Canada, followed by Austin et al. who present a nitrogen budget for Argentina, one of the most important regions of cereal production in the world. Filoso et al. for Brazil, Baisre for Cuba and Ortiz for Puerto Rico synthesize quantitative nitrogen assessments for these whole countries, while Borbor et al. restrict their analyses for the most developed watershed in Equador. After these country-scale assessments, Howarth et al. and Scavia & Bricker illustrate aspects of the nitrogen cycle in important regions of the USA. Howarth et al. illustrate how climate and human activity interact to regulate the nitrogen export from the major 16 watersheds of the Northeastern USA region, one of the most polluted regions of the country. Scavia & Bricker present a comprehensive approach to problems that the coastal regions of the USA are facing due to the large amount of nitrogen that these key areas receive. The last three papers in the issue are mostly a product of the first workshop held in Ubatuba and deal more with nitrogen as a key element in natural systems. Bustamante et al. present nitrogen as one of the most important limiting element in savanna ecosystems throughout the world. Huszar et al. analyze nitrogen limitation in tropical lakes in comparison to temperate lakes. Finally, Ometto et al. present a detailed description of the variability of nitrogen and carbon stable isotopes in tropical forests of the Amazon region. We believe this volume represents a significant advancement in our ability to understand natural processes and human perturbations of the nitrogen cycle in both temperate and tropical systems, as well as to recognize several major differences in nitrogen cycling across several countries and regions of the Americas.

We are in debt with the many institutions and people who helped us to organize our workshops and prepare the manuscripts. Elvira Cuevas and Mercedes Bustamante and their staffs were wonderful hostesses in San Juan and Brasília, respectively. The Ubatuba workshop had financial support from the Fundação de Amparo à Pesquisa do Estado de São Paulo (FAPESP) and the Esso Brasileira de Petróleo company. The San Juan workshop was supported by the Inter American Institute for Global Changes (IAI) and by the University of Puerto Rico through a grant from the US National Science Foundation to Crest (Center for Applied Tropical Ecology and Conservation). The Brasília workshop was mainly supported by the Conselho Nacional de Desenvolvimento e Pesquisas (CNPq), through its Pro-Sul program, and also by the IAI and the University of Brasília. We also thank Suzanne Mekking, the publisher of *Biogeochemistry*, for her flexibility and support in preparing this issue.

Luiz A. Martinelli
CENA, Laboratory of Isotope Ecology
University of São Paulo
Piracicaba, SP, Brazil
e-mail: martinelli@cena.usp.br

Robert W. Howarth
Department of Ecology Evolutionary Biology
Cornell University
Ithaca, USA

Biogeochemistry (2006) 79: 3–24
DOI 10.1007/s10533-006-9000-3

Sources of reactive nitrogen affecting ecosystems in Latin America and the Caribbean: current trends and future perspectives

LUIZ A. MARTINELLI[1,*], ROBERT W. HOWARTH[2], ELVIRA CUEVAS[3], SOLANGE FILOSO[2], AMY T. AUSTIN[4], LORETA DONOSO[5], VERA HUSZAR[6], DENNIS KEENEY[7,8], LUCIENE L. LARA[1], CARLOS LLERENA[9], GEORGE McISSAC[10], ERNESTO MEDINA[11], JORGE ORTIZ-ZAYAS[12], DONALD SCAVIA[13], DAVID W. SCHINDLER[14], DORIS SOTO[15] and ALAN TOWNSEND[16]

[1]*CENA, Av. Centenário 303, 13416-000, Piracicaba, SP, Brazil;* [2]*Ecology & Environmental Biology Department, Cornell University, E311 Corson Hall, Cornell, Ithaca, NY 14853, USA;* [3]*Department of Biology, College of Natural Sciences, University of Puerto Rico, PO Box 23360San Juan, Puerto Rico 00931-3360;* [4]*Faculty of Agronomy, University of Buenos Aires and IFEVA-CONICET, Avenida San Martín 4453, C1417DSE, Buenos Aires, Argentina;* [5]*Instituto Venezuelano de Investigaciones Científicas – IVIC, Centro de Ecología, IVIC, Aptdo. 21827, Caracas 1020-A, Venezuela;* [6]*Departamento de Botânica, Universidade Federal do Rio de Janeiro, Museo Nacional, Quinta da Boa Vista s/n, São Cristóvão – Rio de Janeiro, RJ 20940040, Brasil;* [7]*Agriculture and Biosystems Engineering, Iowa State University, 3402 Eisenhower Ave., Ames, Iowa 50010, USA;* [8]*Institute for Agriculture and Trade Policy, Minneapolis, Minnesota, USA*[9]*Universidad Nacional Agraria La Molina, Apartado 456, Lima, Peru;* [10]*Department of Natural Resources and Environmental Sciences, University of Illinois at Urbana Champaign, w-503 Turner Hall, 1102 South Goodwin Avenue, Urbana, IL 61801, USA;* [11]*Instituto Venezolano de Investigaciones Científicas, Laboratory Plant Physiological Ecology, Centro de Ecología, IVIC, Aptdo. 21827, Caracas, 1020-A, Venezuela;* [12]*Institute for Tropical Ecosystems Studies, University of Puerto Rico, PO Box 23341San Juan, Puerto Rico 00931-3341;* [13]*School of Natural Resources and Environment, University of Michigan, G520 Dana 1115, Ann Arbor, MI, 48109, USA;* [14]*Ecology Department, University of Alberta, Z 811, Biological Sciences Bldg., 114 St – 89 Ave, Edmonton, Alberta, T6G 2E1, Canada;* [15]*Universidad Austral de Chile, Casilla 1327, Puerto Montt, Chile;* [16]*INSTAAR and EPO-Biology, University of Colorado, 1560 30th St., Boulder, CO 80309, USA; *Author for correspondence (e-mail: martinelli@cena.usp.br; phone: +55-19-3429-4074; fax: +55-19-3434-9210)*

Key words: Biomass burning, Caribbean, Latin America, Nitrogen, Reactive agriculture, Urbanization

Abstract. While the amount of reactive nitrogen circulating at the global level has increased markedly in the last century, the effects of this increase are largely seen at the regional level due to interacting ecological and socio-economic factors. In contrast with most other regions of the world, Latin America and the Caribbean (LA-Ca) stand out due to the fact that the major input of reactive nitrogen (Nr) still occurs naturally via biological nitrogen fixation (BNF) in natural ecosystems as opposed to anthropogenic inputs of synthetic fertilizer, fossil fuel combustion and cropping with leguminous species. Largely due to economic reasons, the consumption of fertilizer N in the LA-Ca region is still low in comparison with the average consumption of the world. However, the fertilizer N consumption is increasing at a much faster rate than that in developed regions of the world, like

USA and Canada. The Nr production through BNF in cultivated plants that fix nitrogen (C-BNF) is 5 times lower than that occurring naturally in Latin America, but is still equivalent to 16% of the world C-BNF. The cultivation of nitrogen-fixing crop species in the LA-Ca region is also increasing, almost entirely due to the expansion of soybean fields in the central and northern regions of Brazil and the Pampa region of Argentina. Other anthropogenic activities in the region that contribute to an increase in the circulation of reactive nitrogen include the impact of biomass burning and urbanization. In the last decade, an average of 47,000 km^2 per year of forests was burned in the LA-Ca region. The environmental impact of urban centers in the LA-Ca region has become very important, since an intense urbanization process is occurring in this region, at an intensity that far exceeds urban development in the northern hemisphere. The consequences of increased urbanization include increased emissions of NO_x to the atmosphere due to the fossil fuel combustion, and the lack of sewage treatment facilities in most cities of the LA-Ca result in a large volume of untreated sewage discharged into surface waters, creating serious environmental problems. The combination of rapid urbanization and agricultural intensification in this region suggest that concern is warranted for the potential for increase in the circulation of reactive nitrogen in the very near future. At the same time, the opportunity still exists to mitigate some of the consequences of human impact on the nitrogen cycle in a region that still maintains a large fraction of its natural ecosystems intact.

Introduction

The concentration of reactive nitrogen (Nr) on the Earth has dramatically increased in the last century due mainly to the production of synthetic nitrogen fertilizer through the Haber–Bosch process, fossil fuel combustion, and the cultivation of plants that associate with bacteria that fix atmospheric N (Howarth et al. 2002; Galloway et al. 2004). It has been estimated that in 1860 the production of anthropogenic Nr was 15 Tg N yr^{-1}, and that by the early 1990's it increased by almost 10 times to approximately 140 Tg N yr^{-1} (Galloway et al. 2004). Although these estimates are important in providing a global picture of the alteration of the N cycle by human activity, they do not allow us to propose actions that would help mitigate the deleterious effects of N in the environment and for human health occurring at regional or local scale. Much more detailed spatial information on the production and fate of Nr is needed to propose such actions.

Galloway et al. (2004) investigated geopolitical N budgets and showed that Latin America stands out as the region with highest inputs of Nr by naturally occurring biological nitrogen fixation (BNF) – an amount equivalent to 25% of the world's Nr created in terrestrial ecosystems. The Nr production through BNF in cultivated plants that fix nitrogen (C-BNF) in 1995 was 5 times lower than that occurring naturally in Latin America, but was still equivalent to 16% of the world C-BNF. On the other hand, Latin America was responsible in 1995 for only 6% of the world's Nr created through fossil fuel combustion, and for only 4% of the Nr originated from synthetic N fertilizer (Galloway et al. 2004).

In this paper, we provide an evaluation of the origin of reactive nitrogen in Latin America, and compare it to other areas of the world. The contrasts between impacts at the global and regional scale may help to suggest ways in which we can mitigate the deleterious effects of human impact on the N cycle in the coming decades. Assessing not only the current state of human impact, but the velocity of change in this region highlights the importance of focusing on regions where the potential human impact in the coming decades will be large, and where preventative steps could be taken to minimize the negative impacts of human-induced changes in nitrogen cycling.

The geography and socio-economic aspects of the Latin America and the Caribbean (LA-Ca) regions

Geography

Because statistics are generally categorized by country or political unit in most data bases, we define Latin America and the Caribbean, including their countries and territories, as shown in Table 1 (http://www.cia.gov/cia/publications/factbook/geos/cj.html).

The most typical definition of Latin America is the one that considers the countries of South America, Central America, the Caribbean, and southern North America that speak a Romance language, including Spanish, Portuguese, and French (Table 1). The Caribbean is defined as a group of islands located in the Caribbean Sea, organized in 25 distinct political units, including states and territories (Table 1). Geographical or economical and social indexes are available for most independent countries and, in some cases, for political units as well. For example, Puerto Rico is a commonwealth of the United States (US), and as a result, its data are sometimes aggregated to that the US statistics, while in other cases, they are reported as an independent political unit. As most production and use of Nr is linked to economic indicators, such as use of fossil fuel, fertilizer use, and cultivation of commercially important plants that have N-fixing symbioses, it is useful to provide Latin American economic and social indicators for context.

Social and economical indicators

The Gross Domestic Product (GDP) index (Human Development Report 2003), (defined as the total value of goods and services produced by a nation and expressed on a per capita basis) varied from 0.48 to 0.83 for LA-Ca countries in 2003 (Figure 1).

Approximately 40% of the LA-Ca countries had a GDP index higher than the overall average for developing countries (0.70). Bolivia and Nicaragua had indexes only slightly higher than 0.5, while Haiti was the only country with an index lower than the average of the least developing countries (0.5). (Figure 1). The highest indexes were observed in the Bahamas and Barbados. However, both values were significantly lower than the average of the richest countries of

Table 1. List of countries in Latin America (LA) and the Carribean.

Political unity	Region	Status
1-Anguilla	Caribbean	Territory of UK
2-Antigua and Barbuda	Caribbean	Independent country
3-Aruba	Caribbean	Territory of the Netherlands
4-Bahamas	Caribbean	Independent country
5-Barbados	Caribbean	Independent country
6-British Virgin Islands	Caribbean	Territory of UK
7-Cayman Islands	Caribbean	Territory of UK
8-Cuba	Caribbean (LA)	Independent country
9-Dominica	Caribbean	Independent country
10-Dominican Republic	Caribbean (LA)	Independent country
11-Grenada	Caribbean	Independent country
12-Guadeloupe	Caribbean (LA)	Territory of France
13-Haiti	Caribbean (LA)	Independent country
14-Jamaica	Caribbean	Independent country
15-Martinique	Caribbean (LA)	Territory of France
16-Montserrat	Caribbean	Territory of UK
17-Netherlands Antilles	Caribbean	Territory of the Netherlands
18-Puerto Rico	Caribbean (LA)	Commonwealth USA
19-Saint Kitts and Nevis	Caribbean	Independent country
20-Saint Lucia	Caribbean	Independent country
21-Saint Vincent/Grenadines	Caribbean	Independent country
22-SouthGeorgia/Sandwich Is	Caribbean	Territory of UK
23-Trinidad and Tobago	Caribbean	Independent country
24-Turks and Caicos Is	Caribbean	Territory of UK
25-Virgin Islands – US	Caribbean	Territory of US
26-Belize	Central America	Independent country
27-Costa Rica	Central America (LA)	Independent country
28-El Salvador	Central America (LA)	Independent country
29-Guatemala	Central America (LA)	Independent country
30-Honduras	Central America (LA)	Independent country
31-Nicaragua	Central America (LA)	Independent country
32-Panama	Central America (LA)	Independent country
33-Mexico	North America (LA)	Independent country
34-Argentina	South America (LA)	Independent country
35-Bolivia	South America (LA)	Independent country
36-Brazil	South America (LA)	Independent country
37-Chile	South America (LA)	Independent country
38-Colombia	South America (LA)	Independent country
39-Ecuador	South America (LA)	Independent country
40-Falkland Is (Malvinas)	South America	Territory of UK
41-French Guiana	South America (LA)	Territory of France
42-Guyana	South America	Independent country
43-Paraguay	South America (LA)	Independent country
44-Peru	South America (LA)	Independent country
45-Suriname	South America	Independent country
46-Uruguay	South America (LA)	Independent country
47-Venezuela	South America (LA)	Independent country

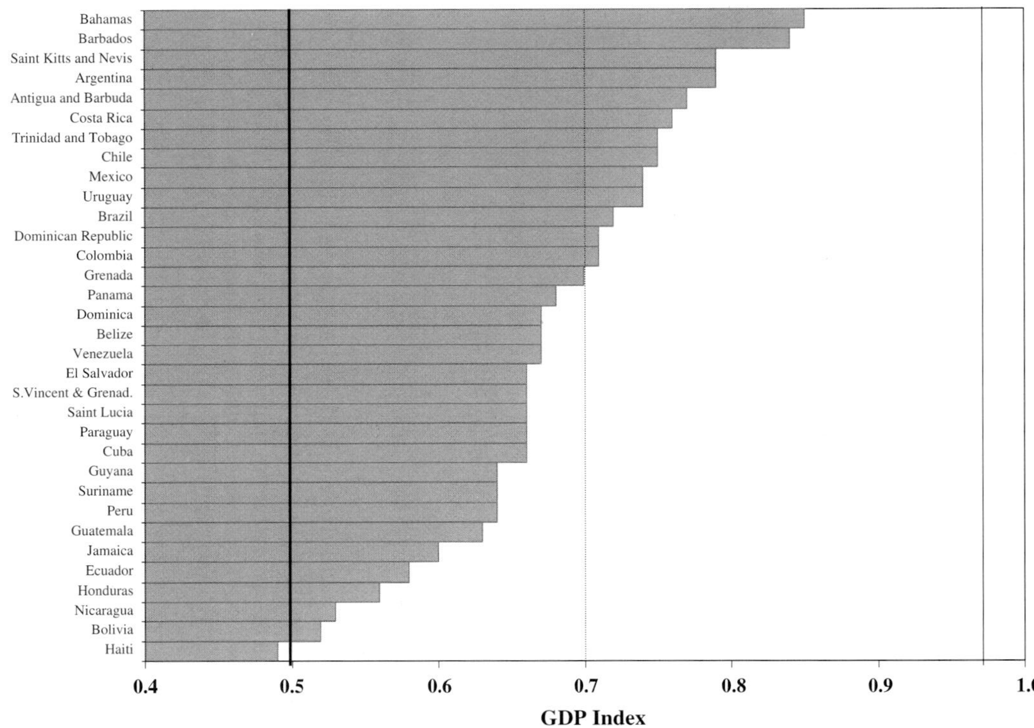

Figure 1. The Gross Domestic Product (GDP) index of LA-Ca countries and political unities. The tick continuous line is the average GDP index for the least developed countries, the fine continuous line is the average for the most developed countries and the dashed line is the average for the developing countries.

the world (Figure 1). The Human Development Index (HDI – Human Development Report 2003) is a weighted index which includes life expectancy, level of education, and GDP per capita. For the LA-Ca region, the HDI varied from 0.45 to 0.88 (Figure 2), while the average HDI for developing countries is 0.6. With exception of Haiti, all LA-Ca countries had indexes above that value. Barbados and Argentina had the highest HDI in the region but, again, the values were lower than those for the richest countries in the world.

Considering these two indexes together, as well as an index for poverty not shown here (Human Development Report 2003), it is clear that LA-Ca has an intermediate position in relation to the richest developed and the least developed countries of the world.

Finally, it is useful to assess the status of sanitation in LA-Ca, because lack of proper treatment of domestic sewage is an important issue in alteration of the N cycle. Sewage is processes at some level in most LA-Ca countries (Figure 3); however, 9 of 33 countries, of which two are the most populated countries in the region (Mexico and Brazil) have an index of sanitation (the percentage of the population with access to adequate excreta disposal facilities, such as a connection to a sewer or septic tank system, a pour-flush latrine, a simple pit latrine or a ventilated improved pit latrine – Human Development Report 2003), lower than the average for developed countries, indicating that a large fraction of the sewage is discharged untreated to waterways.

Land use in Latin America and the Caribbean

Land use

The main land use in LA-Ca is for pasture. Total world pasture area in 2002 was 3.5 billion ha, with approximately 18% (0.62 billion ha) of this in the LA-Ca region (FAOSTAT 2004). In South America, 80% of the area devoted to agriculture areas is under management for pasture, followed by 70% in Central America, and 50% in the Caribbean. Five countries are responsible for about 80% of pasture area in LA-Ca – Brazil (32%), Argentina (23%), Mexico (12%), Colombia (7%), and Bolívia (6%). Most pasture land in LA-Ca are used for cattle grazing, which are typically C_4 grasses, and which are rarely fertilized with nitrogen. Almost 30% of the world's domestic cattle are located in the LA-Ca region. The largest stocks are located in Brazil (50%), Argentina (14%), Mexico (8%), and Colombia (7%).

The second most important LA-Ca land use is for cereal crops. The LA-Ca region is responsible for 8% of the world cereal production. In Central America, 40% of the arable land is used for cereals, followed by 33% in South America, and 26% in the Caribbean. Brazil (38%), Mexico (21%), and Argentina (21%) are responsible for 80% of the cereal harvested in the area in 2003. In Brazil and Mexico, corn is byfar the most important cereal, while wheat is the main cereal cultivated in Argentina. Rice is the second most

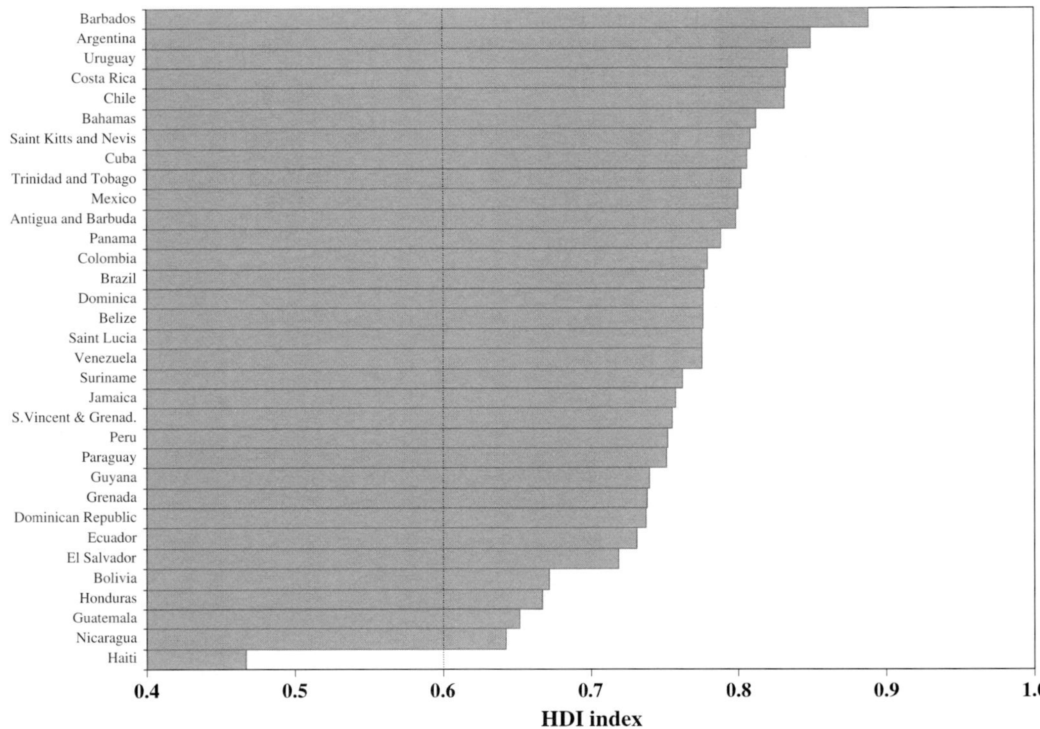

Figure 2. The Human Development Index (HDI) of LA-Ca countries and political unities. The dashed line is the average for the developing countries.

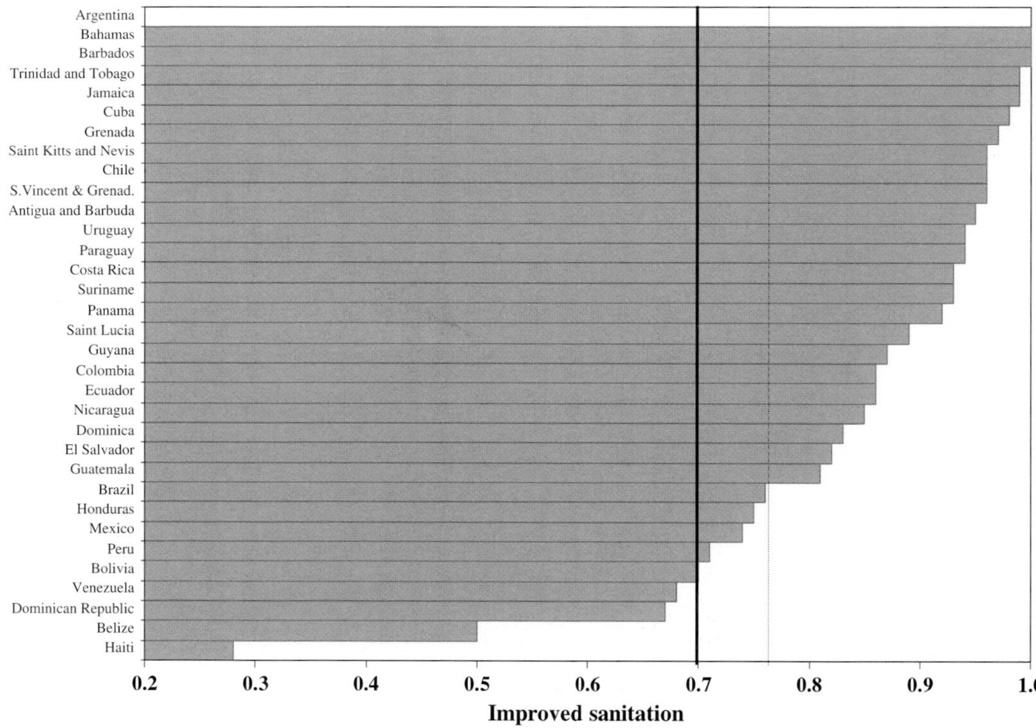

Figure 3. Percentage of houses with improved sanitation in countries and political unities of LA-Ca region. The tick continuous line is the average percentage of houses with improved sanitation for the least developed countries, and the dashed line is the average for the developing countries.

important in Brazil, and wheat is the second most important in Mexico, and corn is the second most important in Argentina.

Fertilizer N consumption in Latin America and the Caribbean

In 2002 the LA-Ca region consumed approximately 5 million metric tons of fertilizer N, equivalent to 6% of the global consumption (FAOSTAT 2004). Three countries – Brazil, Mexico and Argentina – are responsible for almost 70% of the total fertilizer N consumption in the LA-Ca region (Figure 4).

The average application of fertilizer N per hectare of arable land in the LA-Ca region is approximately 34 kg N ha^{-1} (see definition at FAOSTAT home page at http://www.apps.fao.org/default.jsp). Central America has the highest rate (46 kg N ha^{-1} yr^{-1}), followed by South America and the Caribbean, with approximately 32 kg N ha^{-1} yr^{-1}. These rates are below the world average of 60 kg N ha^{-1}, and well below rates for countries like China and the Netherlands where rates in 2002 were approximately 180 and 310 kg N ha^{-1}, respectively (FAOSTAT 2004). It is important to note that some countries in the LA-Ca region, such as Haiti and Bolivia,the rates are even lower than the average rate in the least developed countries (13 kg N ha^{-1} in 2002) (Figure 5). On the other hand, small Central American countries (Costa Rica) and some islands of the Caribbean (Martinique and Guadeloupe) have high usage of fertilizer N (Figure 5); even higher than rates in the Netherlands. In these three countries the main land use are fruits and sugar cane, with banana and plantains the most important crops in terms of harvested area. Costa Rica also produces a variety of other fruits, especially citrus, pineapples, mangos and melons (FAOSTAT 2004). These rates of fertilizer application in the various countries represent the interaction among socio-economic factors such as availability and feasibility of fertilizer use as well as ecological factors including natural levels of fertility, crop selection and relative nutrient limitation by nitrogen vs. other nutrients.

Another way to characterize fertilizer use and to make it comparable with other countries is by reviewing per capita fertilizer consumption. The average value for LA-Ca is approximately 9 kg N per capita, almost 5 times higher than the average for the least developed countries of the world, but below the average for the whole world (14 kg N per capita) and for developing countries (11 kg N per capita). Some small countries (St.Kits and Neives, Guadeloupe) and territories (Virgin Islands) of the Caribbean, and the Guyana in South America showed the highest consumption of fertilizer N per capita, with rates similar to those in China (19 kg N per capita), but well below the USA (38 kg N per capita) or Canada (53 kg N per capita) (Figure 6). This relatively low per capita use of N fertilizer is mainly due to economic constraints, since most of the LA-Ca countries are net importers of fertilizer (FAOSTAT 2004). Venezuela is an exception, importing only 20% of the amount consumed.

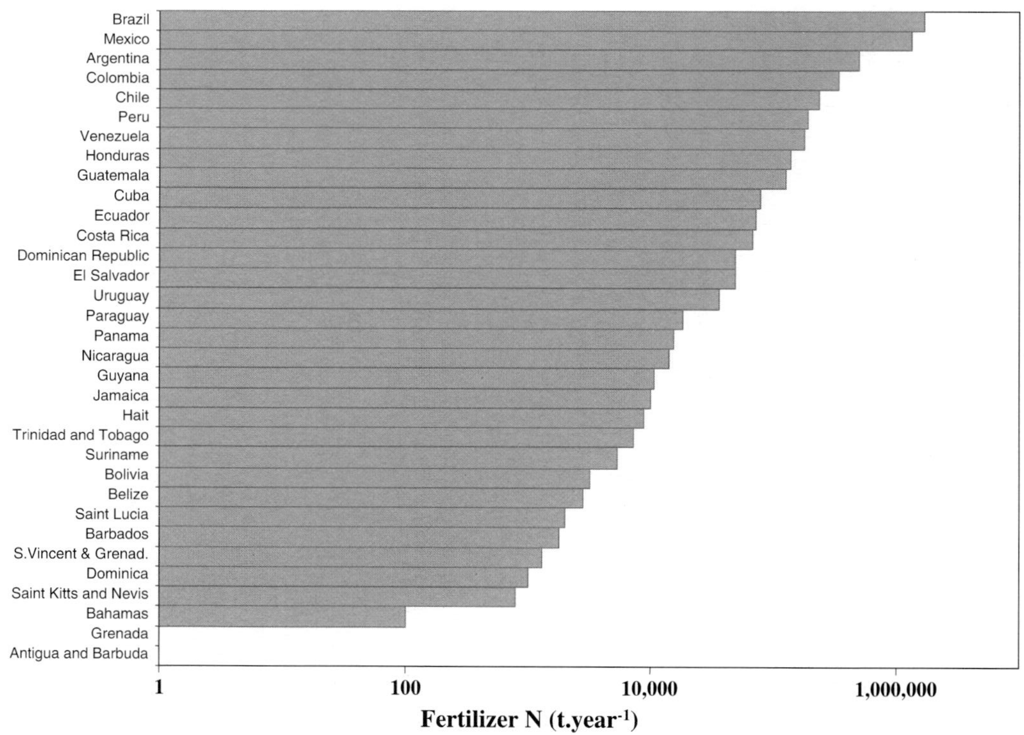

Figure 4. Consumption of fertilizer N in 2002 by countries and political unities of LA-Ca region.

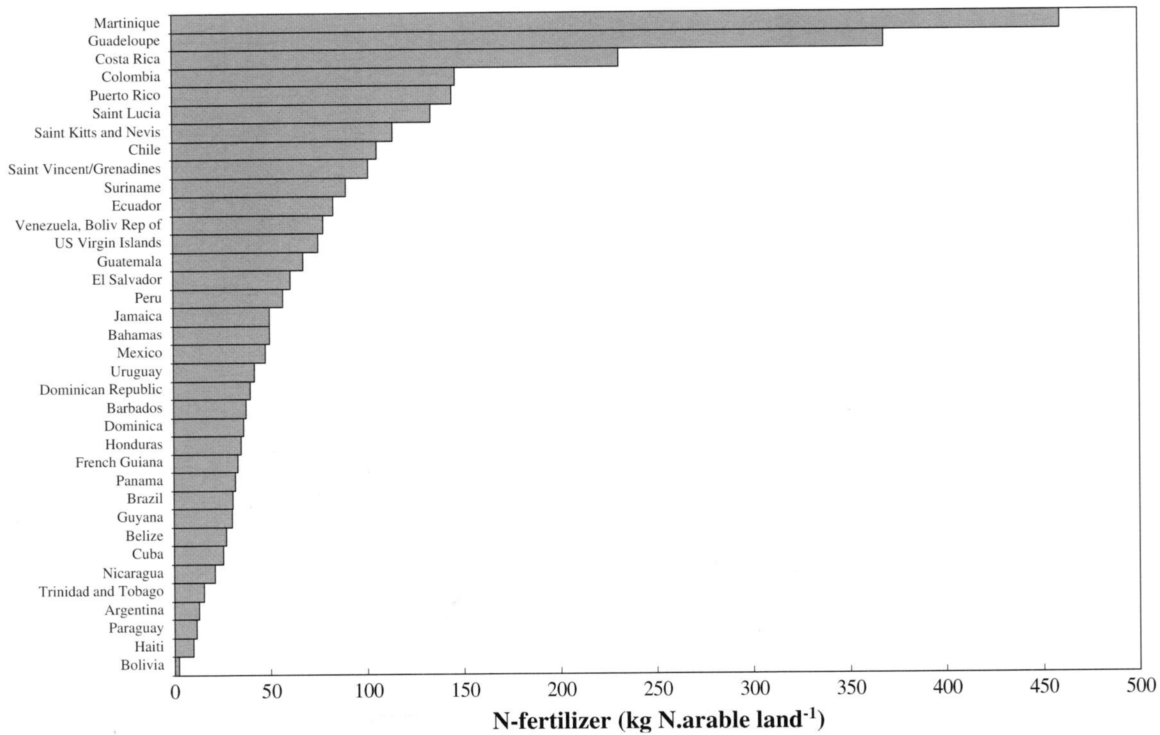

Figure 5. Consumption of fertilizer N per hectare of arable land in 2002 by countries and political unities of LA-Ca region.

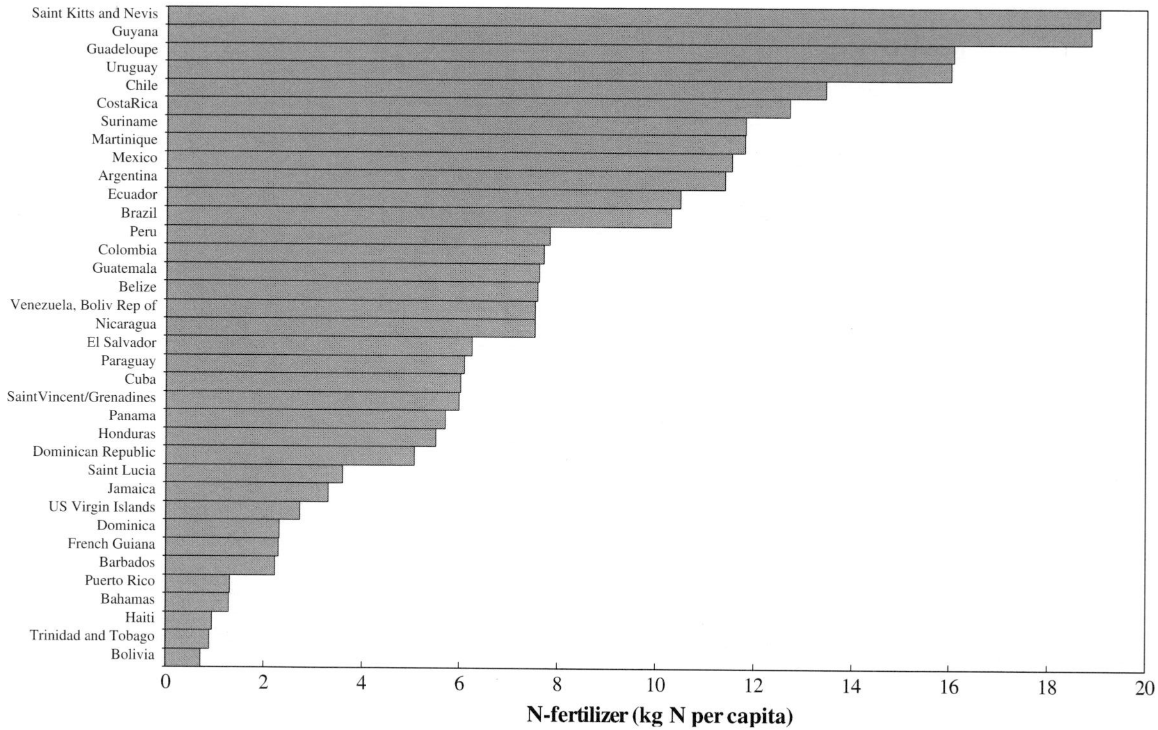

Figure 6. Consumption of fertilizer N per capita in 2002 by countries and political unities of LA-Ca region.

Brazil and Mexico import approximately 70% of their consumption; Argentina and Cuba importing approximately 50% (FAOSTAT 2004).

Although fertilizer N use in LA-Ca is still relatively low, its trends are similar to the progressively increasing trends in Asia, in contrast with Europe and USA (Figure 7). The historical increase in fertilizer N consumption in the three major Latin American consumers – Brazil, Mexico, and Argentina – has shown different trends (Figure 8). While Brazil had a constant and progressive increase since 1961, Argentina experienced a major increase in consumption only in the early 90's (Austin et al., this volume). Major growth in consumption in Mexico was achieved in the end of the 80's, but since then no further increase has been observed (Figure 8).

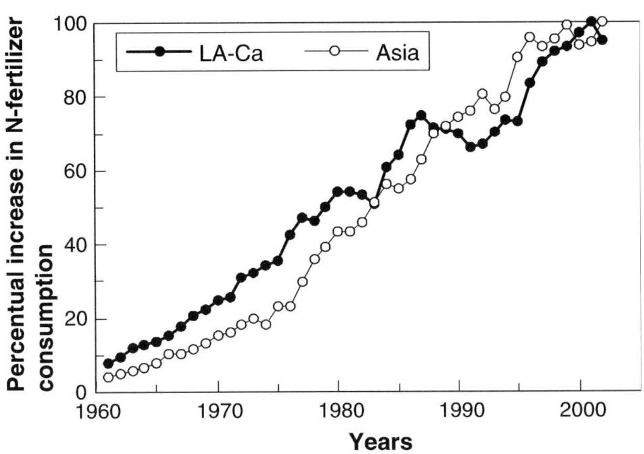

Figure 7. Percentage increase of fertilizer N consumption from 1960 to 2002 in LA-Ca, Asia, Europe and North America.

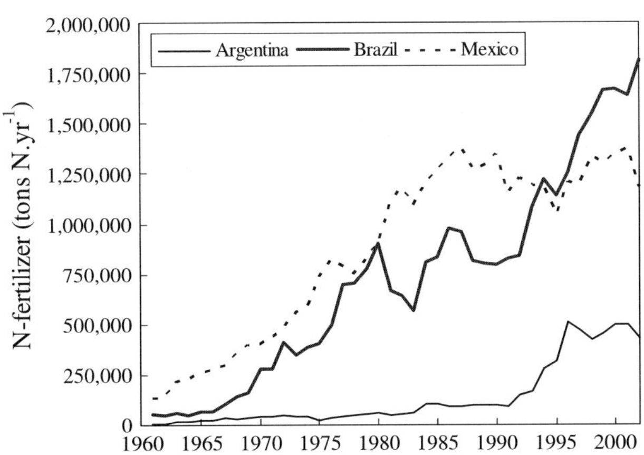

Figure 8. Consumption of fertilizer N from 1960 to 2002 in Argentina, Brazil, and Mexico.

C-BNF historical trends in Latin America and the Caribbean

Latin America and the Caribbean were responsible in 1995 for approximately 16% of the world's N fixation through cultivation of plants that have associations with bacterial N-fixers (C-BNF, Galloway et al. 2004). Three main landcovers that contribute to C-BNF are soybean, pastures, and sugar cane. In 2003, approximately 33 million ha of soybean was harvested in LA-Ca, which is equivalent to 40% of the world production of this crop (FAOSTAT 2004). In 2002, the area harvested in LA-Ca exceeded the area harvested in the North America (Figure 9). Major producers of soybean in LA-Ca are Brazil (55% of the harvested area), Argentina (38%), Paraguay (5%), and Bolivia (2%). In Brazil, soybeans are able to obtain most of their required nitrogen from biological fixation (Boddey et al. 1991; Alves et al. 2003; Filoso et al., this volume); however, this is not the case in Argentina (Austin et al., this volume). Therefore an increase in the area cultivated for soybean in Latin America or in the Caribbean not necessarily implies an addition of newly fixed nitrogen through biological fixation.

As mentioned above, pasture is the main land use in LA-Ca, and the area covered with pasture showed an average increase of 2.6 million ha per year since 1961, in contrast to North America where pasture area has remained practically constant for the last four decades (Figure 10). Most of the nitrogen fixation in grasses like *Brachiaria* spp., *Panicum maximum*, and species of *Pennisetum*, that are extensively cultivated in South America, is derived from non-symbiotic associations with endophytic bacteria or from bacteria in the rizosphere (Miranda and Bodey 1987; Boddey et al. 2004; Filoso et al., this volume).

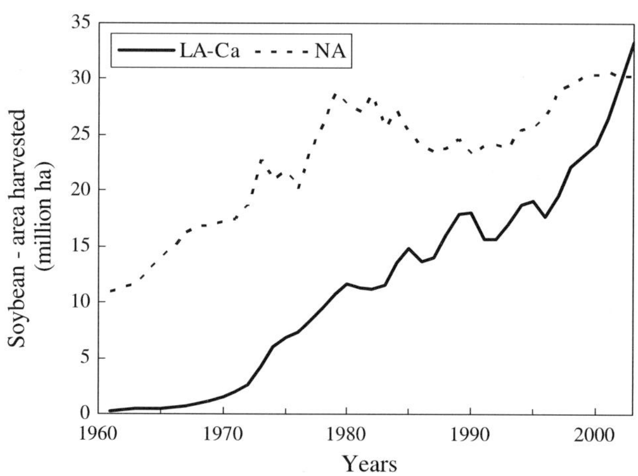

Figure 9. Increase in the area harvested with soybean in the LA-Ca and North America from 1960 to 2002.

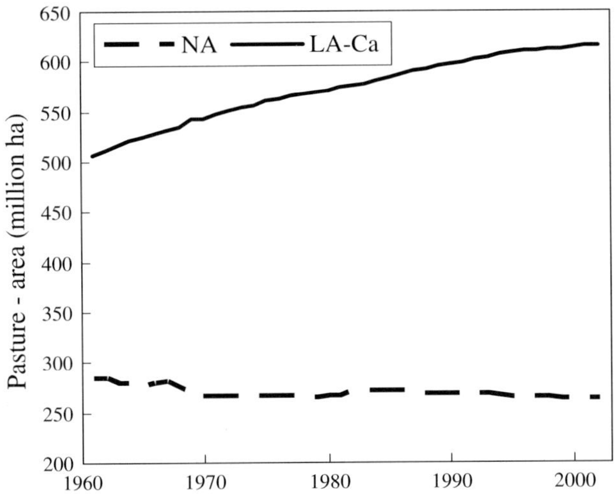

Figure 10. Increase in the pasture area in the LA-Ca and North America from 1960 to 2002.

The third crop that may contribute to C-BNF in the LA-Ca region is sugarcane (Boddey et al. 1991, 2003), which has been a traditional crop in this region (FAOSTAT 2004). Nearly 42% of the harvested area of sugar cane globally comes from Brazil, Cuba, Mexico and Colombia, and in the LA-Ca region alone, these countries account for 80% of the sugar cane production. The area devoted to sugar cane production increased markedly in the LA-Ca region between the beginnings of the 70's until the middle of the 80's (Figure 11), an increase that was led by Brazil, which is currently responsible for 62% of the

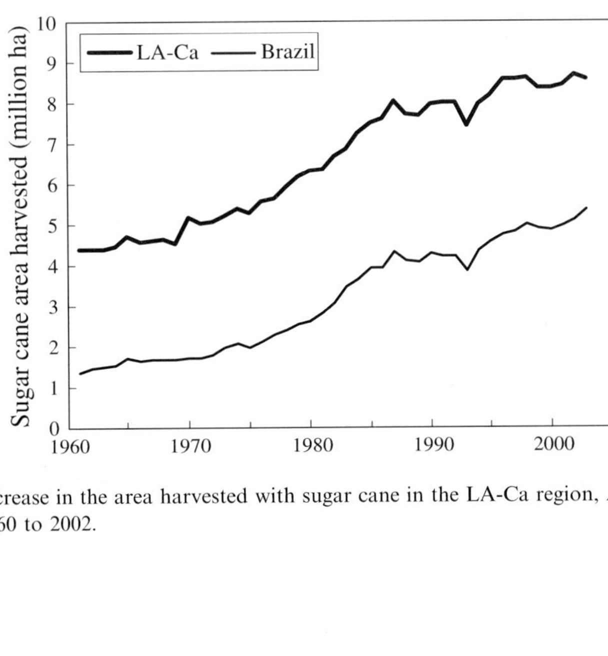

Figure 11. Increase in the area harvested with sugar cane in the LA-Ca region, Asia, Africa and Brazil from 1960 to 2002.

harvested area in the LA-Ca region. The rapid increase in sugar cane cropping stemmed from a series of fiscal incentives to increase sugar cane production for ethanol, a fuel used by Brazilian automobiles, has caused major increases in Brazilian production during this period (Oliveira et al. 2005).

Using current estimates for area in Latin America cultivated with soybean, pasture and sugar cane (shown above), and using equivalent biological nitrogen fixation rates and assumptions from Filoso et al. (this volume), we estimated C-BNF in Latin America for 2003 as approximately 10.5 Tg yr^{-1}. This value is practically the double of the amount estimated by Galloway et al. (2004) for Latin America using land use data from 1995. We believe that this marked increase in the estimated amount of C-BNF for 2003 stems from the extensification in the cultivated area of soybean (Figure 9), pasture (Figure 10) and sugar cane (Figure 11) from 1995 to 2003.

Biomass burning (source of N to the atmosphere)

Deforestation in LA-Ca between 1990 and 2000 averaged approximately 47,000 km^2 yr^{-1} (FRA 2000); Brazil alone was responsible for approximately half of that total (23,000 km^2 yr^{-1}) (Figure 12). In Brazil, most of the deforestation is taking place in the Amazon region, where the average deforestation rates from 1988 to 2002 was approximately 18,000 km^2 per year (INPE 2004). In the last 2 years deforestation rates increased to approximately 25,000 and 30,000 km^2 per year, respectively.

In the Amazon region, most of the forest is converted to pasture for cattle through slash and burn methods during the dry season. Pasture is also burned every 4–5 years as a management method to 'clean the pasture' and eliminate the growth of shrubs and small trees. In southeastern Brazil, fire is also used during the dry season (May–September) prior to manual harvesting of sugar cane (which is the dominant method) to facilitate the process. In Central America and in the Andean countries, fuel wood is the major type of biomass burning. During clearing and burning of tropical rainforest, biomass-associated N is volatilized and a large fraction is emitted to the atmosphere in form of gaseous NH_3 and NO_x (Trebbs et al. 2006).

Most estimates of NO_x emissions for tropical regions have been generated by models due to the lack of measurements. According to some models, the estimated NO_x emissions from biomass burning in LA-Ca ranges from 0.9 to 3.7 Tg NO_x yr^{-1}, while global emissions range between 7.8 and 15.9 Tg NO_x yr^{-1} (Galanter et al. 2000; Shultz 2002; Hoelzemann et al. 2004). On the other hand, Potter et al. (2002) estimated that, in the Amazon region alone, biomass burning produces 3.5 Tg NO_x yr^{-1} emitted to the atmosphere, making this region the source of 20–33% of global NO_x emission fluxes to the atmosphere.

Most of the values predicted by models for N wet deposition in LA-CA ranged from 1 to 3 kg N ha^{-1} yr^{-1} (Dentener and Crutzen 1994; Holland et al. 1999; Galloway and Cowling 2002). However, some few studies in Brazil have

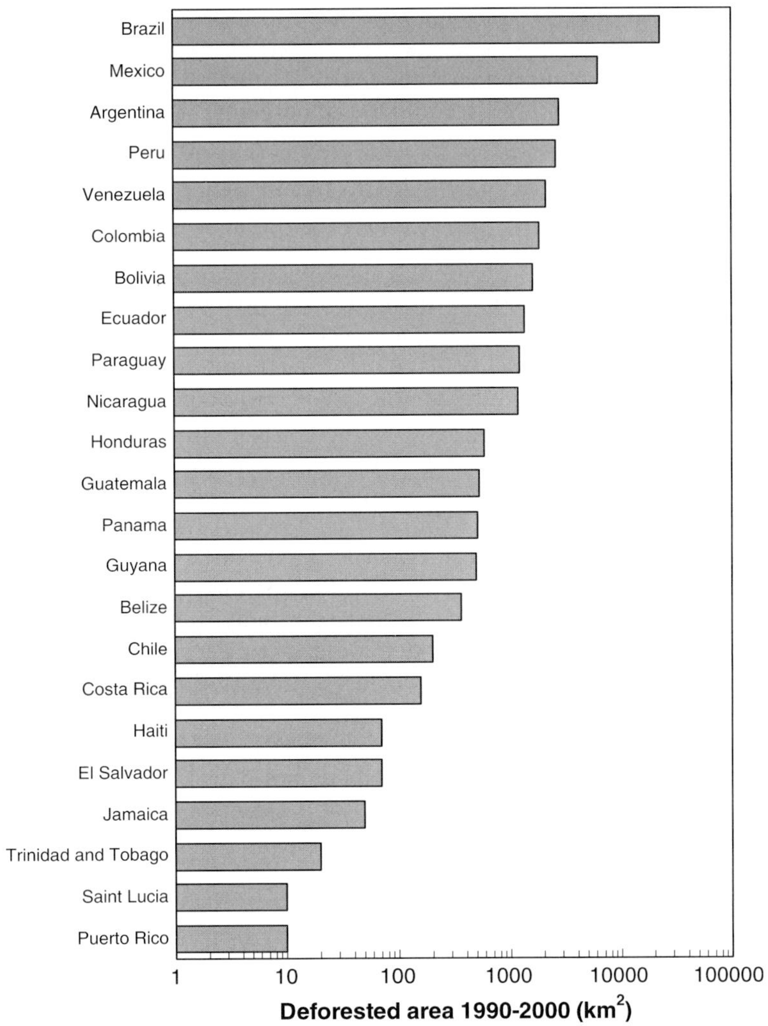

Figure 12. Deforested area from 1990 to 2000 in countries and political unities of the LA-Ca region.

measured a rate of N wet deposition around 1–2 kg N ha^{-1} yr^{-1} in more pristine areas (Williams and Melack 1997; Lara et al., this volume), while in disturbed areas the deposition is almost twice as much (Trebbs et al. 2006). In addition, the burning of biomass in either forests or sugar cane in Brazil is changing the dominant form of nitrogen deposition from nitrate to ammonium (W. Zamboni de Mello, University Federal Fluminense, personal communication). In the atmosphere, ammonium aerosols can reduce visibility, and may have negative effects on human health. Additionally, the formation of ammonium affects the atmospheric transport of SO_2 and NO_x emissions. Once deposited, ammonium releases acidity since the nitrogen is either accumulated in organic form or nitrified and leached as nitrate (Schurkes and Mosello 1988). As biomass burning is one of the major sources of N to the atmosphere, several other regions in LA-Ca could be undergoing similar changes in the pattern of N deposition.

Urbanization (fossil fuel and sewage)

In 2003, approximately 48% of the world population lived in urban centers and urban areas covered approximately 2% of the Earth's land surface. It has been estimated that by 2030, almost 61% of the world population will live in urban centers (World Urbanization Prospects 2003). However, the environmental impact of urban centers extends well beyond their limited spatial area due to the significant 'ecological footprints' of cities (Austin et al. 2003; World Urbanization Prospects 2003).

The Americas, defined here as the sum of North America, Latin America, and the Caribbean, had the highest percentage of their population living in urban centers in 2003 (78%), followed by Europe and Oceania, with 75 and 73%, respectively. Urbanization has increased rapidly. In 1950, the percentage in urban areas was only 53% (World Urbanization Prospects 2003). In addition, 6 out of 20 mega cities of the world (population higher than 10 million people) are located in the Americas.

Although from 1950 to 2003 there was an increase in urbanization in North America, the highest increase was observed in Central and South America (Figure 13). In developing countries of Latin America and the Caribbean, migration was from rural to urban centers, while in North America both rural-to-urban migration and a migration from high density urban centers to low

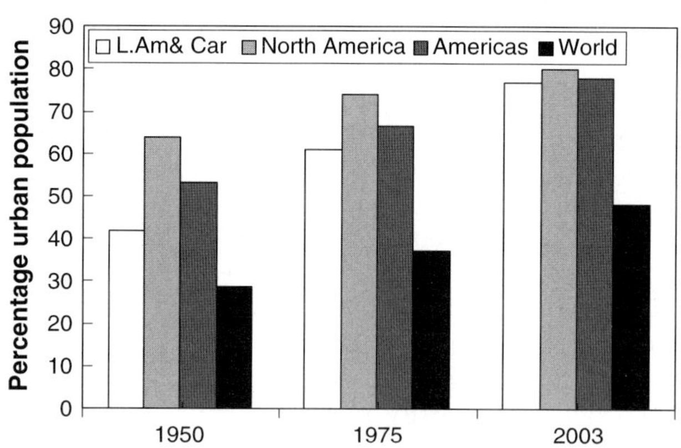

Figure 13. Percentage of urban population in Latin America and Caribbean, North America, Americas and World. Source: World Urbanization Prospects, 2003.

density suburbs surrounding city cores (urban sprawl) occurred. In both cases, urbanization led to the concentration of domestic and industrial solid wastes, air pollution, and problems associated with lack of sewage treatment (Global Environmental Outlook 3 2003). Fewer than 35% of cities in the developing world have their wastewater treated. In the Metropolitan area of São Paulo Brazil (with almost 18 million people) only 10% of the domestic sewage load is treated (Martinelli et al. 2002). In densely populated Caribbean regions, like Puerto Rico, urban sprawl is rapidly transforming the landscape. By 2000, nearly 50% of the 3.8 million inhabitants of Puerto Rico were connected to public sewer systems. Most of the wastewater, however, receives only primary treatment before it is discharged into the ocean. Consequently, a large volume of untreated sewage is discharged into surface waters, creating serious environmental problems. Similar situations are found throughout the developing countries of the Americas. Untreated domestic sewage is particularly rich in NH_4^+ and labile organic matter (Martinelli et al. 1999). The organic matter decomposes, consuming the dissolved oxygen concentrations in water bodies (Daniel et al. 2002). Under anaerobic conditions, NO_3^- is denitrified and lost to the atmosphere while NH_4^+ accumulates in the water (Martinelli et al. 1999). In contrast, in North America 90% of urban sewage is treated (Global Water Supply and Sanitation Assessment 2000). However, most of this treatment is just for the removal of organic matter, and little effort has gone into removal of nitrogen from wastewater streams in most areas in North America, although increasingly such efforts are beginning.

In urban areas, residents rely mainly on fossil fuels for transportation. This is especially true in the United States and Canada due to, in part, to long-distance travel associated with urban sprawl. In 1975, a passenger drove approximately 8000 km per year; but by 1999 this distance increased to almost 15,000 km per passenger-year (Global Environment Outlook 3 2003). Also, in 1975, an average North American traveled about 300 km per year on public transportation, a distance very similar to that observed in 1999 (Global Environment Outlook 3 2003). Thus, over the last 24 years, the use of private transportation has increased drastically, while the use public transportation has remained stable.

Several industries that use fossil fuel as their main source of energy are also located in urban centers. Consequently, urban centers are responsible for increasing emissions of NO_x. In the atmosphere, the dominant sink of NO_x is the oxidation of NO_2 by OH to form HNO_3, which then collects on aerosols or dissolves in precipitation and is subsequently deposited, causing rain acidification (Galloway 2003; Holland and Carroll 2003).

Conclusion

Current trends of economic development and human impact on N cycle in the LA-Ca region demonstrate levels of impact due to agricultural practices and urbanization which are, overall, less than that which is seen in more developed countries in the northern hemisphere. In Latin America, natural biological nitrogen fixation (BNF) is still an important source of Nr to the globe, accounting for 25% of the world's Nr created in terrestrial ecosystems (Galloway et al. 2004, in press). However, the rapid rate of change in almost all indicators of increased reactive nitrogen, from leguminous cropping, synthetic fertilizer use, biomass burning, and fossil fuel combustion merit concern for the potential impact of these indicators in the region in the very near future. For example, the increasing rates of deforestation and extensive conversion of natural ecosystems in Brazil and Argentina to cropping systems of soybean and other crops is a trend which is increasing at an alarming rate and whose consequences for the circulation of reactive nitrogen in the region and at the global level will be seen in the coming decades. While fertilizers are not yet heavily used in the LA-Ca region compared to other regions of the world, its consumption has been increasing constantly in most countries of the region. At the same time, it is important to call attention to the fact that in less developed countries of the LA-Ca region, fertilizer N application currently does not adequately meet the food requirements of an increasing population (Mosier et al. 2004). Fire is extensively used in the LA-Ca region as the cheapest way to remove the natural vegetation for crops, and also to 'clean' pastures from bushes and small trees, and facilitate sugar cane harvesting. During burning, biomass associated N is volatilized and a large fraction is emitted to the atmosphere in form of gaseous NH_3 and NO_x. Part of this emitted nitrogen returns via wet and dry deposition. Increasing emissions of NO_x due to use of fossil fuel have been also observed in urban centers of LA-Ca region. Due to the migration of rural areas to urban centers, most of the cities in LA-Ca region have been growing very rapidly in these last three decades. This increase has not been followed by effective sewage collection and treatment. As a consequence, most of the domestic sewage is dumped without treatment in water bodies, leading to severe pollution problems.

The LA-Ca region, as a region undergoing rapid economic development, has both a challenge and an opportunity to produce food and use energy generated by fossil fuels in a way that keeps disruptions of the natural nitrogen cycle as small as possible. This developing region has the objective to meet the food and energy requirements of a growing population by using their natural resources to achieve such goals. However, the LA-Ca region has a unique opportunity to carry out this task without the same mistakes that have occurred in other areas of the world, where high rates of fertilizer application and mismanagement of reactive nitrogen has resulted in a number of negative consequences both for natural and managed ecosystems.

Acknowledgements

Thanks to the organizers of the various workshops in Ubatuba, Brazil, San Juan, Puerto Rico and Brasilia, Brazil, among others Inter American Institute for Global Changes (CRN-12), Conselho Nacional de Desenvolvimento e

Pesquisa do Brasil (CNPq), Programa PROSUL; University of São Paulo, Cornell University, and Fundación Antorchas of Argentina.

References

Alves B.J.R., Boddey R.M. and Urquiaga S. 2003. The success of BNF in soybean in Brazil. Plant Soil 252: 1–9.

Austin A.T., Howarth R.H., Baron J.S., Chapin III F.S., Christensen T.R., Holland E.A., Ivanov M.V., Lien A.Y., Martinelli L.A., Melillo J.M. and Shang C. 2003. Human disrpution of element interactions: drivers, consequences and trends for the 21st century. In: Melillo J.M., Field C.B. and Moldan B. (eds), Interactions of Major Biogeochemical Cycles: Global Change and Human Impacts, Island Press, Washington DC, pp. 15–46.

Austin A.T., Piñiero G. and Gonzalez Polo M. More is less: agricultural impacts on the N cycle in Argentina. Biogeochemistry (this volume).

Boddey R.M., Urquiaga S., Reis V.M. and Döbereiner J. 1991. Biological nitrogen fixation associated with sugarcane. Plant Soil 137: 111–117.

Boddey R.M., Urquiaga S., Alves B.J.R. and Reis V. 2003. Endophytic nitrogen fixation in sugarcare: present knowledge and future applications. Plant Soil 252: 139–149.

Boddey R.M., Macedo R., Tarre R.M., Ferreira E., de Oliveira O.C., Rezende C.P., Cantarutti R.B., Pereira J.M., Alves B.J.R. and Urquiaga S. 2004. Nitrogen cycling in Brachiaria pastures: the key to understanding the process of pasture decline. Agric. Ecosyst. Environ. 103: 389–403.

Daniel M.H.B., Montebello A.A., Bernardes M.C., Ometto J.P.H.B., Camargo P.B., Krusche A.V., Ballester M.V., Victoria R.L. and Martinell L.A. 2002. Effects of urban sewage on dissolved oxygen, dissolved inorganic carbon and organic carbon, and electrical conductivity of small streams along a gradient of urbanization in the Piracicaba River basin. Water Air Soil Poll. 136: 189–206.

Dentener F.J. and Crutzen P.J. 1994. A three-dimentional model of the global ammonia cycle. J. Atmos. Chem. 19: 331–369.

FAOSTAT. 2004. Food and Agriculture Organization of the United Nations, Statistical Databases [online] URL: http://www.apps.fao.org/.

Filoso S., Martinelli L.A., Howarth R.W., Boyer E.W. and Dentener F. Human activities changing the N cycle in Brazil. Biogeochemistry (this volume).

FRA 2000 – Global Forest Resources Assessment. 2000. Food and Agriculture Organization of the United Nations. Forestry Department.

Galanter M., Levy II H. and Carmichael G.R. 2000. Impacts of biomass burning on tropospheric CO, NO_x, and O_3. J. Geophys. Res. 105(D5): 6633–6653.

Galloway J.N. 2003. Acid deposition: S and N cascades and elemental interactions. In: Melillo J.M., Field C.B. and Moldan B. (eds), Interactions of the Major Biogeochemical Cycles. Global Changes and Human Impacts, Island Press, Washington DC, pp. 259–272.

Galloway J.N. and Cowling E.B. 2002. Reactive nitrogen and the world: 200 years of change. Ambio 31: 64–71.

Galloway J.N., Dentener F.J., Capone D.G., Boyer E.W., Howarth R.W., Seitzinger S.P., Asner G.P., Cleveland C., Green P., Holland E., Karl D.M., Michaels A.F., Porter J.H., Townsend A. and Vorösmarty C. 2004. Nitrogen cycles: past, present and future. Biogeochemistry 70(2): 153–226.

Global Environment Outlook 3. 2003. Past, present and future perspectives. United Nations Environmental Programme.

Global Water Supply and Sanitation Assessment. 2000. World Health Organization – UNICEF Joint Monitoring Programme for Water Supply and Sanitation.

Holland E.A. and Carroll M.A. 2003. Linking the bio-atmosphere carbon and nitrogen cycles and atmospheric chemistry. In: Melillo J.M., Field C.B. and Moldan B. (eds), Interactions of the Major Biogeochemical Cyles. Global Changes and Human Impacts, Island Press, Washington DC, pp. 273–292.

Holland E.A., Dentener F.J., Brasswell B.H. and & Sulzman J.M. 1999. Contemporary and pre-industrial global reactive nitrogen budgets. Biogeochemistry 46: 7–43.

Howarth R.W., Boyer E.W., Pabich W.J. and Galloway J.N. 2002. Nitrogen use in the United States from 1961–2000 and potential future trends. Ambio 31: 88–96.

Hoelzemann J.J., Schultz M.G., Brasseur G.P., Granier C. and Simon M. 2004. Global Wildland Fire Emission Model (GWEM): Evaluating the use of global area burnt satellite data. J. Geophys. Res. 109, D14S04, doi:10.1029/2003JD003666.

Human Development Report. 2003. Millennium Development Goals: a compact among nations to end human poverty (2003). United Nations Development Programme (UNDP), Oxford University Press, New York.

INPE. 2004. Monitoramento da floresta Amazonica brasileira por satelite – Projeto PRODES. Instituto Nacional de Pesquisas Espaciais, [online] URL: http://www.obt.inpe.br/prodes/.

Lara L.L., Holland E.A., Artaxo P., Camargo P.B. and Martinelli L.A. Linking nitrogen deposition patterns and land use change in tropical areas. Biogeochemistry (this volume).

Martinelli L.A., Krusche A.V., Victoria R.L., Camargo P.B., Bernardes M.C., Ferraz E.S., Moraes J.M. and Ballester M.V. 1999. Effects of sewage on the chemical composition of Piracicaba River, Brazil. Water Air Soil Poll. 110: 67–79.

Martinelli L.A., Silva A.M., Camargo P.B., Moretti L.R., Tomazelli A.C., Silva D.M.L., Fischer E.G., Sonoda K.C. and Salomão M.S.M.B. 2002. Levantamento das cargas orgânicas lançadas nos rios do Estado de São Paulo. Biota Neotrópica 2: 1–18.

Miranda C.H.B. and Boddey R.M. 1987. Estimation of biological nitrogen fixation with 11 ecotypes of *Panicum maximum* grown in nitrogen-15 labeled soil. Agron. J. 79: 558–563.

Mosier A.R., Syers J.K. and Freney J.R. 2004. Nitrogen fertilizer: an essential component of increased food, feed and fiber production. In: Mosier A.R., Keith Syers J. and Freney J.R. (eds), Agriculture and the Nitrogen Cycle, Assessing the Impacts of Fertilizer Use on Food Production and the Environment SCOPE 65, Island Press, Washington DC, pp. 3–18.

Oliveira M.E.D., Vaughan B.E. and Rykiel Jr. E.J. 2005. Ethanol as fuel: energy, carbon dioxide balances, and ecological footprint. Bioscience 55: 593–602.

Potter C., Genovese V.B., Klooster S. and Torregrosa A. 2002. Biomass burning emissions of reactive gases estimated from satellite data analysis and ecosystem modeling for the Brazilian Amazon region. J. Geophys. Res. 107((D20)): Art. No. 8056.

Schultz M.G. 2002. On the use of ATSR fire count data to estimate the seasonal and interannual variability of vegetation fire emissions. Atmos. Chem. Phys. 2: 387–395.

Schurkes J.A. and Mosello R. 1988. The role of external ammonium inputs in freshwater acidification. Schweiz. Zeit. Hydrol. 88: 71–86.

Trebbs I., Lara L. L., Zeri L. M. M., Gatti L. V., Artaxo P., Dlugi R., Slanina J., Andreae M. O. and Meixner F. X. 2006. Dry and wet deposition of inorganic nitrogen compounds to a tropical pasture site (Rondônia, Brazil). Atmos. Chem. Phys. 6: 447–469.

Williams M.R. and Melack J.M. 1997. Solute export from forested and partially deforested catchments in the central Amazon. Biogeochemistry 38: 67–102.

World Urbanization Prospects. 2003. The 2003 Revision (2004) Department of Economic and Social Affairs. Population Division, United Nations, New York.

Biogeochemistry (2006) 79: 25–44
DOI 10.1007/s10533-006-9001-2

A review of anthropogenic sources of nitrogen and their effects on Canadian aquatic ecosystems

DAVID W. SCHINDLER[1,*], PETER J. DILLON[2] and HANS SCHREIER[3]
*[1]Department of Biological Sciences, University of Alberta, Edmonton, AB, T6G 2E9, Canada; [2]Environmental and Resource Studies, Trent University, 1600 West Bank Drive, Peterborough, ON, K9J 7B8, Canada; [3]Institute for Resources, Environment and Sustainability, University of British Columbia, 2202 Main Mall, Vancouver, BC, V6T 1Z4, Canada; *Author for correspondence (e-mail: d.schindler@ualberta.ca; phone: +780-492-1291; fax: +780-492-9234)*

Key words: Acid rain, Agriculture, Eutrophication, Water pollution

Abstract. Nitrogen releases to air and water are low in most of Canada, but in southern areas with rapid development there are telltale signs of the problems from releases to air and water that are described elsewhere in this volume. These include higher nitrogen in water and releases to the atmosphere from urban areas, industry and agriculture. As a result, in parts of Ontario and Quebec underlain by Precambrian geology, nitrogen deposition is near the critical loads found for geologically similar areas of Europe. In particular, combined inputs of sulphuric and nitric acids are causing base cation depletion in forest soils and keeping some lakes at pH values too low to allow the recovery of biological communities. In southern Ontario, Alberta and British Columbia, rapidly expanding human populations, industry and agriculture are causing high concentrations of nitrate in surface and groundwaters. At present, there is little sign of estuarine eutrophication in Canada, but it appears to be imminent on the Pacific coast, as the result of expanding human populations and intensifying agriculture in the lower Fraser Valley and Puget Sound. Steps should be taken now to prevent the widespread problems caused by nitrogen pollution that have occurred in Europe, the USA, and other populous and industrialized regions.

Introduction

Most areas of Canada still have low human population densities, and are relatively free from industrial activity. As a result, much of the country has escaped the problems associated with nitrogen contamination that have plagued more industrialized countries. These include the contamination of precipitation, surface and ground water, soil saturation with nitrate, acidification of receiving waters, smog formation and pollution of surface and groundwaters to beyond guidelines to protect human health and ecosystem integrity. The problem of managing nitrogen in most of Canada is thus still largely one of prevention, ideally profiting from the mistakes made in more populous and industrialized countries.

Most Canadian freshwaters still contain low concentrations of inorganic nitrogen. In a survey of provincial databases in eastern Canada, Jeffries (1995) found that almost all lakes contained less than 10 μeq l^{-1} of NO_3. Provincial mean concentrations for nitrate in lakes were all less than 10 μeq l^{-1}, except for Prince Edward Island, where four of six lakes examined had over 100 μeq l^{-1} and New Brunswick, where about 10 percent of lakes had nitrate concentrations of 10–40 μeq l^{-1}. Much of the land in these two small Atlantic provinces is used for agriculture, and most of the soil is sandy. Heavily developed parts of southern Ontario, Quebec, southern Alberta and coastal southern British Columbia are other areas with higher than background nitrogen. Lakes and streams show the expected symptoms of overfertilization with nutrients, including nitrogen. In the remainder of this review, we will focus largely on these areas.

Canada-wide trends in airborne emissions of nitrogen

Chambers et al. (2001) estimate that human activities discharge 1.4 million tonnes of nitrogen to the atmosphere in Canada, with ammonia releases from fertilizer and manure being the biggest source. Release of N_2O was evenly split among transportation, industry and agriculture. NO_x emissions from industry and transportation were also important. No NO_x data were available for agriculture, but it was assumed to contribute about as much as industry. More detailed information is given below.

Emissions of NO_x in Canada nearly doubled from 1970–1985. From 1985 to 2000 emissions were relatively constant. A 17% decrease in eastern Canada caused by regulation of emissions from vehicles and smelters was approximately balanced by a 29% increase from electrical generation, vehicles and oil and gas industries in western Canada (Figure 1). Country-wide emissions are projected to decrease 17% by 2020, a balance between a 39% reduction in eastern Canada and a 5% increase in the west, where increased emissions from oil and gas extraction, and coal-fired electrical power are expected.

Ammonia emissions increased by 9% between 1995 and 2000, largely in agricultural areas of the west. They are expected to increase by 50% between 2000 and 2020 largely as the result of the increasing intensity of livestock and poultry production, commercial fertilizer and pesticide manufacturing (Figure 1, Vet et al. 2005).

American emissions of NO_x and ammonia are also of concern, because prevailing winds carry the gases from areas of intensive electricity generation and agriculture in the US Midwest over southeastern Canada. US and eastern Canadian NO_x emissions have declined slightly and a downward trend is expected to continue. In contrast, ammonia emissions increased by 15% between 1990 and 2000. They are expected to continue to increase slightly until 2020 (Vet et al. 2005).

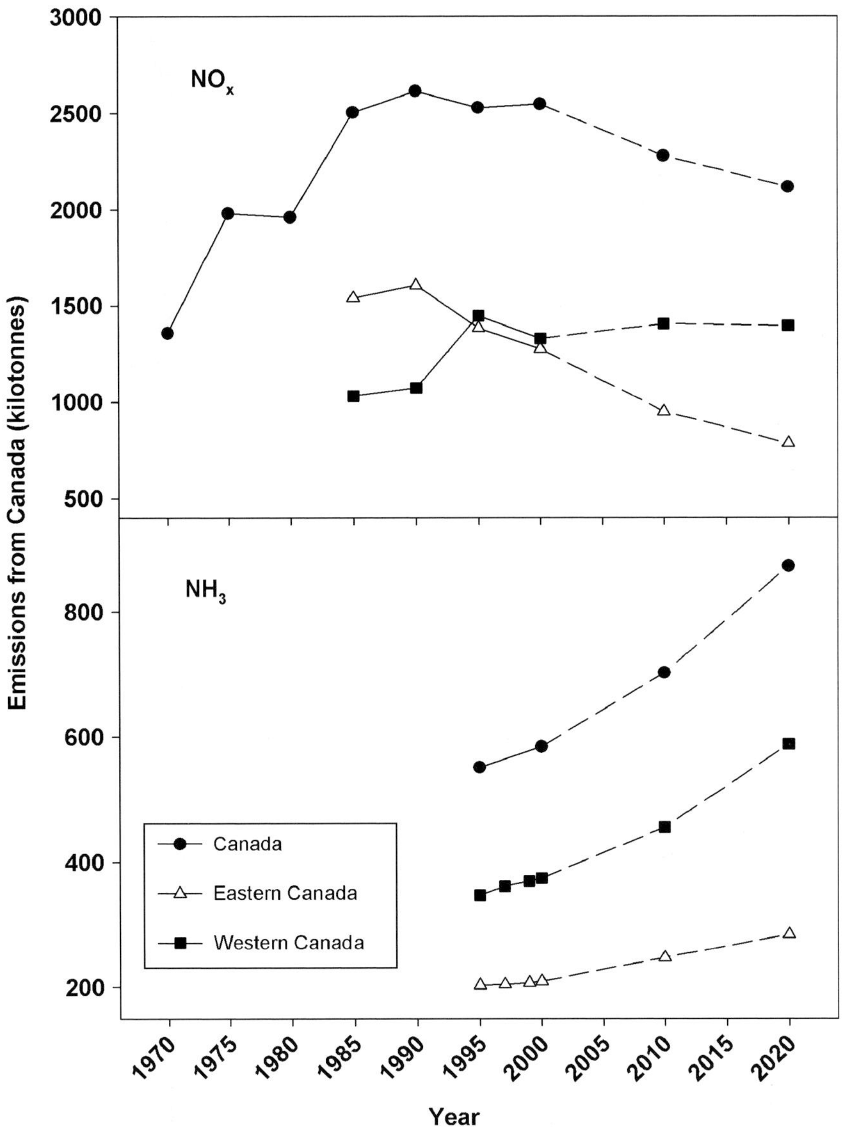

Figure 1. Recent trends in nitrogen emissions in Canada. Dashed lines indicate projected levels. Data from Vet et al. (2005).

The sources of NO_x in Canada are somewhat different from more urbanized and industrialized countries, with off-road vehicles contributing almost as much NO_x as on-road vehicles. In total, vehicles emit almost 50% of NO_x in Canada. The oil and gas industries also contribute disproportionately to Canadian emissions. In contrast, electrical power generation is a smaller percentage than in many countries, as a result of heavy reliance in most of Canada on hydroelectricity rather than fossil fuel burning. Agriculture contributes most of the ammonia emissions. Livestock culture is the predominant source, followed by manufacturing of commercial fertilizer (Table 1). Both gases contribute to soil acidification, water pollution with nitrate and ammonium, and urban smog.

Atmospheric deposition of nitrogen and its impact on ecosystems

As a land area, Canada is only slightly smaller than all of Europe. Most of the population, industry and agriculture are in a band a few hundred kilometres wide along the US border. Very few studies of deposition have been done north of 60° N. latitude.

Eastern Canada has been studied much more thoroughly than the west, due to concerns about the effects of acidifying deposition on geologically-sensitive terrain. For convenience, we will treat eastern and western Canada separately, dividing the country at the Ontario–Manitoba border.

Eastern Canada

Annual wet nitrate deposition in most of eastern Canada north of 52° N. latitude and east of the Manitoba–Ontario border is <5 kg NO_3 ha^{-1} y^{-1} (Figure 2). Only in populous industrial regions of southern Ontario and Quebec does nitrate deposition exceed 10 kg NO_3 ha^{-1} y^{-1} over large areas. An even smaller area along the US border with those provinces has nitrate deposition of 20–25 kg NO_3 ha^{-1} y^{-1} (Figure 2; Vet et al. 2005). The elevated nitrate concentrations in these regions are the result of NO_x emissions from industrial and automotive sources in southern Canada and the northern USA (Whelpdale and Galloway 1994; Vet et al. 2005). Fortunately, the highest

Table 1. Percentage contribution of major emitting sectors to total nitrogen emissions in Canada. Data from Vet et al. 2005.

Sector	NO_x	NH_3
Non-ferrous mining and smelting	–	–
Electrical power generation	11.4	–
Upstream oil and gas	13	–
On-road vehicles	32.6	3.1
Off-road vehicles	26.9	–
Industrial fuel combustion	–	–
Other fuel combustion	–	–
Agriculture (animals)	–	55
Pesticides and fertilizer	–	34.6
Chemicals and products	–	–

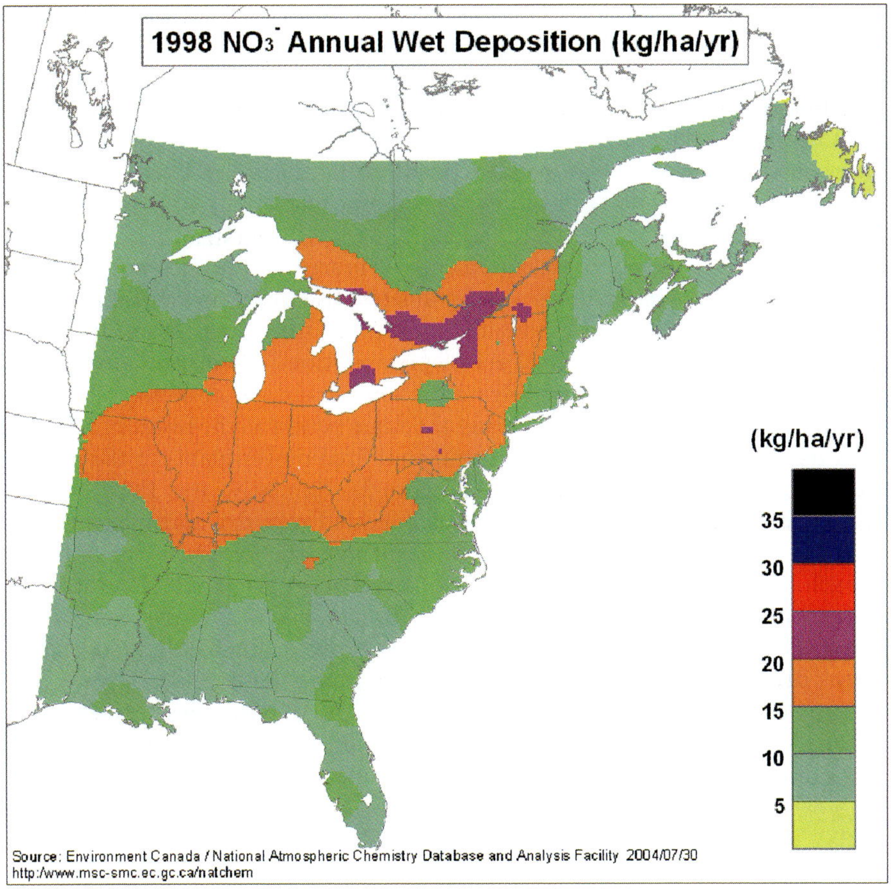

Figure 2. A map of nitrate deposition in southeastern Canada and the eastern USA for 1998. Reprinted with permission from Vet et al. 2005. Note that in this and subsequent figures, deposition values have been left in their original units. For convenience in converting, 1.00 mg l^{-1} $NO_3 = 0.224$ mg l^{-1} N = 16 μeq l^{-1}.

nitrate deposition occurs in areas where soils and water are well buffered. Moreover, deposition has remained relatively constant for some time.

Preliminary wet ammonium deposition maps of eastern Canada show a somewhat different pattern. Here, the highest values observed, of 5–6 kg NH_3 $ha^{-1}y^{-1}$ extend through southern parts of far western Ontario, near the Manitoba border (R. Vet, pers. comm.). There is little population, agriculture or industry in the area, which is largely underlain by Precambrian rock with little overburden. The elevated values probably reflect the proximity to agricultural areas of the US Midwest and southern Manitoba. North of 52° N latitude, values return to a near-pristine < 2.5 kg NH_3 ha^{-1} y^{-1}.

Dry deposition is significant at the eight sites where it has been measured, all south of 52° N. It ranges from 17 to 41 percent of total nitrogen deposition. It is highest near industrial areas along the US border. Combining with wet deposition to obtain total N deposition, values range from 2.6 kg N ha^{-1} y^{-1} at the western edge of Ontario to 7.5 kg N ha^{-1} y^{-1} at several sites in the East.

Dry deposition values have only been measured for 5 years, so there are no detectable long-term trends. Peroxyacetyl nitrate (PAN) and NO_2 are not included in dry deposition, so total deposition estimates are likely underestimated by as much as 40% in industrial regions, and about 10% for most of Canada (Vet et al. 2005).

Wet and bulk deposition measurements of inorganic nitrogen species have been made since the early 1980s at several sites. In Dorset, in central Ontario, near the center of the steep gradient from high southern deposition values to near-pristine northern ones, 26 years of precipitation record show no statistically-significant trend in either nitrate or ammonium. In contrast, sulfate in rainfall has declined by 45–50% during the same period as the result of reduced sulfur oxide emissions from smelters and coal-fired power plants in the region, and recent reductions in the USA (Figure 3). Similarly, the Experimental Lakes Area in northwestern Ontario has collected wet, dry and bulk precipitation for many years. There are no significant trends in deposition of sulfur or nitrogen, although there is a suggestion of a recent upturn in wet deposition of nitrogen (Linsey et al. 1987; Vet et al. 2005). The major source areas for the site are the Midwestern USA, followed by western Canada. Total nitrogen and sulfur deposition at the ELA site are only 2.6 kg N ha^{-1} y^{-1} and 2.9 kg S ha^{-1} y^{-1} (Vet et al. 2005). Management of nitrogen deposition in eastern Canada cannot be done in isolation from the USA. Canada emits only 7 percent of the combined nitrogen emissions in eastern Canada and the USA,

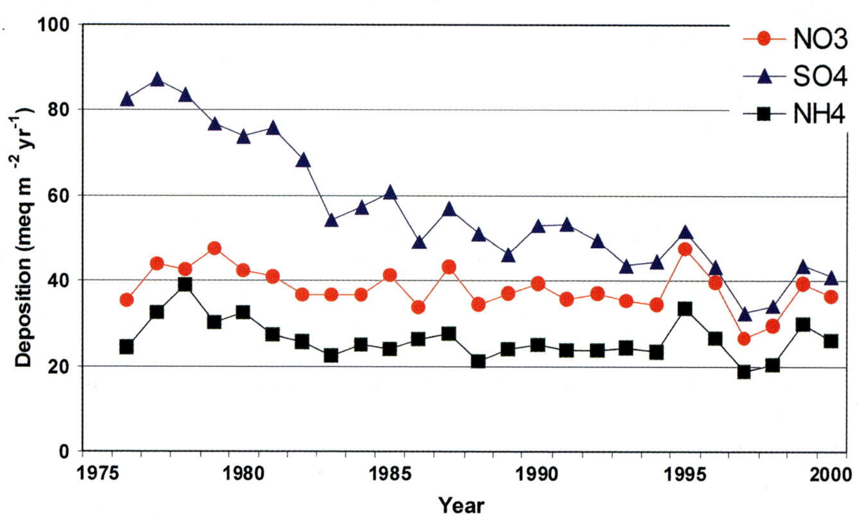

Figure 3. Twenty five years of sulfate, nitrate and ammonium concentrations in bulk deposition at Dorset Ontario.

but receives 32 percent of the resulting wet deposition of nitrogen, because of being downwind of areas of high NO_x production in the American Midwest.

Impact of high nitrogen deposition on ecosystems

Compounds of both sulfur and nitrogen contribute to the acidification of soils and aquatic systems, and with respect to the effects of acidification, the elements cannot be treated in isolation. Various models have been used to predict whether the capacity of terrestrial and aquatic ecosystems to tolerate acidifying substances have been exceeded in eastern Canada. Critical load models combining the deposition of acidifying sulfur and nitrogen compounds similar to those used in Europe have generally been used for these assessments (Critical load is defined here as in the UN-ECE as "a quantitative estimate of an exposure to one or more pollutants below which significant harmful effects on specified sensitive elements of the environment do not occur according to present knowledge" following Nilsson and Grennfelt 1988).

One concern in base-poor soils such as those overlying Precambrian rock has been that continued loading of acidifying substances is leaching calcium in excess of the rate that it is being replenished by deposition and weathering (Likens et al. 1996). Vet et al. (2005) used critical load models for soils and freshwaters to project the extent of acidification damage under "best case" and "worst case" emissions control scenarios for Canada and the USA. It was estimated that between 0.5 and 1.8 million km^2 or 21 to 75 percent of eastern Canada continues to receive acid deposition in excess of critical loads. The effect of high acid deposition is particularly severe for forest soils, which continue to export base cations in excess of inputs from weathering and deposition. Calcium is particularly badly affected, with soil pools in south-Central Ontario depleted by up to 30% since the early 1980s (Watmough and Dillon 2003a, b). Declines in sugar maple growth (Watmough 2002) and hardwood forest production (Oimet et al. 2001) have been attributed to calcium depletion caused by exceedance of critical loads.

An additional concern is the combination of acid deposition and forest harvesting. Watmough and Dillon (2003b) calculate that the combination of calcium removal by forest harvesting and leaching could leave soils unable to support new forest growth within a few decades because of calcium deficiency. In brief, it appears that controlling acid deposition will be an important consideration in sustainable forestry in eastern Canada.

Aquatic critical load models also indicate widespread acidification damage. Vet et al. (2005) estimate that about 15% of lakes in eastern Canada south of 52° N that historically had pH values > 6 (a critical threshold for many aquatic species) will be incapable of maintaining such pH values under the current acid deposition regime. The total number of lakes in eastern Canada south of 52° N that are affected according to this criterion is 500,000 to 600,000 (Vet et al. 2005). As acidifying emissions continue, the continuing depletion of base cations in catchment soils will require greater and greater reductions in emissions of sulfur and nitrogen to reach target pH values in receiving waters. Climate warming and drought have also caused the reoxidation of sulfur deposited in catchments during previous periods of high deposition, exacerbating the acidification problem (Bayley et al. 1992; Lazerte 1993).

In the Dorset, Ontario area, where long-term chemical records are available for several lakes and their inflow streams, Henriksen et al. (2002) SSWC steady-state model indicates that the proportion of lakes where critical loads are exceeded has declined from >90 percent in the late 1970s to <40 percent in the late 1990s, largely the result of reductions in sulfur oxide emissions. As the result of these reductions, total inorganic nitrogen deposition (nitrate plus ammonium) in equivalent units is now about 50 percent higher than sulfur deposition (Figure 3). Despite the documented depletion of soil nutrients and nitrogen deposition that exceeds long-term critical loads in analogous parts of Europe, there are no detectable trends in stream or lake concentrations of inorganic nitrogen in central Ontario during the last two decades. Nitrogen export varies considerably between catchments and years, ranging from 0.1 to 1.5 kg N ha^{-1} y^{-1} (Watmough and Dillon 2003a).

Dean Jeffries (pers. comm.) has also analyzed 1081 lakes in eastern Ontario using Henriksen's SSWC model. Current deposition exceeds critical loads in 45 percent of the lakes. The nitrogen component of the exceedances is, however, very small.

An exception to the above is Lake Superior, which receives little phosphorus from either the atmosphere or via inflow streams, but receives moderate loading of airborne nitrogen from industries and agriculture in the USA. As a result, nitrate has nearly doubled in the latter half of the 20th century, from 200 μg l^{-1} in 1950 to almost 360 μg l^{-1} in 2001 (Hugh Dobson, National Water Research Institute, pers. comm.).

Watmough et al. (2004) analyzed trends over a 16-year period in sixteen forested catchments in Ontario. They found that there were no uni-directional long-term trends in stream nitrate concentration, but that there were common long-term patterns. Two stream groupings were identified. Those with shallow soils, moderate slopes and large wetlands had low nitrate concentrations, while those with deeper soils, steeper slopes and little wetland area had high nitrate concentrations. These differences were attributed to various climate factors including greater impact of droughts in the latter group. The importance of the watersheds' characteristics in determining nitrate concentration is streams was also noted by Schiff et al. (2002), who observed nitrate exports differing by a factor of 10 from two neighbouring streams with different slopes and wetland components.

Nitrate plays a role in lowering pH during spring snowmelt in eastern Canada. During snowmelt, concentrations of nitrate and hydrogen ion in streams and lakes can increase several-fold. At some sites, nitrate can exceed 20 μeq l^{-1} for short periods in the spring, enough to contribute to depression of pH and alkalinity in lakes where alkalinity is less than 100 μeq l^{-1}. Alkalinity and pH depressions also occur in the autumn in streams and the littoral zones of lakes, particularly following dry summers. Such events are also a major problem for the biota (Dillon et al. 1997; Eimers and Dillon 2002). Although sulphate was the dominant factor in most watersheds in these events, in one of nine watersheds an increase in nitrate concentration was the cause of the decline in alkalinity (Laudon et al. 2004).

Kaste and Dillon (2003) compared input and output for several terrestrial catchments in eastern Ontario and Norway, and so far, deposited nitrogen has been more efficiently retained in Canadian forested ecosystems than in similar European catchments. The eight Ontario sites are in an area that receives near-maximum nitrogen deposition, averaging 9 kg TIN ha^{-1} y^{-1} for 19–23 years. They all retained over 90% of input. In contrast, four Norwegian catchments that had deposition roughly twice as high retained only 43–71% of input. An exception was Langtjern, which had a 97% TIN retention, but a deposition equal to the Canadian sites. From these data it appears that there is a major breakpoint in terrestrial saturation in northern forests, occurring somewhere between 9 and 18 kg TIN ha^{-1} y^{-1}, above which leakage from catchments increases rather rapidly. This is within the range estimated by Grennfelt and Hultberg (1986) for nitrate leaching at 10–15 kg N ha^{-1} y^{-1} (Reviewed by Jeffries 1995).

At the Turkey Lakes Watershed (TLW), a few hundred km northwest of Kaste and Dillon's sites in Ontario, nitrogen retention was only 65%, at average TIN deposition of about 9 kg ha^{-1} y^{-1}. There is some evidence that the source of the nitrate in outflow is nitrification of organic nitrogen in forest soils (Spoelstra et al. 2001), and there has been little change in the efficiency of retention over time. Hazlett and Foster (2002) showed that nitrate concentrations in subsurface throughflow draining toward Little Turkey Lake in the TLW varied greatly in relation to slope, reflecting the soil horizons through which the water had passed. In summary, there appears to be stable, and usually high retention of nitrogen at the forested monitoring sites in eastern Canada, even in those that receive the highest nitrogen deposition.

Wetlands are ubiquitous in the boreal ecozone. Beaver ponds (Devito and Dillon 1993a) and *Sphagnum*-conifer swamps (Devito and Dillon 1993b), the two dominant types of wetlands, are effective sinks for inorganic N, although they release an equivalent amount of organic N. They therefore contribute substantially to the effectiveness of landscapes in retaining nitrate and ammonium, although they are in balance with respect to total N.

However, it is still unclear how close to critical loads these systems may be. The FAB model (Posch et al. 1997) has been used in Ontario to calculate the combined critical load of S and N (Aherne et al. 2004). This model takes into account the major biogeochemical processes that involve nitrogen in a lake and its terrestrial catchment, including uptake by vegetation, immobilisation in the soil and denitrification, and in-lake retention. The model assumes that at least some of the N leaks from catchments. Results suggest that the combined S and N deposition exceeds the critical load to a higher proportion of lakes than if critical loads based on only S were considered. In Europe, there seems to be a several year lag before breakthrough begins at intermediate nitrogen deposition, even at constant rate. A long lag in reversal is also observed at the one site where nitrogen deposition was experimentally decreased (Wright et al. 1994). It would be wise to err on the side of caution, regulating nitrogen emissions to keep deposition in Canada at current values or lower.

Kaste and Dillon (2003) also found that the terrestrial catchments of lakes were the predominant sink for nitrogen because they tended to be large relative to the area of the lake to which they drained (drainage area/lake area ratios of 3–49). On the other hand, lakes were more efficient sinks per unit area, retaining 42–70% of deposition compared to 20–27% for terrestrial catchments. The most important sink in oligotrophic boreal lakes is believed to be denitrification in lake sediments (Kelly et al. 1990; Molot and Dillon 1993).

Western Canada

So far, deposition of nitrogen has been relatively low in most of western Canada. Deposition values are <2 kg NO_3 ha^{-1} y^{-1} for most of the area west of Manitoba (Jeffries et al. 2003), but these are based on relatively few stations. Esther, Alberta, near the center of the western prairies, has a total deposition of only 1.8 kg N ha^{-1} y^{-1}. One exception is the lower Fraser Valley in B.C., where rates of more than 10 kg N ha^{-1} y^{-1} have been reported (Belzer et al. 1998; Schreier et al. 1999). NO_x emissions have increased by 29 percent in western Canada between 1985 and 2000, in contrast to the declines in eastern Canada. As of 2000, NO_x emissions in the west exceeded those in eastern Canada for the first time. Emissions are predicted to increase by another 5 percent between 2000 and 2020. This estimate may be low, due to rapid population growth and massive developments in the Alberta Oil Sands. As described later, these developments are expected to cause enormous increases in nitrogen emissions to the atmosphere (Allen 2004).

There are only measurements of dry deposition at two sites in western Canada, an area half the size of Europe. The climate of this region varies greatly, making it is impossible to generalize about either dry or total deposition. Given rapid increases in population, agriculture and industry in the west, more deposition monitoring stations must be considered as an urgent priority.

There is a slight suggestion of recently increased deposition of nitrate in southern Alberta, at the only site in the Rocky Mountains with long term data. Although annual values are scattered, deposition appears to have doubled in about 15 years, from approximately 2 to about 4 kg NO_3 ha^{-1} y^{-1} (Alberta Environment, unpublished data). The region is not immediately downwind of any large urban or industrial development. However, rapidly-growing industrial areas near Calgary Alberta occupy a position on the eastern slopes of the Rocky Mountains similar to those of Colorado, where nitrogen deposition has increased rapidly as industry and population have expanded (Williams et al. 1996). Dated ice samples taken in 1995 from glaciers at the highest point in the Canadian Rockies showed low concentrations and no trend in ammonia and nitrate deposition in 50 years (D.W. Schindler, unpublished data), but this situation may have changed in the ensuing decade. There is some evidence that N deposition is increasing downwind of major population centers, as described in the next section.

Increasing humans, livestock and commercial fertilizer as sources of atmospheric nitrogen

Rapidly increasing populations of humans, industrial development, livestock, fertilizer manufacturing, and increasing use of manure and commercial fertilizer in southern British Columbia and Alberta are causing increased atmospheric concentrations of NO_x and ammonia (Schreier et al. 2003; Schreier and Brown 2004). Vet et al. (2005) estimate that ammonia emissions in western Canada increased by 8 percent between 1985 and 2000. Most of the emissions were in southern parts of the prairie provinces and the lower Fraser River watershed, where agriculture is concentrated. Highest values were in central Alberta and coastal British Columbia, where both agriculture and fossil fuel combustion are greatest. Ammonia emissions in western Canada are projected to increase by 57 percent between 2000 and 2020, largely the result of livestock and poultry culture and other agricultural activities. Nitrogen released by centers of human population is causing increased airborne emissions and deposition of nitrogen well outside the geographical bounds of cities. Kochy and Wilson (2001) found that deposition of nitrogen was higher in national parks downwind of prairie cities than in more remote parks of the prairies. Cheng (1994) predicted such results from modelling studies in Alberta. His models predicted deposition of up to 4 kg N ha^{-1} y^{-1} near Edmonton and Calgary, cities of approximately 1 million people each. However, these deposition values are lower than those reported by Kochy and Wilson, using resin bags rather than precipitation collections. Their results show deposition of 8 kg N ha^{1} y^{-1} for the most pristine sites in Jasper National Park of the Rocky Mountains, and 22 kg N ha^{-1} y^{-1} for the highest sites downwind of Edmonton. Intercalibration of such methods with direct analysis of precipitation samples is in order before results of the two methods can be realistically compared.

The Athabasca Oil Sands

Another potential new source of nitrogen to the atmosphere is rapid development in the oil sands of northeastern Alberta. So far, the two oil sands plants that have been in operation for almost 30 years have not produced excessive emissions of nitrogen. But with rapidly-increasing oil prices, a shortage of secure sources of oil in the Middle East, and recent hurricane damage to oil rigs and refineries around the Gulf of Mexico, expansion has escalated. A third oil sands plant is now operating, two more are under construction, and several others are in the approval process. Huge trucks capable of carrying hundreds of tonnes each are expected to produce NO_x that will cause nitrogen emissions to increase by 359% over 1998 values in the near future (Environment Canada 2003). Nitrate deposition in the mid-1990s was generally less than 2 kg ha^{-1} y^{-1} (Anonymous 2000). Near the center of oil sands activity, nitrogen deposition is projected to be as high as 65 kg N ha^{-1} y^{-1} (Allen 2004), and deposition greater than 20 kg N ha^{-1} y^{-1} is expected over thousands of square kilometres (Anonymous 2000). Fortunately, most of the region around the oil sands is not geologically-susceptible to acidification. The nearest areas of concern from acidification is in northern Saskatchewan, over 100 km east of the oil sands, and in a few parts of the Birch Mountains, about 100 km to the northwest. Prevailing winds in the area are from the east, followed by the west. So far, modelled emission and deposition scenarios have not been verified with precipitation measurements, and the establishment of a precipitation monitoring network should be given high priority. Emissions of CO_2 and SO_2 are also expected to increase substantially as the result of oil sands development.

In summary, little information is available on sources or deposition of atmospheric nitrogen in much of western Canada. There have been no analyses of where critical loads of acidifying substances may be exceeded at present or in the future. Such studies are urgently required in order to plan for ecologically sustainable development in several areas of rapid population and industrial growth, including the lower Fraser River Valley of British Columbia, the corridor from Calgary to Edmonton in Alberta, and the Athabasca Oil Sands in northern Alberta.

Agriculture and urbanization as direct sources of nitrogen to water

The most rapidly increasing source of nitrogen in Canada is from agriculture. Much of the increase in dissolved nitrogen results from direct movement of nitrogen to water, rather than via the atmosphere. Both the use of commercial fertilizers and the production of manure from livestock husbandry have increased very rapidly. The increases in nitrogen are much more rapid than those in phosphorus and potassium, the other two main elements in commercial fertilizer. Since records began in 1950, nitrogen fertilizer production in Canada has increased about 75-fold, while use of phosphorus and potassium have increased by only 5 and 7-fold, respectively (Korol and Rattray 1999).

In some cases it is difficult to distinguish losses of nitrogen by agriculture to water from losses to air. Releases can be as N_2O, a potent greenhouse gas and precursor of urban smog, or as ammonia, which produces odors at high concentrations, is a NO_x precursor, and can both acidify sensitive soils and act as a fertilizer.

Increased demand for meat, particularly in the USA and the Orient, and decreased financial returns per animal have caused an explosive increase in livestock culture in Canada. In several provinces, populations of cattle, hogs and poultry have increased several fold since the mid-20th century. In the five years 1996–2001, Canadian cattle increased by 4.4%, hogs by 26.4%, and chickens by 23.4%, despite a human population growth of only 4.0% (Stats-Can, 2001). Alberta currently has the most cattle, 6.4 million. They increased by 11.3% in the 1996–2001 period. The stated goal of the Alberta government is to double that number. Hog increases have been greatest in Quebec and Manitoba, but numbers in Alberta, Saskatchewan and other provinces have also increased substantially. Chicken number have increased most rapidly in British Columbia, by 37% over the same 5 year time period.

Chambers et al. (2001) calculate that 4.3 million tonnes of nitrogen were added to Canadian farm land in 1996 from anthropogenic sources. Of this, 1.97 million tonnes were applied to agricultural land as manure, fertilizer and biosolids, 0.77 million tonnes were fixed by domestic legumes, and 0.043 million tonnes as atmospheric deposition of nitrate and ammonium. Only 58 percent of the amount added to agricultural lands was removed in the form of crops. A total of 0.3 million tonnes of nitrogen were estimated to enter fresh, ground and coastal waters as a result of human activity.

Most of the reported exceedances of groundwater standards for human consumption of nitrate in Canada were in agricultural areas. Seventeen percent of agricultural land in Ontario, 6% in Quebec, and 3% in Atlantic provinces would be expected to produce runoff or seepage with >14 mg l^{-1} N, largely as nitrate (Chambers et al. 2001). In Ontario, a significant proportion of both shallow and deep wells have nitrate in excess of drinking water standards of 10 mg NO_3 l^{-1} (Goss et al. 1998; Rudolph et al. 1998).

According to Canadian statistics, hogs are particularly large sources of phosphorus and nitrogen, with approximately 10 and 5 times the per capita output of humans, respectively (Chambers et al. 2001). The application of manure to land is high in parts of Canada that have high populations of livestock. Most notably, southern Alberta, southern Ontario and Quebec, and the lower Fraser Valley and Okanagan regions of southern B.C. have manure applications of 1000 kg N ha^{-1} and more.

Eighty percent of nitrogen consumed by farm animals is excreted or egested, after which 4–95% is lost to the atmosphere. Agriculture emitted an estimated 132 thousand tonnes of N_2O–N in Canada in 1996, with the largest sources in Alberta and Saskatchewan. The analogous figure for NH_3–N was 849 thousand tonnes from agriculture (1990 values). Fifty-five percent of the emissions were from fertilizer. Anhydrous ammonia is the most popular fertilizer in Canada, followed by urea. Both can have high losses to the atmosphere and acidify the soils. Fifty five percent of the nitrogen emissions from livestock are emitted by cattle (Desjardins and Keng 1999). Some of the highest atmospheric ammonia deposition has been measured in the most intensively used agricultural area in the lower Fraser Valley in B.C. where annual total atmospheric ammonia deposition rates of 9.5 kg N ha^{-1} y^{-1} were measured in 1997 (Schreier et al. 1999).

The lower Fraser River Valley in British Columbia is one of the most rapidly developing areas of Canada, with human populations and intensive agriculture increasing very rapidly (Schreier et al. 2003). The region has the highest density of livestock in Canada. Livestock culture is also changing rapidly from traditional methods to industrial confined feeding operations (CFOs). Nitrogen applications of both manure and commercial fertilizer are high, with over 50% supplied by manure in many of the subregions. Total applications in most areas are 200 to more than 300 kg N ha^{-1} y^{-1}. In most cases, these applications are in excess of plant needs, and excesses of application/demand are typically over 100 kg N ha^{-1} y^{-1}, with calculated excesses in some areas exceeding 300 kg N ha^{-1} y^{-1}. While populations of cattle, hogs and horses appear to be relatively stable, there has been an explosion in poultry culture, with chicken populations nearly doubling in between 1991 and 2001 (Schreier et al. 2003).

Commercial fertilizer use in Canada has also increased, by 20–30% every 5 years, according to Statistics Canada data (Statscan 2001). Using a nutrient budget modelling approach, Schreier et al. (2003) calculated that average annual surplus applications for the arable area of the lower Fraser Valley were more than 50 kg N ha^{-1} y^{-1}. Seven of the twenty subregions of the area had application rates exceeding 100 kg N ha^{-1} y^{-1} (Figure 4, Schreier and Brown 2004). In the Sumas River watershed, which is located in the most intensively used agricultural area in the valley, surplus levels of up to 300 kg N ha^{-1} y^{-1} were reported by Berka et al. (2001) and Schreier et al. (2003). Berka (1996) showed a significant positive relationship between surplus application rates and ammonia concentrations in surface and groundwater of the Sumas River watershed during the winter season and a significant negative relationship between surplus N and dissolved oxygen. Nitrate concentrations in the river have gone up steadily for over 30 years (Figure 5). Smith (2004) showed positive correlations between animal stocking density and surplus nitrogen applications within stream buffer zones (100 m wide on both sides of the river, and 500 m long stream segments) and ammonia and nitrate values in the Sumas River during the winter. Smith was also able to show how the rapid increase in animal stocking density between 1973 and 2003 has impacted the nitrate levels in the streamwater during the winter, when active uptake by vegetation is low.

Urbanization presents other problems. Decreased infiltration of water, coupled with lawn fertilizer, pet excrement, street dust, etc. can make storm water quality quite similar to that of sewage. Destruction of wetlands (80% in the lower Fraser, 70% in the western prairies), channelization of streams, and destruction of riparian buffer zones lead to the efficient delivery of all chemicals to water.

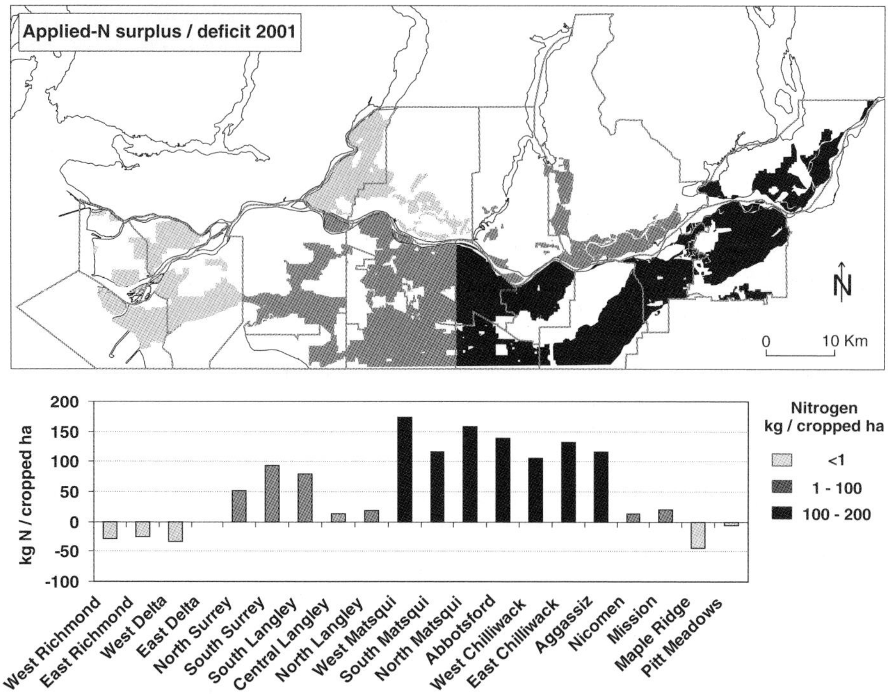

Figure 4. Annual surplus/deficit nitrogen application in 20 enumeration areas in the Lower Fraser Valley in B.C. in 2001. Values are in kg N ha^{-1} y^{-1} and represent application rates above crop nutrient requirements.

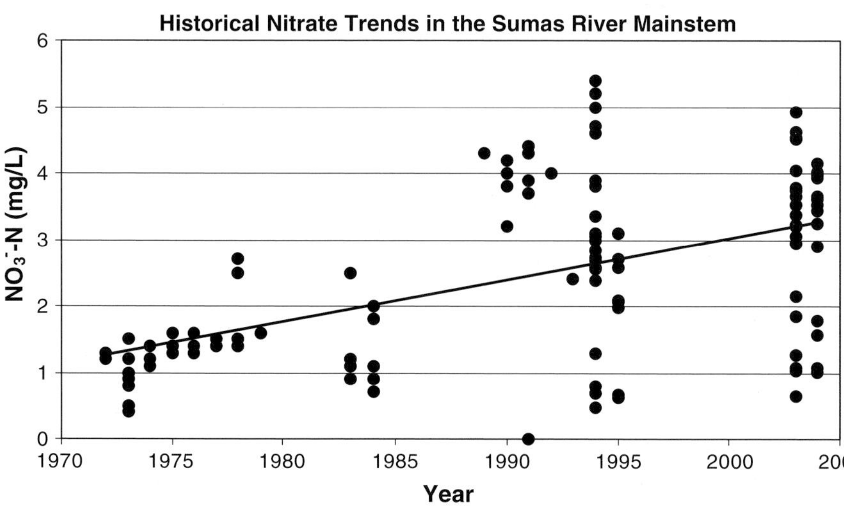

Figure 5. Trends in nitrate in the Sumas River, British Columbia, showing the effects of increasing agricultural intensity between 1972–2004.

In most regions of Canada, regulations for discharge of animal wastes are rudimentary. Most wastes are simply spread or sprayed on the land. Hog operations usually discharge to surface lagoons, which have high ammonia losses to the atmosphere. Timing and rates of manure application are not well-regulated, with application on frozen ground being common. Riparian corridors and wetlands are filled and destroyed in many regions. On average, 70% of wetlands in southern Canada have been destroyed, with little regard for their role as important nitrogen sinks or sites of denitrification. As a result, nitrogen has increased greatly in many surface waters. Phosphorus has also increased, and as most waters in the western prairies are phosphorus rich to begin with, one result is rapidly increasing eutrophication.

Eutrophication of freshwaters and marine ecosystems

Most Canadian freshwaters are phosphorus limited, and in general phosphorus released from human activity has increased even more rapidly than nitrogen (Chambers et al. 2001; Schindler 2001). In Lake Winnipeg, concentrations of total nitrogen have more than doubled in less than 10 years, from about 350 to over 800 $\mu g\ l^{-1}$. Phosphorus and chlorophyll have also increased, from 30 to 55 $\mu g\ l^{-1}$ and 5 to over 20 $\mu g\ l^{-1}$, respectively. The current state of the lake with respect to nutrient concentrations and algal blooms is roughly the same as Lake Erie was at the height of the eutrophication problem in the early 1970s (M.P. Stainton, Freshwater Institute, pers. comm.).

Increasing trends for phosphorus, nitrogen and algal populations are commonly identified in paleoecological studies of western lakes (Blais et al. 2000; Schindler unpublished data). In general, phosphorus has increased more rapidly than nitrogen, resulting in the proliferation of nitrogen-fixing Cyanobacteria. In the west, increasing temperatures (1–3° C in the western prairies), and drought conditions (which have caused huge declines in water flows in rivers and water renewal times in lakes) have exacerbated problems with water quantity and quality (Schindler and W. Donahue, unpublished data).

Land-use change has been responsible for some of the nutrient increases in lakes in Canada. The prairies have been largely converted to agricultural land, so that new lands for agriculture are taken from areas that were originally forested. Studies in Ontario (Dillon and Kirchner 1975) and Alberta (Neufeld and Schindler unpublished) indicate that simply clearing a small parts of forested catchments for agriculture can cause yields of nutrients to increase by at least two-fold. Once cleared, pastures or fertilized croplands usually follow, causing additional increases in nutrient runoff. Many of the lakes in western Canada have undergone moderate to severe eutrophication as a result of these changes. In Alberta, with its rapid population growth, rapid lakeshore development has contributed to eutrophication in many of the same lakes.

So far, there is little sign of coastal eutrophication in Canada even though large coastal cities like Halifax, Victoria, and part of Vancouver discharge sewage directly into coastal waters. Perhaps the most threatened area in the country is the Georgia Basin, which receives the discharge of the Fraser River as mentioned above, but is surrounded by 4.5 million people and their enterprises, if the combined populations of the Vancouver Area, Victoria on Vancouver Island, and the American city of Seattle are included. The population of the area is increasing rapidly, with the 2010 population predicted to be 28% higher than at present. Eutrophication is presently confined to small protected bays, because of high exchange rates with the main Pacific Ocean via the Strait of Juan de Fuca (Chambers et al. 2001).

On the east coast of Canada, Halifax Harbor has received the effluents from the cities of Halifax, Dartmouth and Bedford for many decades. Raw sewage is discharged, and although the human population is much smaller (343000) than the Georgia Basin, the Bedford Basin and Halifax Harbor form a small, protected marine estuary. Fortunately, very high tides produce a rapid exchange with the main Atlantic Ocean. While it is difficult to dissect the effects of nitrogen from other components of raw sewage, dinoflagellate blooms and occasional summer fish kills have been observed (Chambers et al. 2001).

Summary

While Canada has so far escaped major problems with airborne or waterborne nitrogen, many of the problems experienced by other countries are beginning to emerge as human populations and industry increase and livestock culture intensifies. Potential problems with acid rain, increased nitrate in surface and ground water, and eutrophication, especially of marine estuaries, are problems that are expensive and difficult to control once they have reached critical stages. It would be economically astute to limit nitrogen releases before the problem becomes acute.

Acknowledgements

We are grateful to Dean Jeffries for reviewing an early draft of the manuscript and for providing unpublished data. Bob Vet supplied copies of nitrogen deposition maps for eastern Canada. Michael Stainton shared his recent chemical data for Lake Winnipeg. Research by DWS was supported by a Natural Sciences and Engineering Research Council Discovery Grant # 89673.

References

Aherne J., Posch M., Dillon P.J. and Henriksen A. 2004. Critical loads of acidity for surface waters in south-central Ontario, Canada: regional application of the First-order Acidity Balance (FAB) model. Water Air Soil Pollut. Focus 4: 25–36.

Allen E. 2004. Effects of nitrogen deposition on forests and peatlands: A literature review and discussion of the potential impacts of nitrogen deposition in the Alberta Oils Sands Region. Wood Buffalo Environmental Association, Fort McMurray, AB.

Bayley S.E., Schindler D.W., Parker B.R., Stainton M.P. and Beaty K.G. 1992. Effects of forest fire and drought on acidity of a pase-poor boreal forest stream: similarities between climate warming and acidic precipitation. Biogeochemistry 17: 191–204.

Belzer W., Evans C. and Poon A. 1998. Atmospheric Nitrogen Concentrations in the Lower Fraser Valley. Environment Canada, Aquatic and Atmospheric Science Division, Vancouver, B.C. DOE FRAP 1997: 1–23.

Berka C., Schreier H. and Hall K. 2001. Linking Water Quality with agricultural intensification in a rural watershed. Water Air Soil Pollut. 127(1–4): 389–401.

Berka C. 1996. Relationships between agricultural land use and surface water quality using GIS: Sumas River Watershed, Abbotsford, B.C. M.Sc. Thesis. Resource Management and Environmental Studies, University of British Columbia. June 1996, 174 pp.

Blais J.M., Duff K.E., Schindler D.W., Smol J.P., Leavitt P.R. and Agbeti M. 2000. Recent eutrophication histories in Lac Ste Anne and Lake Isle, Alberta, Canada, inferred using paleolimnological methods. Lake and Reserv. Manage. 16: 292–304.

Chambers P.A., Guy M., Roberts E.S., Charlton M.N., Kent R., Gagnon C., Grove G. and Foster N. 2001. Nutrients and their impact on the Canadian environment. Agriculture and Agri-Food Canada, Environment Canada, Fisheries and Oceans Canada, Health Canada and Natural Resources Canada, Ottawa. 241 p.

Cheng L. 1994. Concentration and Deposition of Anthropogenic Pollutants in Alberta: Summary Report. Alberta Research Council, Edmonton, p. 10.

Desjardins R.L. and Keng J.C. 1999. Nitrous oxide emissions from agricultural sources in Canada. In: Desjardins R.L. Keng J.C. and Haugen-Kozyra K. (eds) Proceedings of the International Workshop on Reducing nitrous Oxide Emissions from Agroecosystems, Banff, Alberta March 3–5 1999. Conservation and Development Branch, Alberta Agriculture, Food and Rural Development, Agriculture and Agri-Food Canada, Research Branch, Edmonton, Alberta.

Devito K.J. and Dillon P.J. 1993a. The importance of runoff and winter anoxia to the P and N dynamics of a beaver pond. Can. J. Fish. Aquat. Sci. 50: 2222–2234.

Devito K.J. and Dillon P.J. 1993b. The influence of hydrologic conditions and peat oxia on the phosphorus and nitrogen dynamics of a conifer swamp. Wat. Resourc. Res. 29: 2675–2685.

Dillon P.J. and Kirchner W.B. 1975. The effects of geology and land use on the export of phosphorus from watersheds. Water Res. 9: 135–148.

Dillon P.J., Molot L.A. and Futter M. 1997. A note on the effect of El Niño-related drought on the recovery of acidified lakes. Int. J. Environ. Mon. Assess. 46: 105–111.

Eimers M.C. and Dillon P.J. 2002. Climate effects on sulphate flux from forested catchments in south-central Ontario. Biogeochemistry 61: 337–355.

Environment Canada 2003. Submission to the Alberta Energy and Utilities Board. Canadian Natural Resources Ltd. Horizon Oil Sands Mine Application 1273113. Sept. 2003.

Goss M.J., Barry D.A.J. and Rudolph D.L. 1998. Contamination in Ontario farmstead domestic wells and its association with agriculture: 1. Results from drinking water wells. J. Contam. Hydrol. 32: 267–293.

Grennfelt P. and Hultberg H. 1986. Effects of nitrogen deposition on the acidification of terrestrial and aquatic ecosystems. Water Air Soil Pollut. 30: 945–963.

Hazlett P.W. and Foster N.W. 2002. Topographic controls of nitrogen, sulfur and carbon transport from a tolerant hardwood hillslope. Water Air Soil Pollut. Focus 2: 63–80.

Henriksen A., Dillon P.J. and Aherne J. 2002. Critical loads of acidity for surface waters in south-central Ontario, Canada: regional application of the Steady-State Water Chemistry (SSWC) model. Can. J. Fish. Aquat. Sci. 59: 1287–1295.

Jeffries D.S. 1995. Freshwater acidification in Canada caused by atmospheric deposition of nitrogen pollutants: A preliminary assessment of existing information. A draft discussion paper. National Water Research Institute, Burlington, Ontario contribution 95–116, 68 pp.

Jeffries D.S., Clair T.A., Couture S., Dillon P.J., Dupont J., Keller W., McNicol D.K., Turner M.A., Vet R. and Weber R. 2003. Assessing the recovery of lakes in southeastern Canada from the effects of acid deposition. Ambio 32: 176–182.

Kaste O. and Dillon P.J. 2003. Inorganic nitrogen retention in acid-sensitive lakes in southern Norway and southern Ontario, Canada-a comparison of mass balance data with an empirical N retention model. Hydrol. Processes 17: 2393–2407.

Kelly C.A., Rudd J.W.M. and Schindler D.W. 1990. Acidification by nitric acid – future considerations. Water Air Soil Pollut. 50: 49–61.

Kochy M. and Wilson S.D. 2001. Nitrogen deposition and forest expansion in the northern Great Plains. J. Ecol. 89: 807–817.

Korol M. and Rattray G. 1999. Canadian Fertilizer Consumption, Shipments and Trade, 1997–1998. Agri-Foods and Agriculture Canada, Ottawa.

Laudon H., Dillon P.J., Eimers M.C., Semkin R.G. and Jeffries D.S. 2004. Climate induced episodic acidification of streams in central Ontario. Environ. Sci. Technol. 38: 6009–6015.

Lazerte B.D. 1993. The impact of drought and acidification on the chemical exports from a minerotrophic conifer swamp. Biogeochemistry 18: 153–175.

Likens G.E., Driscoll C.T. and Buso D.C. 1996. Long-term effects of acid rain: response and recovery of a forest ecosystem. Science 272: 244–246.

Linsey G.A., Schindler D.W. and Stainton M.P. 1987. Atmospheric deposition of nutrients and major ions at the Experimental Lakes Area in northwestern Ontario. Can. J. Fish. Aquat. Sci. 44(Suppl. 1): 206–214.

Molot L.A. and Dillon P.J. 1993. Nitrogen mass balances and denitrification rates in central Ontario Lakes. Biogeochemistry 20: 195–212.

Nilsson J. and Grennfelt P. 1988. Critical Loads for Sulfur and Nitrogen. Nordic Council of Ministers, Copenhagen, Denmark, p. 31.

Oimet R., Duschesne D., Houle D. and Arp P.J. 2001. Critical load and exceedances of acid deposition and associated forest growth in the northern hardwood and boreal coniferous forests in Quebec, Canada. Water Air Soil Pollut. Focus 1: 119–134.

Posch M., Kamari J., Forsius M., Henriksen A. and Wilander A. 1997. Exceedance of critical loads for lakes in Finland, Norway and Sweden: Reduction requirements for acidifying nitrogen and sulfur deposition. Environ. Manage. 21: 291–304.

Rudolph D.L., Barry D.A.J. and Goss M.J. 1998. Contamination in Ontario farmstead domestic wells and its association with agriculture: 2. Results from multilevel monitoring well installations. J. Contam. Hydrol. 32: 295–311.

Schiff S.L., Devito K.J., Elgood R.J., McCrindle P.M., Spoelstra J. and Dillon P.J. 2002. Two adjacent catchments: dramatically different NO_3 export. Wat. Resourc. Res. 38: 1292, doi: 10.1029/2000WR000170.

Schindler D.W. 2001. The cumulative effects of climate warming and other human stresses on Canadian freshwaters in the new millennium. Can. J. Fish. Aquat. Sci. 58: 18–29.

Schreier H., Bestbier R. and Derksen G. 2003. Agricultural nutrient management trends in the Lower Fraser Valley, BC (Based on Agricultural Census 1991–2001). (Compact Disc) IRES, University of British Columbia.

Schreier H. and Brown S. 2004. Multiscale approaches to water management: land-use impacts on nutrient and sediment dynamics. In: Tchiguirinskaia I., Bonell M. and Hubert P. (eds), Scales in Hydrology and Water Management. International Association of Hydrological Sciences (IAHS), Wallingord, GB, pp. 61–75 (IAHS publication 287).

Schreier H., Hall K., Brown S.J., Wernick B. and Berka C. 1999. Agriculture: an important non-point source of pollution. In: Gray C. and Tuominen T. (eds), Health of the Fraser River Aquatic Ecosystem, Vol. II. Environment Canada, DOE-FRAP, Vancouver, 83–99.

Smith I. (2004) Cumulative effects of agricultural intensification on nutrients and trace metal pollution in the Sumas River watershed. M.Sc. Thesis, Resources Management and Environmental Studies, University of British Columbia, Vancouver, 275 pp.

Spoelstra J., Schiff S.L., Elgood R.J., Semkin R.G. and Jeffries D.S. 2001. Tracing the sources of exported nitrate in the Turkey Lakes Watershed using $^{15}N/^{14}N$ and $^{18}O/^{16}O$ isotopic ratios. Ecosystems 4: 536–544.

StatsCan 2001. Agricultural Census Data. Statistics Canada, Ottawa, On.

Vet R., Brook J., Ro C., Shaw Narayan J., Zhang L., Moran M. and Lusis M. 2005. Atmospheric response to past emission control programs. In: Canadian Acid Rain Science Assessment 2004, Chapter 3. Environment Canada, Ottawa Ontario, pp. 15–98.

Watmough S.A. 2002. A dendrochemical survey of sugar maple in south-central Ontario. Water Air Soil Pollut. 136: 165–187.

Watmough S.A. and Dillon P.J. 2003a. Base cation and nitrogen budgets for seven forested catchments in central Ontario 1983–1999. Forest Ecol. Manag. 177: 155–177.

Watmough S.A. and Dillon P.J. 2003b. Calcium losses from a forested catchment in south-central Ontario, Canada. Environ. Sci. Technol. 37: 3085–3089.

Watmough S.A., Eimers M.C., Aherne J. and Dillon P.J. 2004. Climate effects on nitrate export from forested catchments in south central Ontario. Environ. Sci. Technol. 38: 2383–2388.

Whelpdale D.M. and Galloway J.N. 1994. Sulfur and reactive nitrogen oxide fluxes in the North Atlantic atmosphere. Global Biogeochem. Cycles 8: 481–493.

Williams M.W., Baron J., Caine N., Summerfeld R. and Sanford R. 1996. Nitrogen saturation in the Rocky Mountains. Environ. Sci. Technol. 30: 640–646.

Wright R.F., Cosby B.J., Ferrier R.C., Jenkins A., Bulger A.J. and Harriman R. 1994. Changes in acidification of lochs in Galloway, southwestern Scotland, 1979–1988: The MAGIC model used to evaluate the role of afforestation, calculate critical loads, and predict fish status. J. Hydrol. 161: 257–285.

Biogeochemistry (2006) 79: 45–60
DOI 10.1007/s10533-006-9002-1

More is less: agricultural impacts on the N cycle in Argentina

AMY T. AUSTIN, GERVASIO PIÑEIRO and MARINA GONZALEZ-POLO
*Instituto de Investigaciones Fisiológicas y Ecológicas Vinculadas a la Agricultura (IFEVA) and Department of Ecology, Faculty of Agronomy, University of Buenos Aires and CONICET, Avenida San Martín 4453, Buenos Aires, C1417DSE, Argentina; *Author for correspondence (e-mail: austin@ifeva.edu.ar)*

Key words: Agriculture, Argentina, Biological nitrogen fixation (BNF), Human impact, Pampa, South America, Soybean

Abstract. Human impact on nitrogen cycling, in particular the introduction of reactive nitrogen in terrestrial and aquatic ecosystems, can be examined at multiple scales, from the global impact on atmospheric chemistry to the impact of human activities on soil organic matter and fertility at the scale of square meters. Nevertheless, anthropogenic loading of nitrogen cycling in natural and managed ecosystems can be seen most directly at the regional scale, where concentrated human activity results in disruption of the nitrogen balance, with consequences for biogeochemical cycling and their interactions. Differences in land-use and agricultural practices between North and South America, and the importance of economic drivers that determine the fate of new reactive nitrogen demonstrate a contrasting picture of human impact on N cycling when the consequences are considered at the global vs. the regional scale. In particular, in the Pampa region of Argentina, the central agricultural zone of the country, the expansion of soybean cultivation in the last 20 years and the use of synthetic fertilizers have resulted in an influx of reactive nitrogen into these systems, with unexpected consequences for the nitrogen balance. A mass balance of nitrogen for soybean demonstrates that increased nitrogen inputs from biological fixation do not compensate for losses due to seed export, such that most areas under soybean cultivation are currently experiencing a substantive net loss of nitrogen. In addition, other crops that are currently being fertilized still show a net loss of nitrogen also due to the effect of primary exports from these agroecosystems. These simple models demonstrate that socioeconomic factors in large part drive the contrasting effects of anthropogenic impact on nitrogen cycling at global vs. regional scales. The future impact on nitrogen cycling in the Americas requires an integration of both ecological factors and socioeconomic drivers that will ultimately determine human disruption of the nitrogen cycle.

Introduction

The globalization of agriculture in the last century has resulted in marked changes in the movement of materials, goods and services at local, regional and global scales (Vitousek et al. 1997b). At present, estimates range from 20 to 30% of global net primary production (NPP) that is appropriated for human use, the vast majority stemming from agricultural practices (Vitousek et al.

1986; Rojstaczer et al. 2001; Imhoff et al. 2004). These patterns of appropriation are, however, region-specific with estimates for human consumption of net primary production in North America a much larger fraction than for South America (Imhoff et al. 2004). Beyond the biological import of agricultural practices affecting both managed and natural ecosystems, the importance of market and political forces driving changes in land use and manipulation of natural ecosystems for food production highlights the importance of understanding the socioeconomic framework of agriculture as a global change.

In particular, the increase in the use of nitrogen, both from synthetic fertilizers and by cropping of leguminous species has been estimated to have doubled the circulation of reactive nitrogen (e.g. Galloway and Cowling 2002; Galloway et al. 2004), with a consequent disruption of the N cycle and interaction with other biogeochemical cycles (Vitousek et al. 1997a; Austin et al. 2003). At the regional scale, anthropogenic loading of nitrogen from agriculture varies widely. For example, there are a number of differences in agricultural practices between North and South America, which largely stem from the use of synthetic fertilizer and the intensity of land use conversion in North America. Capital-intensive agriculture, which relies on large inputs of energy, machinery and synthetic fertilizers result in large outputs of inorganic nutrients to adjacent aquatic and estuarine systems (Howarth et al. 1996, 2002; Berman et al. 2005). However, the vast extent of land in the tropical zones of South America often have experienced a different land-use history, which includes slash-and-burn and conversion of forest ecosystems to transient agriculture, and low-input agriculture (Viglizzo et al. 1997a, b). As such, examining these impacts at a global scale may mask important contrasts with the effects of anthropogenic loading of nitrogen at regional or even local scales.

There is a general recognition that increased yields in the last 50 years have resulted in large part from the intensification of cereal crops, rather than the extensification of agriculture to marginal lands (Cassman 1999). It is estimated that agricultural intenstification will accelerate in the Latin American region in the next 50 years (Tilman et al. 2002). It has yet to be seen what the consequences will be for nitrogen loading in particular, as Latin American countries with substantial agricultural production such as Argentina and Brazil increase the use of fertilizers in an effort to meet food demand and maintain a competitive presence in the world agricultural market.

Current and future agricultural practices in Latin America are and will be critical in determining the human impact on the N cycle, due to the importance of the agricultural sector in the economies of most of the region (Martinelli et al. 2006, this volume). At the same time, differences both in climatic and edaphic properties as well as socioeconomic policies make generalization difficult in terms of predicting the consequences of human activity in this region. The objective of this review is to examine aspects of the impact of agriculture on nitrogen cycling in Argentina, principally the impact of soybean cultivation and synthetic fertilizer use, in an effort to understand the important vectors of human-induced global change on the N cycle at the regional scale and in a global context.

Agriculture in Argentina: the breadbasket of South America

Argentina is currently considered one of the major agricultural regions of the world, with most activity centered in the Pampa region of converted natural grasslands of Buenos Aires, Córdoba and Santa Fe provinces (Hall et al. 1992). This extensive region of 52 million hectares ranges in precipitation from more humid systems up to 1100 mm precipitation in the east to semiarid systems of 600 mm rainfall. Cereal grain agriculture began in the 1870s with rapid expansion until 1937, after which agriculture activity diminished. Mechanized agriculture grew in importance starting in the 1970s, but lagged behind other countries due to the access to technology and large-scale international capital (Viglizzo et al. 1997a, b). While land-use changes in the Pampa region are more recent, comparative analysis between North and South America suggest that the impact of agriculture has similar effects on carbon uptake and radiation capture in both continents (Guerschman and Paruelo 2005).

In the main agricultural region of Argentina, the Pampa, the major crops are in descending order of cultivated area, soybean (*Glycine max* (L.) Merr.), wheat (*Triticum aestivum* L.), maize (*Zea mays* L.) and sunflower (*Helianthus annuus* L.) (Figure 1, FAO 2004; SAGPyA 2005). While soybean is the crop with the most dedicated area in the region, there is very little nitrogen fertilizer used in this crop due to the large contribution of biological nitrogen

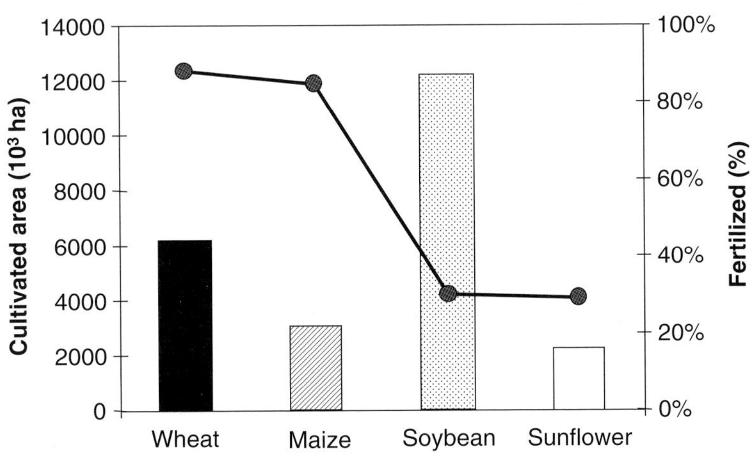

Figure 1. Cultivated area of major cereal crops and percentage of fertilizer use in the Pampa region of Argentina for 2002/2003. The bars represent the area cultivated for each crop; the line indicates that of the cropped area, what percentage is fertilized. Data from FAO (2004), Oliverio et al. (2004), SAGPyA (2005).

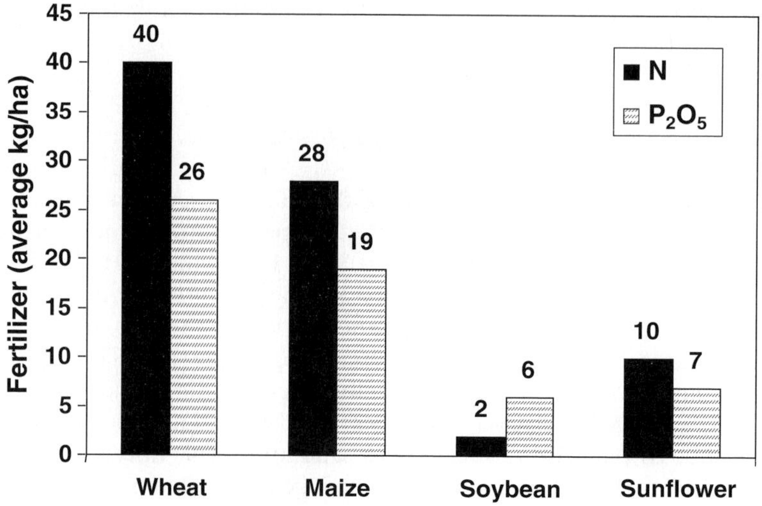

Figure 2. Average fertilizer use of nitrogen and phosphorus in the major cereal crops in the Pampa region of Argentina. Black bars represent synthetic N fertilizer (primarily urea) while striped bars represent phosphorus-based fertilizers. Data from (FAO 2004; Oliverio et al. 2004).

fixation (BNF) associated with its symbiosis with the bacteria *Rhizobium*. The minor application of nitrogen stems from the combined fertilizer use with diammonium phosphate. At the same time, synthetic fertilizer, almost exclusively in the form of urea, is now applied widely to both wheat and maize crops, and a large fraction of the area receives some fertilizer during the crop rotation (Figures 1 and 2). It is important to note that in almost all cases, the amount of nitrogen applied as fertilizer or gained through crop BNF is not sufficient to compensate for losses associated with current agricultural practices in the region (Viglizzo et al. 2001; Díaz-Zorita et al. 2002). We will address these two agriculture practices in the following sections, in an effort to examine the impact on nitrogen cycling and balance in the Pampa region.

More N from legumes? The case of soybean cultivation in Argentina

The expansion of cultivation of leguminous crops at the global scale is currently adding approximately 40 Tg of nitrogen annually, and is one of the principal pathways of reactive nitrogen entering terrestrial ecosystems (Vitousek et al. 1997a). This anthropogenic loading is especially relevant in Argentina, where both the expansion of soybean cropping in marginal land and the intensification of soybean cropping in currently cultivated land have increased markedly in the last 20 years (Figure 3).

In theory, legume-based cropping (or intercropping with other crops such as maize) can reduce carbon and nitrogen losses from cultivated land. The

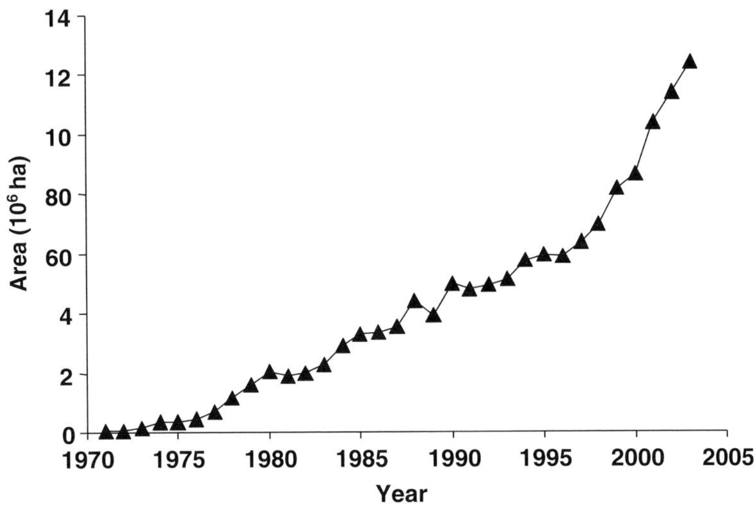

Figure 3. Area of soybean cultivation in Argentina, 1970–2004. Data from (SAGPyA 2005).

incorporation of low carbon:nitrogen litter in soil organic matter from leguminous crop residues, and timing of cropping has been shown to increase carbon and nitrogen retention in temperate agroecosystems (Drinkwater et al. 1998). In the region, areas dedicated to grazing by domestic livestock are frequently seeded with leguminous species such as alfalfa, which increase soil fertility and nitrogen retention (García-Perchác et al. 2004). In the case of crop species, however, the observed increase in soil C and N are compared to losses resulting from traditional cropping methods, generally single crop cultivation. The nitrogen balance from a cropping system must incorporate the various inputs and perhaps more importantly the outputs from the system in question. Our question focused on when natural ecosystems are converted to cropped systems, albeit with leguminous species, what is the effect on the net balance of nitrogen?

Soybean cultivation in Argentina has expanded markedly in the last 30 years in Argentina, with current estimates for 2004 of over more than 14,500,000 ha under cultivation (Figure 3), and the leading crop in the Pampa region (Figure 1). The introduction of glyphosate-tolerant soybean in Argentina, which was genetically modified for herbicide resistance, transformed the way in which this crop was cultivated, and over a very short time period. In 1996, glyphosate-tolerant soybean occupied less than 1% of the planted crop area; by 2002, over 90% of cultivated soybean was of the genetically modified strain, a rate of adoption that exceeded the United States (Trigo and Cap 2003). This was due, in part, to the release of patents on both the soybean seed source and the glysophosphate herbicide that resulted in rapid development of local varieties of glyphosate-tolerant soybean and relatively inexpensive access to the herbicide. A part of the technological package of planting of glyphosate-tolerant soybean was no-till cultivation, which increased in parallel with the expansion of soybean cultivation. As of 2002, over 90% of the area of soybean cultivation in the region is under no-till agricultural practices, with very low fertilizer inputs of both nitrogen and phosphorus (Figure 2).

We calculated a simple mass balance for nitrogen of a soybean crop based on the nitrogen content of the different plant compartments and biological nitrogen fixation, along a gradient of increasing grain yield from 2000 to 6000 kg/ha. We considered novel inputs for nitrogen only from biological fixation from the crop, while losses due to leaching, volatilization or topsoil erosion were not included in the calculations. Aboveground standing biomass was estimated based on grain yield (GY) and harvest index (HI–the ratio of grain biomass to aboveground biomass) from 53 studies from Argentina using local varieties of soybean (Dardanelli et al. 1991; Andrade 1995; Scheiner et al. 1997; Weilenmann de Tau and Lúquez 2000; Sadras and Calvino 2001; Di Ciocco et al. 2004). Our analysis showed that HI did not vary with grain yield ($HI = 1\times10^{-6}*GY + 0.3801$, $r^2 = 0.0012$, n.s.), which has also been shown in studies of soybean varieties from temperate North America (Schapaugh and Wilcox 1980). We assigned a mean value of 0.387 across the yield gradient based on the local studies. Soybean litter decomposed previously to the harvest, was set at 30% of aboveground standing biomass at harvest an upper bound for return of aboveground biomass to the soil. Data for root biomass was almost inexistent, but was estimated based on published studies of root to shoot ratio (root/aboveground biomass including seed), and we used the average value of 0.20 from these studies (Allmaras et al. 1975; Hudak and Patterson 1995; Scheiner et al. 1997).

Nitrogen concentration in soybean grain, aboveground biomass, litter and roots was estimated based on reported values of N content in each plant compartment from local studies and when data was not available, from other temperate ecosystems (Álvarez et al. 1995; Peoples 1995; Di Ciocco et al. 2004). Finally, the percentage of biological nitrogen fixation (BNF) was estimated for the crop as a whole, as well as the allocation to different plant compartments. We used a range of values from 20 to 50% of nitrogen in seed derived from BNF, based on ^{15}N isotpe pool dilution experiments from the region (Álvarez et al. 1995; Di Ciocco et al. 2004). We found no local studies that reported the fraction of BNF allocated to litter and roots, and as such, we estimated BNF in these compartments using simulated values of a widely used crop model in the region, CropGro.

With increasing crop yield, our model showed increasing amount of nitrogen derived from BNF, and increased return of nitrogen to the soil (Figure 4). Interestingly, a larger fraction of BNF was allocated to seed mass than other plant compartments (Álvarez et al. 1995; Di Ciocco et al. 2004) such that a larger fraction of the fixed N was destined for export. At the same time, increasing yield and N exports in grain estimated from the above data varied from 104 to 313 kg/ha (Figure 4). Across all yields, there was a net loss of nitrogen due to seed exports, with N deficits ranging from −42 to −126 kg/ha

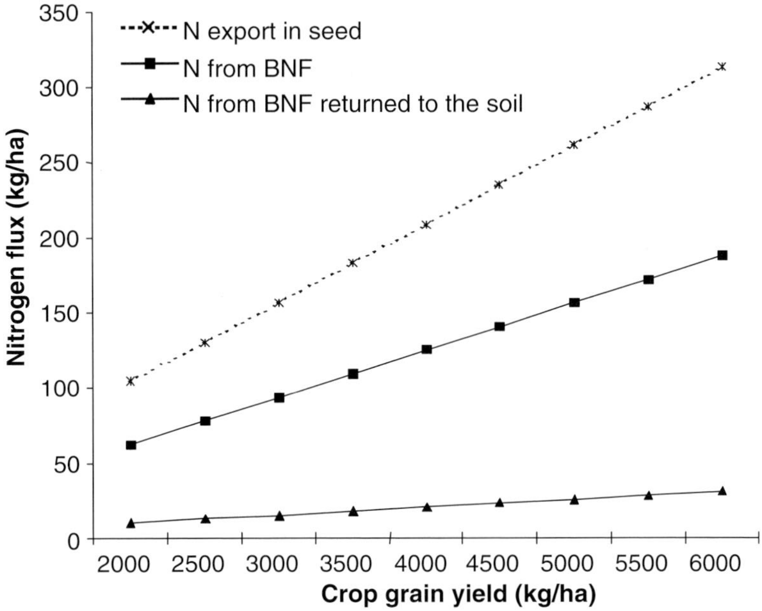

Figure 4. Mass balance of nitrogen for soybean cultivation in Argentina in a range of yields (2000–6000 kg/ha). Nitrogen inputs from fixation, N returned to soil and N exported in grain for different soybean yields (see text for details of model calculations).

(Figure 5), while the percentage of loss across all yields was 23% of the total crop N.

The potential for BNF of soybean varies from 0 to 95% (Unkovich and Pate 2000), and has been shown to vary inversely with soil nitrate concentration at

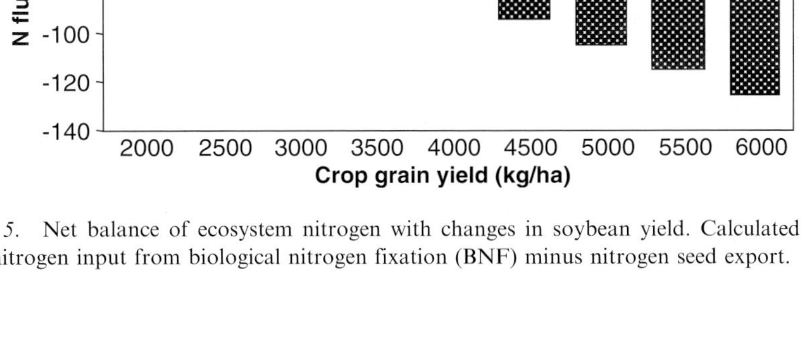

Figure 5. Net balance of ecosystem nitrogen with changes in soybean yield. Calculated deficit from nitrogen input from biological nitrogen fixation (BNF) minus nitrogen seed export.

sowing (Peoples 1995; Peoples et al. 1995; Singh et al. 2003). In addition, the BNF from soybean is reduced with moderate doses of fertilizer (Gan et al. 2002). Furthermore, studies from tropical or infertile soils typically show a much higher percentage of nitrogen in the soybean crop coming from BNF, which can be as high as 80% of biomass nitrogen (Maskey et al. 2001; Alves et al. 2003). The paradox of human impact of soybean cultivation on the nitrogen cycle in Argentina thus stems from the fact that these highly fertile soils may inhibit the capacity for nitrogen fixation, reducing the possible compensation of BNF for seed export of nitrogen. Application of N fertilizer works against the BNF efficiency, such that inorganic N amendments to the soil do not increase yield due to the inhibition of nodulation. Thus the current practice of single rotation soybean actually accentuates nitrogen losses due to the lack of economic and ecological motivation for adding nitrogenous fertilizer. Extrapolating these simple calculations to the Pampa region, for an average yield of 3000 kg/ha, this model estimates that total nitrogen losses from the region at 756,000 tons for 2002, not including possible losses due to leaching, volatilization, or soil erosion. Thus, in spite of a no-till cultivation and a leguminous crop, the example of soybean in Argentina demonstrates that human impact on the nitrogen cycle can result in a substantial net loss of nitrogen at the regional scale, and that current agricultural practices are essentially ,mining' the nutrient capital of this region.

Why fertilize? The case of synthetic fertilizer use in Argentina

The development of the Haber–Bosch process in 1913 to synthetically produce nitrogen fertilizer revolutionized agriculture (Smil 2001; Galloway and Cowling 2002). A readily available source of nitrogenous fertilizer made it possible to increase yields in many crops, and was particularly important as a part of the technological package of the ,green revolution' during the 1960s and 70s (Matson et al. 1997; Tilman et al. 2002). Globally, nitrogen fertilizer use has increased eight-fold in the last 50 years, and is expected to more than double by 2050 (Tilman et al. 2001). The increased use of nitrogen fertilizer globally has doubled food production in the last 35 years but coupled with increased yields have been adverse effects on adjacent downstream and natural ecosystems, including eutrophication of estuarine zones (Rabalais et al. 1996; Howarth et al. 2002). At the same time, there is great deal of regional variation within Latin America in the intensity of nitrogen fertilizer use, with Argentina being on the lower end of the range of fertilizer application (Figure 6a). Because of the highly fertile Pampa soils, however, Argentina has a fertilizer efficiency (kg grain/kg fertilizer applied) that is one of the highest in the world (Figure 6b).

Outside of North America and Europe, ecological consequences of nitrogen use in agriculture have been less evident. Nevertheless, intensive agricultural practices in the Yaqui Valley of México, have resulted in extensive phytoplankton blooms in nitrogen-limited areas of the Pacific Ocean very recently

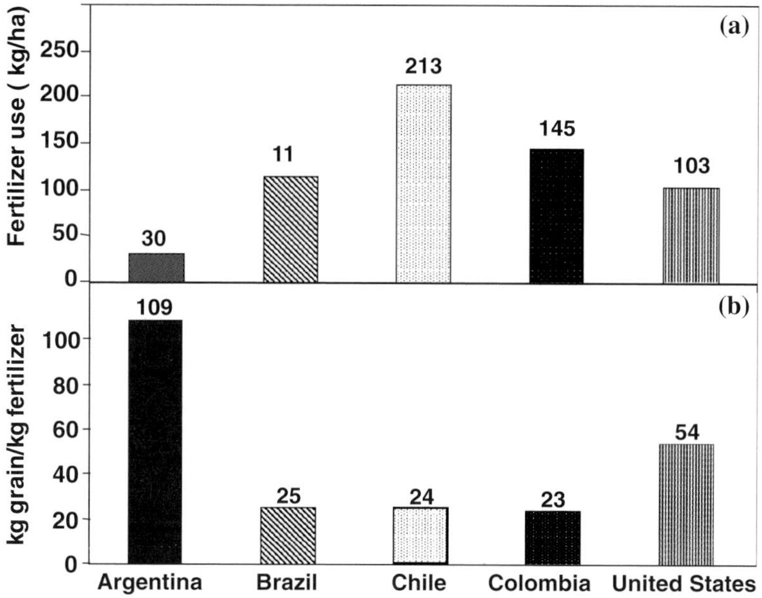

Figure 6. Fertilizer use and fertilizer use efficiency in Argentina and other selected countries of the Americas. A) Total fertilizer use per hectare of arable land and B) grain produced per kilogram of fertilizer applied (Data from FAO 2002).

(Berman et al. 2005). In general, the use of fertilizers in agriculture in South America and in particular in Argentina, has had a much shorter history than North America and Europe. It was not until the 1970s with the introduction of improved crop varieties that the benefits of nitrogen fertilizer could be seen and hence promoted the practice of fertilizer application. In Argentina, the introduction of a government-subsidized credit plan in 1973 that allowed the purchase of fertilizers against future profits from the harvested crop resulted in the growing use of urea, and a stabilization of the relationship in prices between fertilizers and wheat and corn. Much more important for fertilizer use in agriculture, however, was the election of a government in 1989 which was based on a free market economy and open trade policies (Trigo and Cap 2003).

Two consequences of government policy had a large impact on the increase in the use of fertilizers between 1992 and 2001 (Figure 7). The first was the removal of the export tax, which caused a large increase in the profit margin for agricultural exports and an incentive for higher yields. Second, the ‚stabilization' of the Argentine currency by linking it 1:1 with the dollar made it possible to purchase previously unavailable products such as fertilizer and herbicides. The resulting change in the political administration caused a dramatic increase in fertilizer application until 2001, almost entirely from imported sources (Figure 8). After the severe economic crisis of 2001–2002 and the devaluation of the Argentinean peso, the national production of fertilizers increased drastically, to the point where Argentina now is an exporter of

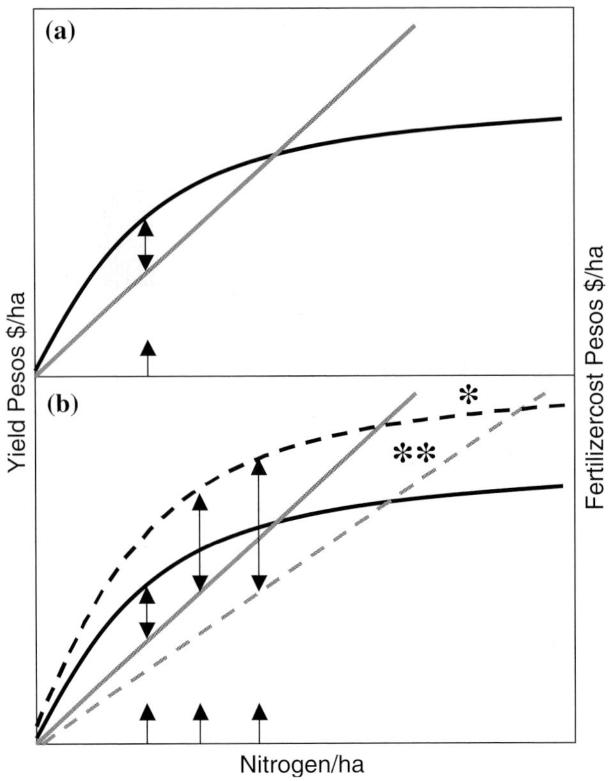

Figure 7. Conceptual diagram of how economic policies in Argentina altered fertilizer use since 1989. A) Simple model of diminishing returns on yield with application of nitrogen fertilizer in a static economic situation. The choice of fertilizer dose, marked with the arrow, is made based on the maximum difference between fertilizer cost and change in yield. B) The removal of the export tax (*) increased the value of the harvested crop, shifting the yield curve upward, and the stability and convertibility of the peso (**) resulted in a decrease in the real cost of fertilizer. The net result is a strong economic incentive to increase fertilizer use as the dose for maximum profit is shifted to the right on the axis of fertilization (marked with arrows).

fertilizers (Figure 8). This trend of increased fertilizer use has been observed in other parts of Latin America, particularly in Brazil, where similar changes in government policy have stimulated economic growth in the agricultural sector (Martinelli et al. 2006, but see Baisre 2006, this volume).

The increased use of nitrogen fertilizer has not compensated for the export of elements in grain from all cropping systems. Analysis of agricultural impact in the last century suggests that low-input agriculture has reduced the nitrogen capital of the Pampa region (Viglizzo et al. 1997a, b), but estimates with current fertilization rates also demonstrate a net loss of N, P, K and S in all major crops (Figure 9, García et al. 2005). In particular, the net balance of nitrogen use in the Pampa region of Argentina for 2002 ranges from 25 to 100 kg/ha deficit of N in this region, due to the export of nitrogen in grain and seed

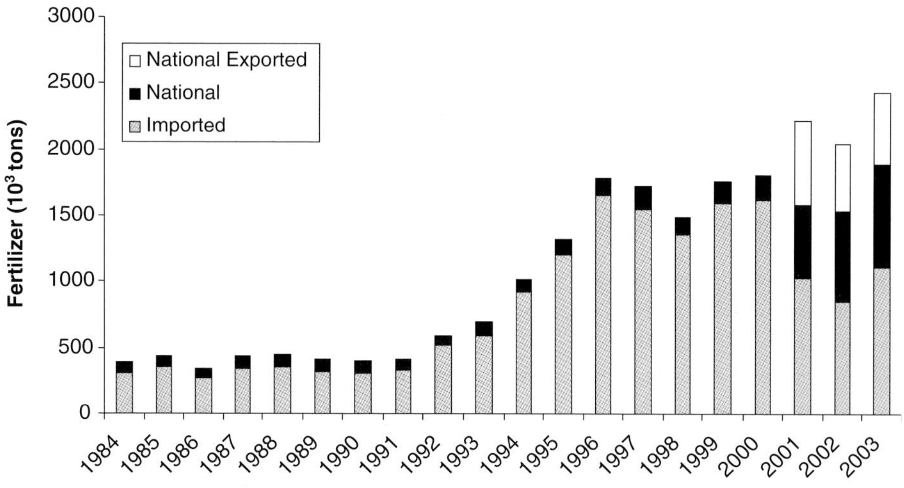

Figure 8. Synthetic fertilizer use in Argentina, 1984–2003, from national and imported sources. Data from (Conde Prat and De Simone 2004 ; Oliverio et al. 2004; SACPyA 2005).

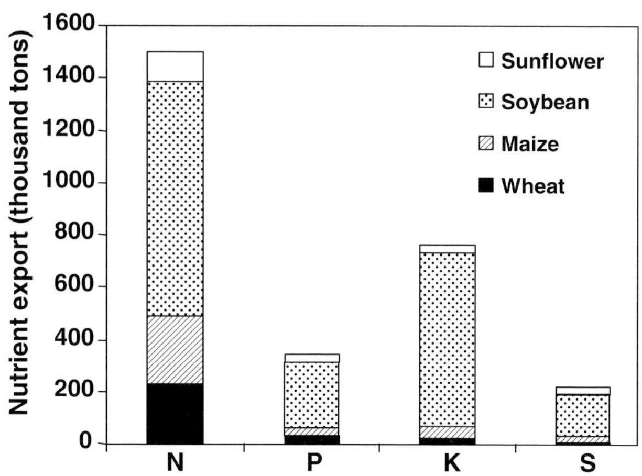

Figure 9. Nutrient losses (N, P, K and S) from the Pampa region of Argentina, based on calculations of fertilizer application and grain export. Source: García et al. (2005) and http://www.inpofos.org.

(Figure 10, García et al. 2005). Again, the relative low input of fertilizer use in this region and the high fertility of the soils results in an export of nutrient capital, and a negative balance for the region, not only for nitrogen but for other elements as well.

It is clear from these two examples of the effect of agricultural practices on nitrogen cycling in Argentina that understanding the present and future human impact on N cycling must include the effects of political and economic policies

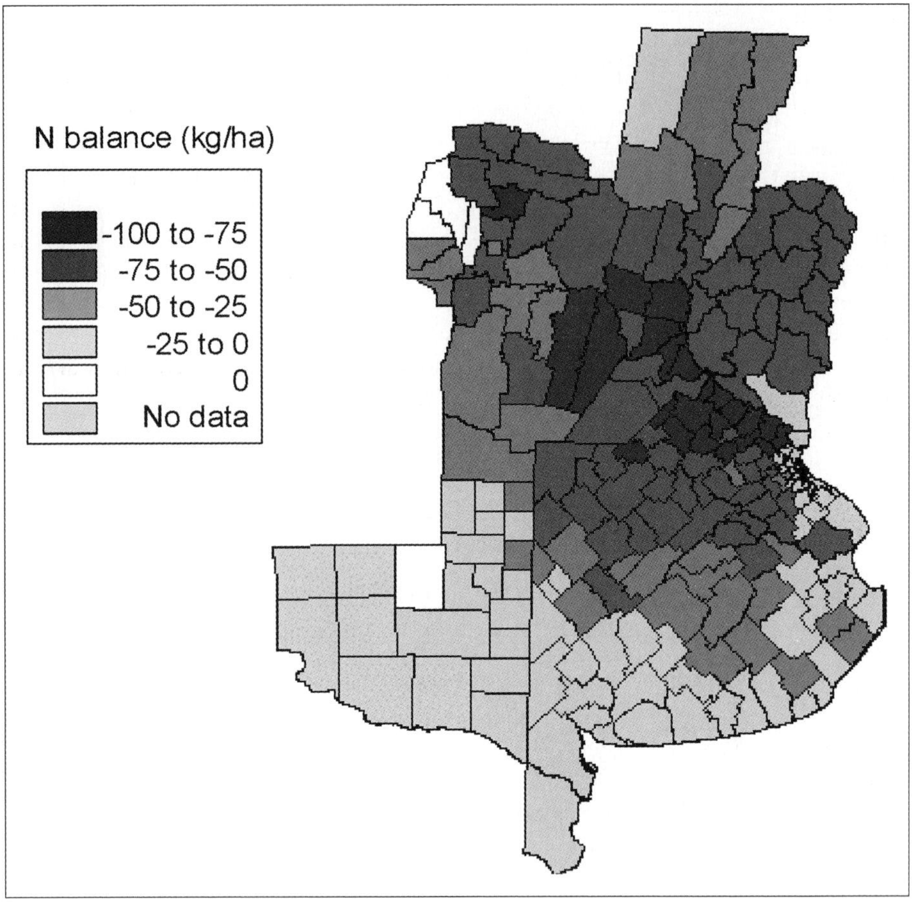

Figure 10. Spatially explicit nitrogen balance in the Argentinean Pampa, 2003 Source: García et al. (2005) and http://www.inpofos.org.

on nitrogen use, cycling and movement across regional boundaries, both in developed and developing economies. In addition, the challenge of integrating global and regional effects of anthropogenic N loading will require region-specific responses due to the dynamic political and economic environment of much of Latin America. Most regions of the northern hemisphere such as Canada and the United States continue to increase the amount of reactive nitrogen in the form of agricultural inputs (Schindler et al. 2006, this volume), while other regions, such as Cuba, have followed a different trajectory of decreasing nitrogen inputs, due to elimination of fertilizer subsidies (Baisre 2006, this volume). Argentina, with its rapid changes in free-market economic policies in the last 20 years coupled with the importance of the agricultural sector serves as an example of a hybrid between the importance of economic and ecological drivers affecting human impact on nitrogen cycling. The central conclusion for Argentina is that in spite of increased cultivation with leguminous crops and a small but dramatic increase in nitrogen fertilizer use, the net effect on N cycling for this region is **negative**, with reductions in N in soil organic matter and net loss of nitrogen from these agroecosystems. These models reinforce other analyses from the region, such that the challenge is to find a crop rotation that combines high BNF efficiency with lower nitrogen losses to begin to compensate for the current negative balance of soybean cultivation (Diaz Zorita and Duarte 2004). These losses, however, are likely to increase in the next 15 years due to growing demand for cereal grain production and the stated goal of 100 million metric tonnes of cereal production in Argentina. The dynamic nature of political change and socioeconomic drivers affecting agricultural practices must be incorporated into our understanding of the consequences of human activity on nitrogen cycling at all scales.

Acknowledgements

This manuscript was a result of collaboration in a series of workshops on the impact of human activity on nitrogen cycling in the Americas, supported by the Inter-American Institute for Global Change Research, CNPq – PROSUL of Brazil, University of Brasilia and Cornell University. We are grateful to Bob Howarth and Luiz Martinelli for leading the effort to bring together a diverse group of people to discuss issues of global change and nitrogen cycling, and to many others including Elvira Cuevas and Mercedes Bustamante for hosting local workshops. Access to maps of nitrogen losses in the Pampa region were kindly provided by INPOFOS ConoSur. We thank Ernesto Medina, Roberto Álvarez, Segundo Urquiaga and an anonymous reviewer for helpful comments on earlier drafts of this manuscript. Additional support came from the Inter-American Institute for Global Change Research (CRN-012) and the Fundación Antorchas of Argentina, and a graduate research fellowship to GP and MGP from CONICET of Argentina.

References

Allmaras R.R., Nelson W.W. and Voorhees W.B. 1975. Soybean and corn rooting in southwestern Minnesota. 2. Root distributions and related water inflow. Soil Sci. Soc. Am. Proc. 38: 771–777.

Álvarez R., Lemcoff J.H. and Merzari A.H. 1995. Balance de nitrógeno en un suelo cultivado con soja. Ciencia del Suelo 13: 38–40.

Alves B.J.R., Boddey R.M. and Urquiaga S. 2003. The success of BNF in soybean in Brazil. Plant Soil 252: 1–9.

Andrade F.H. 1995. Analysis of growth and yield of maize, sunflower and soybean grown at Balcarce, Argentina. Field Crops Res. 41: 1–12.

Austin A.T., Howarth R.W., Baron J.S., Chapin F.S. III, Christensen T.R., Holland E.A., Ivanov M.V., Lien A.Y., Martinelli L.A., Melillo J.M. and Shang C. 2003. Human disruption of element interactions: drivers, consequences and trends for the 21st century. In: Melillo J.M., Field C.B. and Moldan B. (eds.), Interactions of Major Biogeochemical Cycles: Global Change and Human Impacts, Island Press, Washington DC, pp. 15–46.

Baisre J.A. 2006. Assessment of nitrogen flows into the Cuban landscape. Biogeochemistry.

Berman J.M., Arrigo K.R. and Matson P.A. 2005. Agricultural runoff fuels large phytoplankton blooms in vulnerable areas of the ocean. Nature 434: 211–214.

Cassman K.G. 1999. Ecological intensification of cereal production systems: yield potential, soil quality and precision agriculture. Proc. Natl. Acad. Sci. 96: 5952–5959.

Conde Prat M. and De Simone C. 2004. Insumos Agrícolas: Fertilizantes y Terapéuticos. In Foro de Perspectivas Agropecuarias 2004. Edited by S. d. E. A. Área de Análisis Económico de la Dirección de Economía Agropecuaria. Buenos Aires.

Dardanelli J.L., Suero E.E., Andrade F.H. and Andriani J.M. 1991. Water deficits during reproductive growth of soybeans. 2. Water use and water deficiency indicators [water stress, water use efficiency, indicator parameter]. Agronomie 11: 747–756.

Di Ciocco C., Álvarez R., Andrada Y. and Momo F. 2004. Balance de nitrógeno en un cultivo de soja de segunda en La Pampa Ondulada. Ciencia del Suelo 22: 48–51.

Diaz Zorita M. and Duarte G. 2004. Manual práctica para la producción de soja. Hemisferio Sur, Montevideo.

Díaz-Zorita M., Duarte G.A. and Grove J.H. 2002. A review of no-till systems and soil management for sustainable crop production in the subhumid and semiarid Pampas of Argentina. Soil Till. Res. 65: 1–18.

Drinkwater L.E., Wagoner P. and Sarrantonio M. 1998. Legume-based cropping systems have reduced carbon and nitrogen losses. Nature 396: 262–265.

FAO 2002. Food and Agricultural Organization. FAOSTAT database collection., Available at http://www.apps.fao.org.

FAO 2004. Uso de fertilizantes por cultivo en Argentina., 50. Food and Agricultural Organization, Roma.

Galloway J., Dentener F., Capone D., Boyer E., Howarth R., Seitzinger S., Asner G., Cleveland C., Green P., Holland E., Karl D., Michaels A., Porter J., Townsend A. and Vorösmarty C. 2004. Nitrogen cycles: past present and future. Biogeochemistry 70: 153–226.

Galloway J.N. and Cowling E.B. 2002. Reactive nitrogen and the world: 200 years of change. Ambio 31: 64–71.

Gan Y., Stulen I., Posthumus F., van Keulen H. and Pieter J.C. 2002. Effects of N management on growth, N_2 fixation and yield of soybean. Nutr. Cycl. Agroecosyst. 62: 163–174.

García F.O., Oliverio G., Segovia F. and López G. 2005. Fertilizers to sustain production of 100 million metric tons of grain. Better Crops 89: 33–35.

García-Perchác F., Ernst O., Siri-Prieto G. and Terra J.A. 2004. Integrating no-till into crop-pasture rotations in Uruguay. Soil Till. Res. 77: 1–13.

Guerschman J.P. and Paruelo J.M. 2005. Agricultural impacts on ecosystem functioning in temperate areas of North and South America. Global Planet. Change 47: 170–180.

Hall A.J., Rebella C., Ghersa C.M. and Culot J.P. 1992. Field-crop systems of the Pampas. In: Person C.J. (ed.), Field Crop Ecosystems, Ecosystems of the World, Elsevier, Amsterdam.

Howarth R.W., Billen G., Swaney D., Townsend A., Jaworski N., Lajtha K., Downing J.A., Elmgren R., Caraco N., Jordan T., Berendse F., Freney J., Kudeyarov V., Murdoch P. and Zhao-Liang Z. 1996. Regional nitrogen budgets and riverine N and P fluxes for the drainages to the North Atlantic Ocean: natural and human influences. Biogeochemistry 35: 75–139.

Howarth R.W., Walker D. and Sharpley A. 2002. Sources of nitrogen pollution to coastal waters of the United States. Estuaries 25: 656–676.

Hudak C.M. and Patterson R.P. 1995. Vegetative growth analysis of a drought-resistant soybean plan introduction. Crop Sci. 35: 464–471.

Imhoff M.L., Bounoua L., Ricketts T., Loucks C., Harriss R. and Lawrence W.T. 2004. Global patterns in human consumption of net primary production. Nature 429: 870–873.

Martinelli L.A., Howarth R.W., Cuevas E., Filoso S., Austin A.T., Donoso L., Huzsar V., Keeney D., Lara L.L., Llerena C., McIssac G., Medina E., Ortiz-Zayas J., Scavia D., Schindler D.W., Soto D. and Townsend A. 2006. Sources of reactive nitrogen affecting ecosystems in in Latin America and the Caribbean: current trends and future perspectives. Biogeochemistry, in press.

Maskey S.L., Bhattarai S., Peoples M.B. and Herridge D.F. 2001. On-farm measurements of nitrogen fixation by winter and summer legumes in the Hill and Terai regions of Nepal. Field Crops Res. 70: 209–221.

Matson P.A., Parton W.J., Power A.G. and Swift M.J. 1997. Agricultural intensification and ecosystem properties. Science 277: 504–509.

Oliverio G., Segovia F. and López G.M. 2004. Fertilizantes para una Argentina de 100 millones de toneladas., http://www.producirconservando.org.ar/docs/. Buenos Aires: Fundación Producir Conservando.

Peoples M.B. 1995. Enhancing legume N_2 fixation through plant and soil managment. Plant Soil 174: 83–101.

Peoples M.B., Gault R.R., Lean B., Sykes J.D. and Brockwell J. 1995. Nitrogen fixation by soybean in commercial irrigated crops of Central and Southern New South Wales. Soil Biol. Biochem. 27: 553–561.

Rabalais N.N., Turner R.E., Justic D., Dortch Q., Wiseman W.J. and Sen Gupta B.K. 1996. Nutrient changes in the Mississippi River and system responses on the adjacent continental shelf. Estuaries 19: 386–407.

Rojstaczer S., Sterling S.M. and Moore N.J. 2001. Human appropriation of photosynthesis products. Science 294: 2549–2552.

Sadras V.O. and Calvino P.A. 2001. Quantification of grain yield response to soil depth in soybean, maize, sunflower, and wheat. Agron. J. 93: 577–583.

SAGPyA (2005). Secretaría de Agricultura Ganadería Pesca y Alimentos de la Nación Argentina. Estadísticas Agropecuarias., Available at http://www.sagpya.gov.ar.

Schapaugh W.T. Jr. and Wilcox J.R. 1980. Relationships between harvest indices and other plant characteristics in soybeans. Crop Sci. 20: 529–533.

Scheiner J.D., Alvarez Renzi D.F., Lavado R.S. and Torri S.I. 1997. Efecto de la fertilización fosforada y nitrogenada en soja en el centro-oeste bonaerense (Argentina). Ciencia del Suelo 15: 36–38.

Schindler D.W., Dillon P.J. and Schreier H. 2006. Anthropogenic sources of nitrogen in Canada. Biogeochemistry.

Singh A., Carsky R.J., Lucas E.O. and Dashiell K. 2003. Soil N balance as affected by soybean maturity class in the Guinea savanna. Agricult. Ecosyst. Environ. 100: 231–240.

Smil V. 2001. Enriching the Earth: Fritz Haber, Carl Bosch, and the Transformation of World Food Production. MIT Press, Cambridge, Massachusetts.

Tilman D., Cassman K.G., Matson P.A., Naylor R. and Polasky S. 2002. Agricultural sustainability and intensive production practices. Nature 418: 671–677.

Tilman D., Fargione J., Wolff B., D'Antonio C., Dobson A., Howarth R., Schindler D., Schlesinger W.H., Simberloff D. and Swackhamer D. 2001. Forecasting agriculturally driven global environmental change. Science 292: 281–284.

Trigo E.J. and Cap E.J. 2003. The impact of the introduction of transgenic crops in Argentinean agriculture. AgBioForum 6: 87–94.

Unkovich M.J. and Pate J.S. 2000. An appraisal of recent field measurements of symbiotic N_2 fixation by annual legumes. Field Crops Res. 65: 211–228.

Viglizzo E.F., Lértora F., Pordomingo A.J., Bernardos J.N., Roberto Z.E. and Del Valle H. 2001. Ecological lessons and applications from one century of low external-input farming in the pampas of Argentina. Agricult. Ecosyst. Environ. 83: 65–81.

Viglizzo E.F., Roberto Z.E., Lértora F., López Gaya E. and Bernardos J. 1997a. Climate and land-use change in field-crop ecosystems of Argentina. Agricult. Ecosyst. Environ. 66: 61–70.

Viglizzo E.F., Roberto Z.E., Lértora F., López Gaya E. and Bernardos J. 1997b. Climate variability and agroecological change in the Central Pampas of Argentina. Agricult. Ecosyst. Environ. 55: 7–16.

Vitousek P.M., Aber J.D., Howarth R.W., Likens G.E., Matson P.A., Schindler D.W., Schlesinger W.H. and Tilman D. 1997a. Human alteration of the global nitrogen cycle: sources and consequences. Ecol. App. 7: 737–750.

Vitousek P.M., Mooney H.A., Lubchenco J. and Melillo J.M. 1997b. Human domination of Earth's ecosystems. Science 277: 494–499.

Vitousek P.M., Ehrlich P.R., Ehrlich A.H. and Matson P.A. 1986. Human appropriation of the products of photosynthesis. BioScience 36: 368–373.

Weilenmann de Tau M.E. and Lúquez J. 2000. Variations for Biomass, Economic Yield and Harvest Index among Soybean Cultivars of Maturity Groups III and IV in Argentina. Soybean Genetics Newsletter 27, online journal.

Biogeochemistry (2006) 79: 61–89
DOI 10.1007/s10533-006-9003-0

Human activities changing the nitrogen cycle in Brazil

SOLANGE FILOSO[1,*], LUIZ ANTONIO MARTINELLI[2], ROBERT W. HOWARTH[1], ELIZABETH W. BOYER[3] and FRANK DENTENER[4]
[1]*Department of Ecology and Evolutionary Biology, Cornell University, Ithaca, NY 14853 USA;* [2]*Centro de Energia Nuclear na Agricultura, University of São Paulo, Av. Centenário 303, Piracicaba, SP 13416-000 Brazil;* [3]*Department of Environmental Science, Policy and Management, University of California, Berkeley, CA 94720 USA;* [4]*Joint Research Centre, Institute for Environment and Sustainability, Climate Change Unit, TP280, I-21020 Ispra (Va), Italy;* **Author for correspondence (e-mail: sfw6@cornell.edu)*

Key words: Agricultural expansion, Amazon, Brazil, Cerrado, Deforestation, Nitrogen, Nitrogen budget

Abstract. The production of reactive nitrogen worldwide has more than doubled in the last century because of human activities and population growth. Advances in our understanding of the nitrogen cycle and the impacts of anthropogenic activities on regional to global scales is largely hindered by the paucity of information about nitrogen inputs from human activities in fast-developing regions of the world such as the tropics. In this paper, we estimate nitrogen inputs and outputs in Brazil, which is the world's largest tropical country. We determined that the N cycle is increasingly controlled by human activities rather than natural processes. Nitrogen inputs to Brazil from human activities practically doubled from 1995 to 2002, mostly because of nitrogen production through biological fixation in agricultural systems. This is in contrast to industrialized countries of the temperate zone, where fertilizer application and atmospheric deposition are the main sources of anthropogenic nitrogen. In Brazil, the production of soybean crops over an area of less than 20 million ha, was responsible for about 3.2 Tg N or close to one-third of the N inputs from anthropogenic sources in 2002. Moreover, cattle pastures account for almost 70% of the estimated 280×10^6 ha of agricultural land in Brazil and potentially fix significant amounts of N when well managed, further increasing the importance of biological nitrogen fixation in the nitrogen budget. Much of these anthropogenic inputs occur in the Brazilian savannah region (*Cerrado*), while more urbanized regions such as the state of São Paulo also have high rates of nitrogenous fertilizer inputs. In the Amazon, rates of anthropogenic nitrogen inputs are relatively low, but continuing conversion of natural forests into cattle pasture or secondary forests potentially add a significant amount of new nitrogen to Brazil given the vast area of the region. Better measurements of biological fixation rates in Brazil are necessary for improving the nitrogen budgets, especially at a more refined spatial scale.

Introduction

Over the past century, growing human population and increasing human activities related to the production of food and energy have more than doubled the production rate of reactive nitrogen (Nr) on the land surface of the Earth

(Galloway and Cowling 2002; Galloway et al. 2004) and greatly altered the nitrogen cycle globally. Consequently, in many parts of the world the conversion of unreactive N_2 to reactive forms (nitrogen oxides plus other oxidized nitrogen species, NH_3, NH_4, and organic N) became controlled mainly by anthropogenic activities such as fertilizer production, combustion of fossil fuel, and biological fixation in agriculture, instead of being controlled by natural processes such as natural biological nitrogen fixation (BNF) and lightning (Smil 2001; Galloway et al. 2004).

The anthropogenic production of Nr has been especially high in industrialized countries of the temperate zone, where severe eutrophication of estuaries and coastal zones (Howarth et al. 2000; NRC 2000; Rabalais 2002), acidification of lakes and streams (Vitousek and Field 1999), and forest decline (Aber et al. 1995, 2003) have become common environmental problems associated with increasing nitrogen loads to ecosystems. Because the tropics encompass mainly developing countries, the rates of Nr production by anthropogenic activities and inputs in tropical ecosystems have not been an issue commonly addressed in scientific investigations (Matson et al. 1999, 2002). However, important drivers responsible for the increased production of Nr in the temperate zone are increasingly influencing the nitrogen cycle in the tropics and sub-tropics (Matson et al. 1999; Galloway and Cowling 2002), further changing the global cycle. Therefore, advances in our understanding of the nitrogen cycle and the impact of anthropogenic activities at regional to global scales depend on the expansion of scientific studies of fast-developing regions of the world such as the tropics (Galloway et al. 2004).

Vast deforestation, rapid conversion of natural vegetation into agricultural lands accompanied by intensification of agriculture, expansion of nitrogen-fixing crops, increasing rates of fertilizer consumption, population growth and fast urbanization rates are the common drivers altering the nitrogen cycle in the tropics. For instance, deforestation and slash-and-burn practices in the Amazon basin, where forest clearings have reached about 250,000 km^2 between 1990 and 2003 (Laurance et al. 2004), alters the nitrogen cycle by increasing the rates of nitrogen mineralization and mobilization in soils and, consequently, the export of nitrogen in tropical streams (Williams and Melack 1997). After the typical conversion of Amazon forests into pastures, rates of mineralization and nitrification tend to decrease, and reduce the nitrogen availability in soils and delivery rates to water bodies (Neill et al. 1997; Melillo et al. 2001). However, as natural vegetation and aging pastures are increasingly converted to intensive production of export crops such as soybeans, cotton and other lucrative crops, growing rates of both nitrogenous fertilizer consumption and biological nitrogen fixation in agriculture (Boddey et al. 1997) are leading to increased inputs of anthropogenic Nr to the landscape and, eventually, to higher export of nitrogen to surface waters (Downing et al. 1999).

In this paper, we examine the changes in the nitrogen cycle associated with anthropogenic activities in Brazil, the largest tropical country in the world, with 8.5 million km^2 and a wide range of tropical biomes, spanning from humid tropical lowland forests to dry tropical forests, savannas, wetlands and mountain forests. The main large-scale activities changing the nitrogen cycle in Brazil include deforestation of the Amazon, the conversion of vast areas of pastures and natural vegetation to high-intensity agriculture in the central region (*Cerrado*), and high urbanization rates in the southeast region, where urban centers such as Rio de Janeiro and São Paulo have approximately 12 and 18 million people, respectively.

Methods

The effects of human activities on the nitrogen cycle in Brazil are examined using a nitrogen budget approach for large regions (Howarth et al. 1996), and where new net anthropogenic nitrogen inputs (net inputs = inputs − outputs) are quantified and subtracted from outputs and compared with riverine exports (Boyer et al. 2002). We quantify anthropogenic nitrogen inputs for the whole country as well as for some contrasting regions that represent its largest biomes and/or where changes are occurring at the fastest pace. New nitrogen refers to reactive nitrogen (Nr) that is either newly fixed within or transported into a region. New net nitrogen inputs in the budget include NO_y ($NO_y = NO_x$ ($NO + NO_2$) plus other single N species with an oxygen atom), and NH_x ($NH_x = NH_3$ plus aerosol NH_4^+) (Galloway et al. 1995) from atmospheric deposition, nitrogen from fertilizer application, biologically fixed nitrogen in agriculture, and imports of foodstuffs.

In nitrogen budgets constructed for other large countries such as the US (Howarth et al. 2002), it has been assumed that the N status of soils is in steady sate and that the rate of soil-N mineralization equals the rate of nitrogen immobilized on an annual basis, at least on a several year period. In the present study, we also assume that soil N is in steady state. However, because mineralization of nitrogen occurs on such a large scale in Brazil due to deforestation and biomass burning in the Amazon (Williams and Melack 1997) and central region, this assumption needs to be considered with caution because newly mineralized nitrogen from the Amazon forest may function like a new input from anthropogenic activity in the region, especially in aquatic ecosystems.

Data sources

Nitrogen inputs from fixation in agricultural land in Brazil were estimated as the product of the total area of land planted with major crops (i.e., soybeans, common beans, sugarcane) and pasture associated with BNF, and the average fixation rates available in the scientific literature for each one of these agricultural land types, as described below. Land use data for soybeans, pastures, common beans and sugarcane were obtained from the Brazilian Institute of

Geography and Statistics, IBGE (2004). Additional data on land use area of pastures and common beans were obtained from Boddey et al. (2003) and Mostasso et al., (2002), respectively.

Nitrogen fertilizer inputs for Brazil were estimated from several sources, including the Brazilian National Agency for Distribution of Fertilizers (ANDA), the International Potash Institute (POTAFOS-Brasil), and the International Fertilizer Industry Association (IFA). Fertilizer data from the Food and Agriculture Organization of the United Nations (FAO) were excluded from our budgets because their values were consistently low in comparison to those from all the other data sources such as ANDA, POTAFOS, and IFA.

Except for some data on wet deposition for the Central Amazon (Lesack and Melack 1991; Williams and Melack 1997; Filoso et al., 1999) and southeastern Brazil (Lara et al. 2001), N deposition (NH_x and NO_y) in the country is essentially unknown. Therefore deposition of N in Brazil and regions presented here was based on model simulations for global N deposition in the early 1990s as described in Dentener and Crutzen (1994) and Lelieveld and Dentener (2000), and extensively used by Galloway et al. (2004), Rodhe et al. (2002), Seitzinger et al. (2002), and Neff et al. (2002). However, in order to obtain deposition data for Brazil, the model had to be modified to simulate deposition in South America. Some of the uncertainties associated with using the global model for a region of the globe are discussed in Galloway et al. (2004).

Inputs

Inputs from fixation in agriculture

Over the past 40 years, the agricultural area of Brazil has expanded over 100 million ha, at a rate of almost 3 million ha yr^{-1} (Figure 1). Much of this

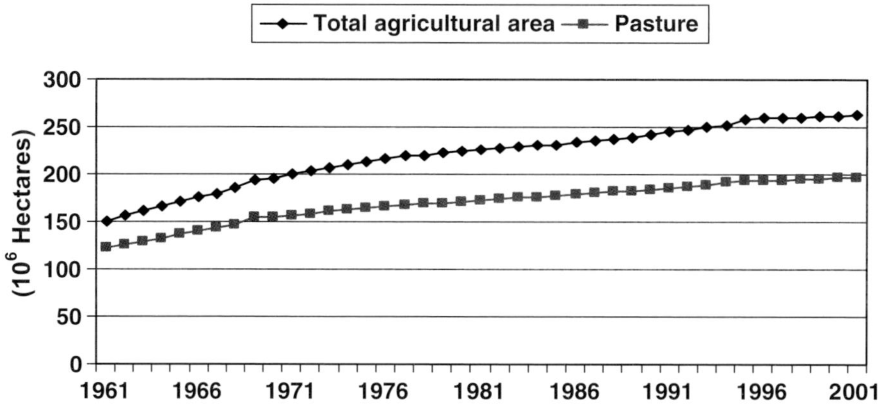

Figure 1. Total agricultural area, and total pasture (natural and introduced) between 1960 and 2002 (FAOSTAT 2004).

expansion occurred due to increasing areas of improved or cultivated pastures for cattle ranching but, in the past decade, the area of annual crops in Brazil has grown at a faster rate than the area of pastures (FAOSTAT 2004). One of the main causes for the rapid expansion of field crops in recent years has been the growth of soybean cultivation in the country, which was partially promoted by advances in the Brazilian soybean-breeding program that lead to the spread of the crop from high to low latitudes and, consequently, to new land entering production in the Brazilian *Cerrado* region (Alves et al. 2003; Machado et al. 2004).

Soybeans need large quantities of nitrogen for plant growth and development, but because Brazilian soybeans are able to obtain between 70 and 85% of the nitrogen required from biological fixation (Alves et al. 2003; Boddey et al. 1991), nitrogen fertilizer is not commonly applied to soybean fields in Brazil. Therefore, the only source of anthropogenic N in Brazil associated with the growing soybean production in the country is biological nitrogen fixation (BNF), as opposed to the U.S. and other countries of the temperate zone, where N fertilizer is often applied in soybean fields.

On average, the productivity of soybeans harvested in Brazil is 2400 kg ha^{-1} yr^{-1} and the BNF rates range between 70 and 250 kg N ha^{-1} yr^{-1} (Alves et al. 2003). Therefore, if we assume an average rate of 170 kg N ha^{-1} yr^{-1}, the cultivation of this crop in Brazil potentially introduced about 2.5 Tg of Nr to the Brazilian and global budgets in 1995, which is equivalent to more than 6% of the amount of nitrogen produced by BNF in agriculture worldwide according to estimates for the mid 1990s by Galloway et al. (2004). In 2004, with 22.9 million ha planted with soybeans in Brazil, the creation of reactive nitrogen via fixation in agriculture potentially reached close to 4 Tg N.

Soybean cultivation is the activity most commonly associated with BNF in agriculture in Brazil. However, two other agricultural activities in the country have been shown to produce significant amounts of reactive nitrogen via BNF; one is the cultivation of sugar-cane (Boddey et al. 1991; Doberheiner 1997) and the other is the cultivation of pasture grasses. BNF in pastures can be especially important for the nitrogen budget in Brazil because of the large extent of this type of land use in the country. According to estimates FAOSTAT (2004), Brazil has about 265 million ha of agricultural land, of which 197 million consists of permanent pastures (Figure 2). Most of these pastures have been formed in the past 30 years, when tropical forests and savannas were replaced with exotic grasses of African origin, especially *Brachiaria* spp. (Boddey et al. 1997). Approximately 80 million ha of *Brachiaria* spp. is now planted in Brazil (Boddey et al. 2004), and several ecotypes of *Brachiaria* spp. and of *Panicum maximum*, another representative type of grass in the region, have been shown to obtain between 40 and 50 kg N ha^{-1} yr^{-1} from plant-associated BNF in field experiments (Boddey and Victoria 1986; Miranda and Boddey 1987), while different species of *Pennisetum* can derive up to 165 kg N ha^{-1} yr^{-1} from BNF. This nitrogen comes mainly from non-symbiotic associations with endophytic bacteria or from bacteria in the rhizophere.

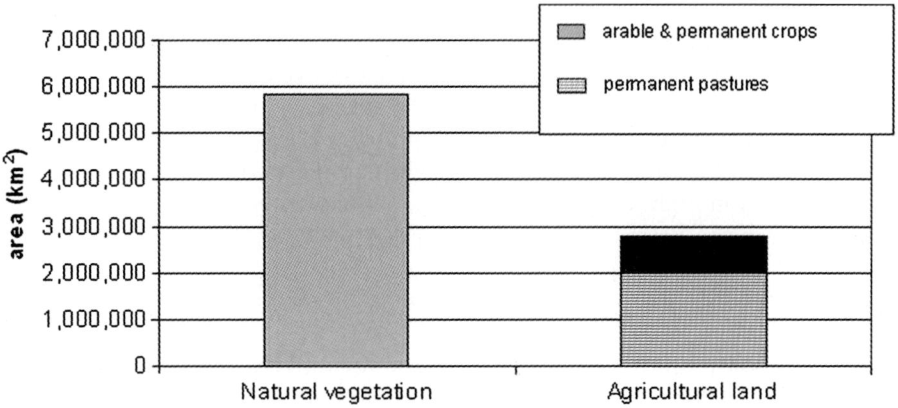

Figure 2. Comparison of natural and agricultural land cover areas in Brazil in 2002.

Fixation rates have not been measured in pastures under grazing, but we assume that they are relatively low (~15–30 kg N ha^{-1} yr^{-1}) even under N limitation conditions, because of limitation of P and other nutrients, as well as because of drought periods during the dry season in regions such as the Brazilian *Cerrado* where much of the pastures are situated (S. Urquiaga, personal communication, EMBRAPA, Brazilian Agency of Agriculture and Pecuary). Another factor that can limit BNF in pastures in Brazil is overgrazing (and consequent degradation), which occurs in about 50% of this land type in the country (Boddey et al. 2004). Therefore, if we assume that BNF is occurring at a conservative rate of 15 kg N ha^{-1} yr^{-1} in 50% of pasture land of Brazil (except for the Amazon where we use a rate of 30 kg N ha^{-1} yr^{-1} as explained below), we can predict that pastures in Brazil potentially create about 3.4 Tg N yr^{-1} in the regional budgets, which is a significant quantity at the global scale (Table 1).

The third most common type of field crop in Brazil associated with BNF is sugar-cane, which accounts for about 5 million ha of the agricultural land of the country. Sugarcane has been cultivated in Brazil since the 16th century under low nitrogen fertilizer inputs, while depletion of soil-nitrogen reserves has not been commonly observed, possibly because of inputs of biologically fixed nitrogen by the sugarcane (Yoneyama et al. 1997; Boddey et al. 2003). Recent field studies with ^{15}N abundance in sugarcane fields planted with commercial varieties have shown that BNF contributes between 0 and 60% of plant N (Boddey et al. 1991, 2003). Sugar-cane in Brazil, commonly, accumulates an average of 110 kg N ha^{-1} (Oliveira et al. 1999), while non-symbiotic N fixation can contribute between 0 and 66 kg N ha^{-1} yr^{-1} (Urquiaga et al. 1992) which is 0 to 0.33 Tg N of the regional and global budgets. While these numbers are relatively low and not significant at a large scale, BNF in sugar-cane can be a very important constituent of nitrogen budgets at the watershed scale where monoculture of sugar-cane is widespread, such as in the Southeast Brazil (Filoso et al. 2003).

Table 1. Nitrogen inputs in Brazil from natural and anthropogenic sources during the pre-colonization, in 1995 and 2002.

Nr Sources	Brazil			Latin America	North America	Global
Natural	1500	1995	2002	1995	1995	1995
Natural BNF	17.1	11.6	10.9*	26.5	11.9	107
Lightning	0.5	0.5	0.5	1.4	0.2	5.4
Total	17.6	12.1	11.4	27.9	12.1	112
Anthropogenic						
Fertilizer use	0	1.2	2.5	3.9	12.7	77.6
BNF in agriculture	0	3.5	7.3	5.0	6.0	31.5
Fossil fuel combustion	0	0.4	0.7	1.3	7.3	24.5
Net imports of foodstuffs	0	−0.1	−1.1	−0.9	−3.0	
Total	0	5.0	9.4	9.3	23.0	134

Values for Latin America, North America and Global are adapted from Galloway et al. 2004. Nitrogen inputs are estimated in Tg yr^{-1}.
*Value for natural BNF in 2002 was estimated using the same data for land cover of natural vegetation in Brazil for the mid 90s minus the amount fixed in the area lost to agricultural lands in the *Cerrado* and Amazon during the period between mid 90s and 2000, according to estimates from INPE (2004) for the Amazon, and from Machado et al. (2004) for the *Cerrado*.

Inputs from fertilizer use

Although much of the expansion of intensive agriculture in Brazil in the last decade has been associated with soybean crops in the *Cerrado* region, other crops such as cotton, maize, sunflower and sorgum have been expanding together with soybeans, especially in the past few years (Machado et al. 2004). The *Cerrado* is composed of highly leached, acidic soils, and have low levels of P and N availability (Bustamante et al. 2004). Therefore, large quantities of lime and fertilizer, including nitrogenous, are required for crops not associated with BNF. Consequently, the agricultural expansion that has occurred in Brazil in the past few years has been accompanied also by the growth of nitrogen fertilizer consumption.

The consumption of nitrogen fertilizer in Brazil increased about 30 times from 1960 to 2002 (Figure 3), from about 0.07 to about 2.5 Tg N yr^{-1} (ANDA 2003; POTAFOS-Brasil 2004). But one of the largest and most rapid increases in the use of nitrogenous fertilizer in the country began in the late 1990s, which also coincides with a period of great agricultural expansion in the *Cerrado* region. According to predictions for the next few decades, the area of intensive agricultural crop production in Brazil, not including the Amazon region (~60% of the Brazilian territory), will expand by about 170 million ha, and much of this is projected to take place in the *Cerrado* region (USDA 2003). For the N budget in Brazil, this expansion in the nutrient-poor soils of the *Cerrado* means a considerable increase of nitrogen fertilizer consumption and, possibly, higher delivery rates of reactive N to aquatic systems.

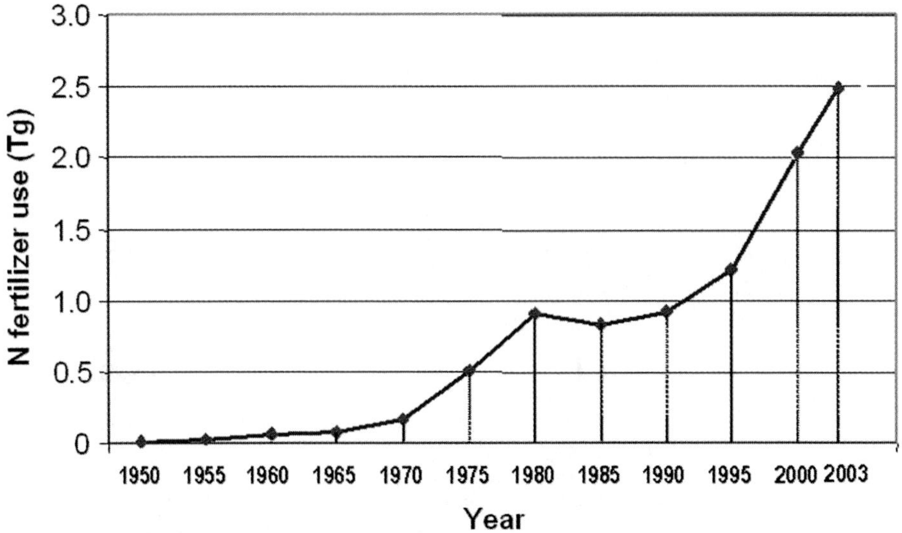

Figure 3. Consumption of nitrogen fertilizer along a period of 40 years in Brazil (Source: SIACES/POTAFOS-Brasil 2003; ANDA 2003).

Inputs from atmospheric deposition

As stated above, deposition of N in Brazil presented here is based on model simulations of global N deposition in the early 1990s described in Dentener and Crutzen (1994) and modified for South America. According to the model simulations for the 1990s, NO_y deposition in most of Brazil varies between 100 and 250 mg m^{-2} yr^{-1} (Figure 4), but along the coast on the North and Northeast region, deposition is lower, varying between 50 and 100 mg m^{-2} yr^{-1}. A few areas with deposition rates up to 500 mg m^{-2} yr^{-1} are clearly associated with highly urbanized and industrialized regions of the country such as São Paulo state, in agreement with data collected in the region in the late 90s (Lara et al. 2001). Overall, assuming an intermediate deposition rate of 175 mg m^{-2} yr^{-1} for the whole country, total deposition of NO_y over Brazil amounts to 1.4 Tg N yr^{-1} (Table 1).

Deposition of NH_x in Brazil is less homogeneous than deposition of NO_y, and a strong gradient can be observed for rates between the southern and northern regions of the country (Figure 5). The highest deposition rates have been estimated for the South and Southeast, where intensive agricultural activities are prevalent. In these regions, deposition of NH_x varied between 500 and 2000 mg m^{-2} yr^{-1} (Figure 5), while in much of the Amazon rates were below 100 mg m^{-2} yr^{-1}. The modeled values fall in the range of the few available measurements of deposition of NH_4 in the Amazon and São Paulo state. We assumed an average deposition rate of 250 mg m^{-2} yr^{-1} for the whole country and estimated that the total deposition of NH_x is on the order of

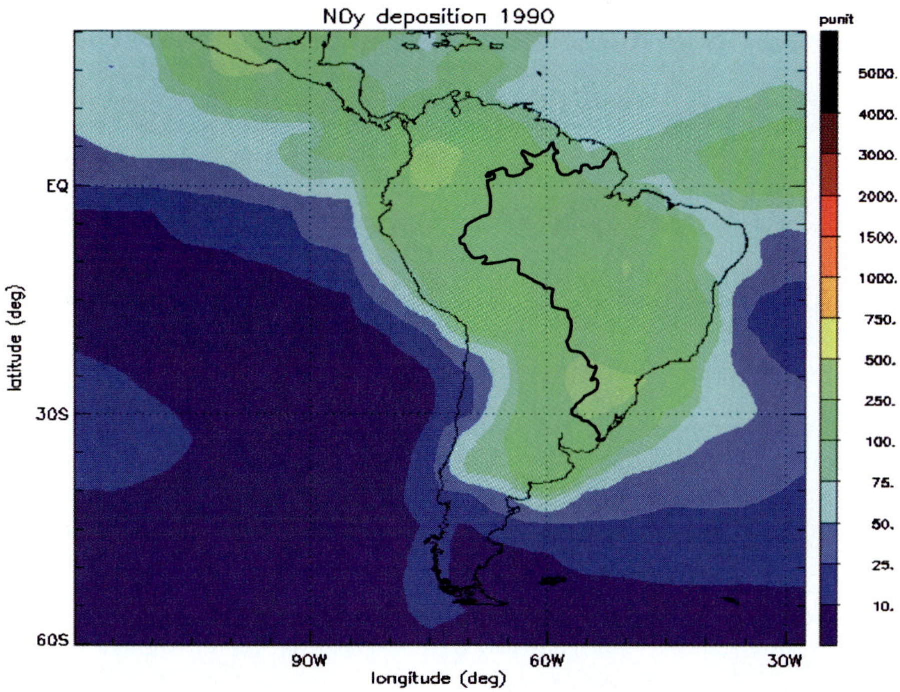

Figure 4. Modeled NO_y deposition in Brazil in the mid 90s. Results were generated by the TM3 global chemistry-transport model of the University of Utrecht on a 5° by 3.75° grid.

2.1 Tg N yr^{-1}, compared to 4.4 Tg N yr^{-1} for South America and 56.7 Tg N yr^{-1} globally (Galloway et al. 2004).

By performing a marked tracer experiment, as described in Lelieveld and Dentener (2000) and Marufu et al. (2000), the origin of these depositions was estimated in relative proportions, and showed that fossil fuel combustion contributed between 20 and 50% of the NO_y emissions along the eastern portion of Brazil, from North to South, while in the central region (*Cerrado*) and the Amazon basin fossil fuel emissions contributed less than 20% of the total NO_x inputs (Figure 6). These estimates are in accordance with population densities in Brazil, which are the highest along the coastal region and the lowest in the Amazon basin. Natural soil emissions contribute the largest percentage of NO_y deposition in the Amazon basin, while biomass burning emissions are becoming increasingly important. In the central region, biomass burning contributes up to 70% of the NO_y deposition probably because of natural and anthropogenic fires that occur mainly in the dry season.

The association of relatively high rates of NO_x emission from fossil fuel with more densely populated areas of the country is partly due to the fact that NO_x emissions in Brazil originate mainly from petroleum products used in on-road vehicles (Figure 7). The consumption of fossil fuel for production of electric

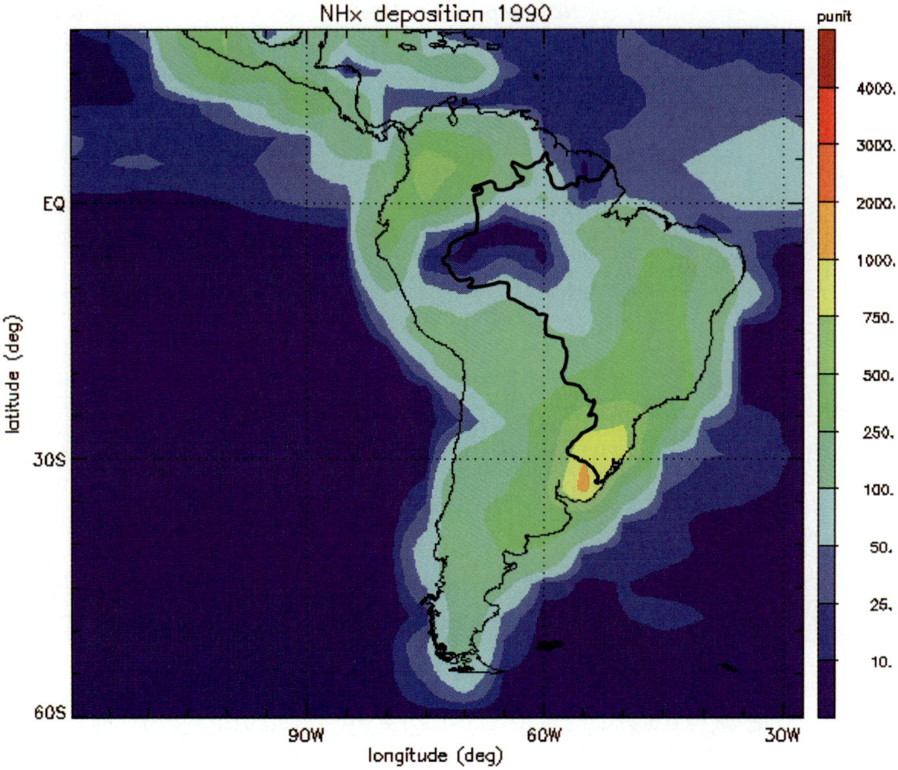

Figure 5. Modeled NH_x deposition in Brazil in the mid 90s. Results were generated by the TM3 global chemistry-transport model of the University of Utrecht on a 5° by 3.75° grid.

energy is negligible. Moreover, emissions from coal burning are insignificant since most of the electrical energy produced in the country comes from hydroelectric power plants.

Anthropogenic sources of NH_x emissions estimated by the model include fertilizer application and animal waste, and together are responsible for 70–90% of the deposition along the eastern portion of the country. In the central and western regions, including the Amazon basin and the central-west, anthropogenic sources contribute, on average, about 30 to 40% of the NH_x deposition (Figure 8a), while emissions from biomass burning are dominant (Figure 8b). Some of the biomass burning is caused by natural fires, especially in the central region of Brazil (Pinto et al. 2002). However, most of the biomass burning is associated with anthropogenic activities such as deforestation and pasture management practices and lead to high amounts of mineralized N in the ecosystems.

Since part of the NH_x deposited in Brazil originates from fertilizer, cattle manure, and biomass burning of agricultural fields such as sugar cane, or even from volatilization of NH_3 from senescing sugar cane plants, it is likely that

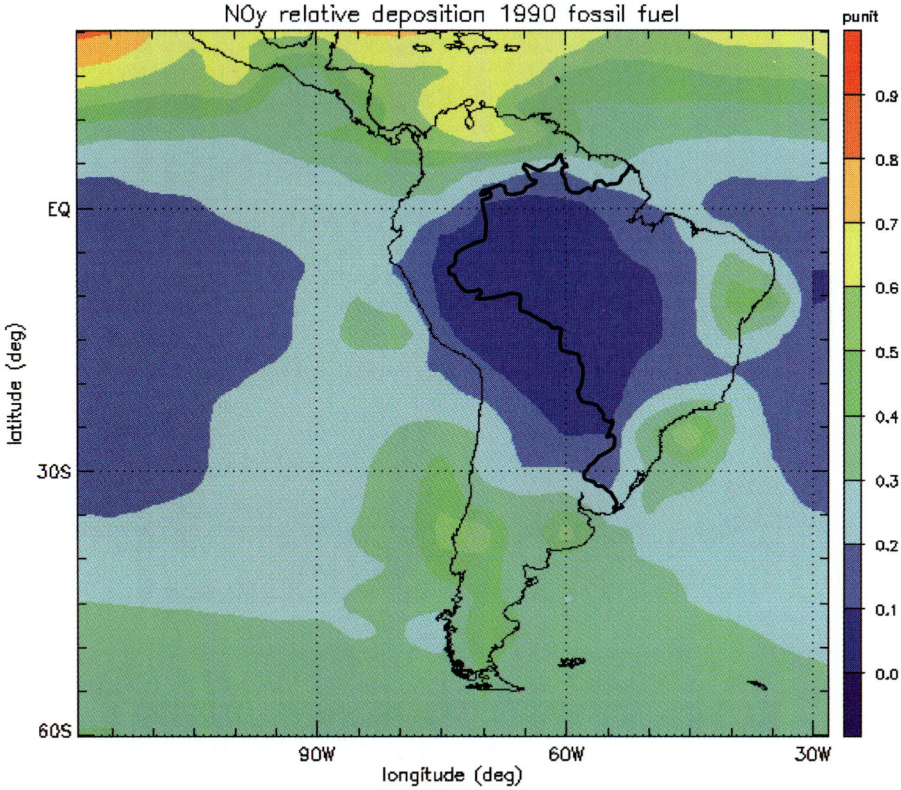

Figure 6. Modeled relative deposition of NO_x in Brazil from fossil fuel emissions in the mid 90s. Results were generated by the TM3 global chemistry-transport model of the University of Utrecht on a 5° by 3.75° grid.

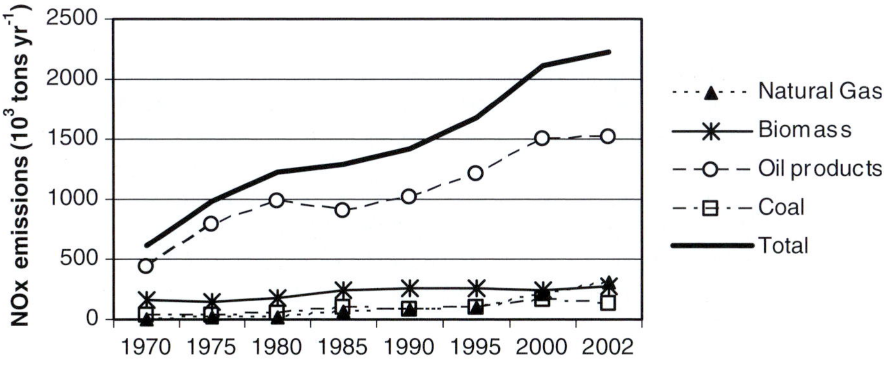

Figure 7. Emissions of NO_x in Brazil from different sources between 1970 and 2002 (Source: Brazilian Ministery of Science and Technology – http://www.mct.gov.br).

(a)

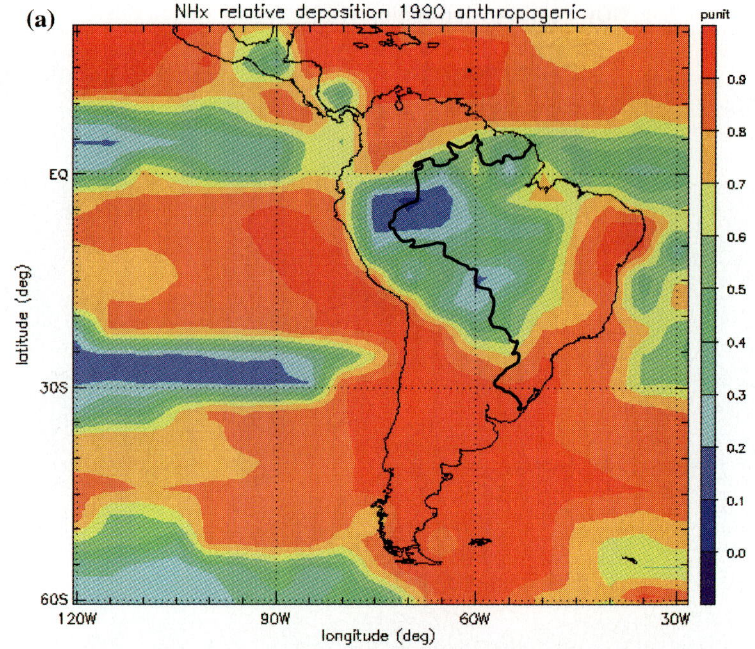
NHx relative deposition 1990 anthropogenic
punit
0.9
0.8
0.7
0.6
0.5
0.4
0.3
0.2
0.1
0.0
EQ
30S
60S
latitude (deg)
120W
90W
60W
30W
longitude (deg)

(b)

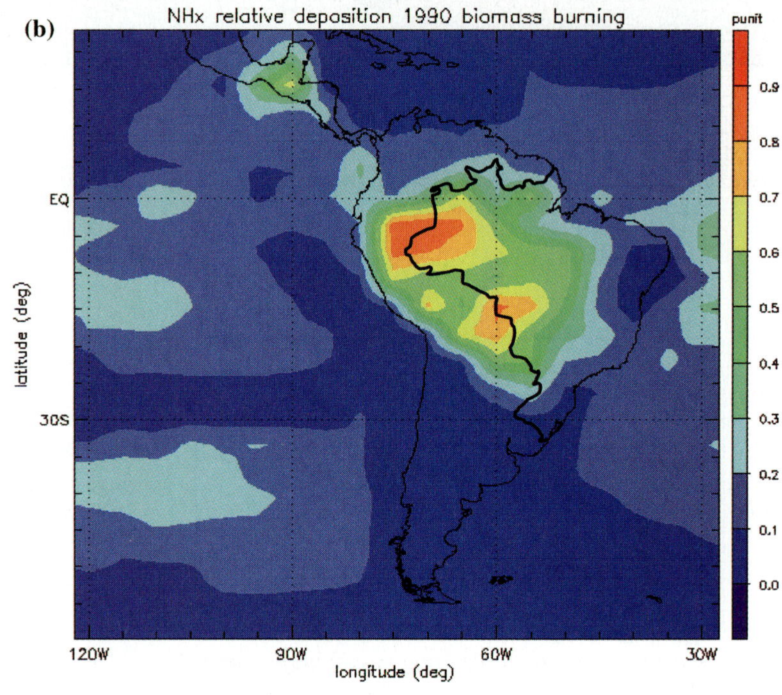
NHx relative deposition 1990 biomass burning
punit
0.9
0.8
0.7
0.6
0.5
0.4
0.3
0.2
0.1
0.0
EQ
30S
latitude (deg)
120W
90W
60W
30W
longitude (deg)

Table 2. Import and export of N in Brazil in foodstuffs for 2002.

	N import Tg yr^{-1}	N export Tg yr^{-1}	Source
Cereals	0.136	0.01	Smil 1999
Starchy roots	0.001	0.0004	Smil 1999
Sugar & sweets	0.000	0.043	Smil 1999
Beans	0.006	0.001	Smil 1999
Treenuts (forage)	0.000	0.000	Smil 1999
Soybeans	0.072	1.099	Alves et al., 2003
Veg/fruits	0.001	0.023	Smil 1999
Coffee	0.013	0.070	Smil 1999
Beef	0.000	0.028	Boddey et al. 2004
Pig meat	0.002	0.017	Smil 1999
Poultry meat	0.000	0.069	Smil 1999
Fish/seafood	0.00	0.00	Smil 1999
Milk	0.002	0.001	http://www.leco.org
Forage	0.005	0.00	Smil 1999
Total	0.24	1.36	

Note: Values of N exported and imported in Brazil through the different food types listed have been indirectly estimated from the sources cited.

some of the N in NH_x deposition may be double counted in the budget. This might also be true for NO_y if some of it comes from the burning of improved pastures which fix N. However, atmospheric deposition contributes small amounts of N to the total budget and, therefore, errors generated from uncertainties associated with N deposition data should be minor.

Outputs

Exports of N in food and feed

Brazil has the largest economy in Latin America, a population of over 174 million people increasing at 1.38% per year, and a consequent positive influence on the demand for food. Overall, most food and feed consumed in Brazil originate in the country itself, although Brazil imports some products such as cereals, oil seeds, some fruits, fish and seafood, and milling products like malt, starch and wheat gluten (Agriculture and Agri-food Canada 2004). Because the amount of these imported products to Brazil is relatively small (FAOSTAT 2004) and/or their N contents low (Smil et al. 1999), the amount of N imported with these food products is unimportant (Table 2).

Figure 8. (a) Relative importance of NH_x deposition in Brazil from anthropogenic sources based on the TM3 global chemistry model-simulated results. (b) Relative importance of biomass burning in the deposition of NO_x in Brazil in the mid 90s. Results are based on simulated values produced by the TM3 global chemistry model.

In contrast, a significant amount of N in agri-food products produced in Brazil is exported to foreign countries, especially to the Netherlands, USA, Russia, Germany and China (Agriculture and Agri-food Canada 2004). The main products exported are soybeans, soybean oil, sugars, fruits, meat, and coffee (FAOSTAT 2004) and the N exported with these products is approximately 1.36 Tg N yr^{-1} (Table 2). Therefore, the net import of N to Brazil from food and feed is negative (Table 1).

Between 1995 and 2002, the net export of N in foodstuffs from Brazil increased about one order of magnitude, from about 0.1 to 1.1 Tg N yr^{-1} (Figure 9). This increase was caused mainly by growing exports of soybeans, while importation of cereals, the largest imported food product in Brazil, remained quite constant throughout the years.

Export of N in wood and other products

The most important forest (wood) products exported in Brazil in terms of quantity are sawnwood and chips/particles, with approximately 2×10^6 m^3 exported annually between 1995 and 2002 (FAOSTAT 2004). Most of the exported material is produced in the Amazon basin, where about 4.5×10^6 m^3 of wood is extracted annually. Assuming that logging in the Amazon basin averages 25–30 m^3 ha^{-1} (Keller et al. 2004), and that the export of N from selective logging is on the order of 200 kg ha^{-1} (Martinelli et al. 2000), we estimate that N export from Brazil in wood products is about 0.36 Gg. In comparison to foodstuffs, the export of N in wood products is significant for

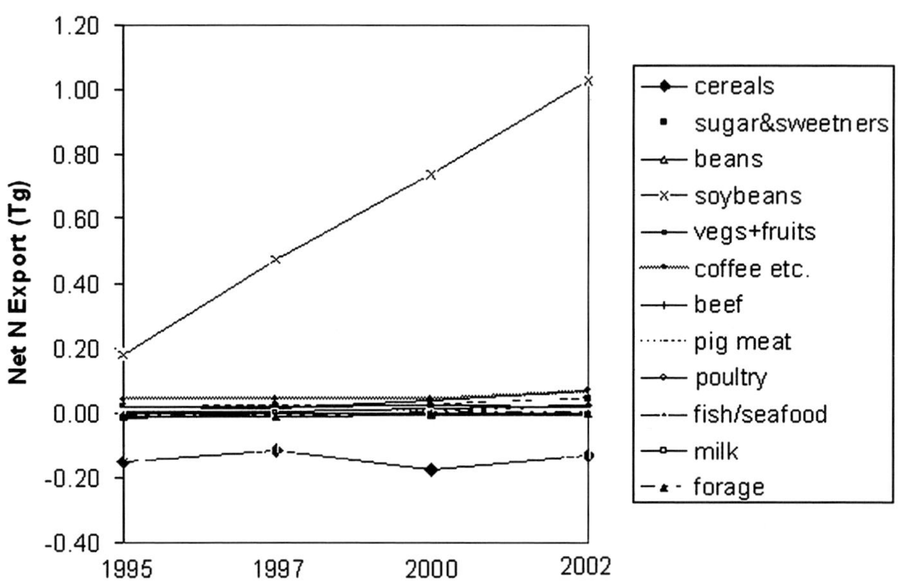

Figure 9. Net export of N from Brazil in foodstuffs in 1995, 1997, 2000, and 2002.

the country as a whole. However, such losses can be significant on a regional scale in the Amazon basin, especially because it is a cumulative loss.

In the South and Southeastern Brazil, the production of wood, mainly from silviculture, is also considerable but the wood produced there is mostly used in the Brazilian industry of paper and cellulose and consumed within Brazil (FAOSTAT 2004). Therefore, we consider that the N export in wood products from other regions of Brazil besides the Amazon is not significant.

Riverine export of Nr to Atlantic Ocean

Annually, world rivers discharge an average of 42×10^3 km^3 of freshwater into oceans (ANA 2004). Rivers in Brazil contribute about 20% of the global discharge, mainly because of the Amazon River, which supplies an average of 6.5×10^3 km^3 yr^{-1} of water to the Atlantic Ocean, and makes Brazil the country with the largest volume of freshwater discharged into oceans ($\sim 8.2 \times 10^3$ km^3 yr^{-1}) (AQUASTAT 2004; ANA 2004). Other major Brazilian watersheds, according to the Brazilian Water Agency (ANA) are the Tocantins, North and Northeast Atlantic, Parnaíba, São Francisco, East Atlantic, Paraguay, Paraná, Uruguay, and South and Southeast Atlantic (Figure 10); they range in size from 178 to 1029 km^2 and generate 41 to 372 km^3 of water annually within Brazil (ANA 2004).

Although Brazil has such an abundance of water resources, more than 73% of the freshwater available in the country is located in the Amazon basin, where it is used by less than 5% of the population (ANA 2004). The remaining

Figure 10. Map of Brazil with the delineation of the major river basins in the country (ANA 2004).

Table 3. Nitrogen export in major rivers in Tg yr^{-1}.

	DIN	TN	Data source
Amazon	0.96	3.0	Lewis et al. (1999)
Tocantins/Araguaia		0.3	Howarth et al. (1996)
Paraná	0.14	0.21	Bonetto et al. (1991)
Paraguay		0.03	Lewis et al. (1999)
TOTAL		3.54	

DIN – dissolved inorganic N, and TN – total N.

27% of the water available is produced in the 11 other major watersheds, and used by 95% of the country's population (ANA 2004) for water supply, agriculture, industry and production of hydroelectric power.

Because of its high water volume, the Amazon River is the single largest source of nitrogen to the Brazilian coast, transferring approximately 3 Tg N yr^{-1} from land to ocean annually (Table 3), with 0.96 Tg N yr^{-1} in dissolved inorganic form. The Amazon basin is still considered pristine, but increasing deforestation and agriculture expansion in the region are likely to cause major impacts on the N-cycle and maybe increase nitrogen transport into aquatic systems (Downing et al. 1999). The Tocantins River watershed is another important drainage system in the Amazon region, with a drainage area of 767,000 km^2 and annual mean discharge of 11,000 m^3 s^{-1} (Costa et al. 2003). However, the export of nitrogen in this river is approximately 0.3 Tg N yr^{-1}, or only 10% of that of the Amazon River, based on estimates presented in Howarth et al. (1996) (Table 3).

The Paraná watershed is the second largest drainage system in Brazil and accounts for 75% of the water discharge at the mouth of the Rio de la Plata. Although only about half of the Paraná River watershed is situated in Brazil, the Brazilian portion drains a region that is densely populated, has vast areas of intensive agriculture, and is highly industrialized. Human impacts in this part of the watershed have severely altered the nutrient dynamics in the entire Paraná River (Bonetto et al. 1991), including the deposition of phosphorus (P) and suspended sediments promoted by man-made reservoirs in the basin, the concomitant decrease in P concentrations coupled with P limitation (Bonetto et al. 1991; Villar et al. 1998), and the increase of N export in the past few decades (Bonetto et al. 1991). Pedrozo and Bonetto (1989) have claimed fertilizer inputs contribute the greatest share of nitrogen inputs into the Paraná basin, although only a small percentage of the inputs are accounted for in the riverine export at a sampling station located approximately 350 km below Iguaçu Falls (Bonetto et al. 1991). At this point in the main stem of the Paraná River, export rates of inorganic N average 188 kg N km^{-2} yr^{-1}, or 0.49 Tg N yr^{-1} (Table 3), and concentrations range between 0.06 and 1.4 mg l^{-1}, with nitrate accounting for 74% of the pool (Bonetto et al. 1991). Closer to the border of Brazil with Paraguay and Argentina, DIN concentrations are on the order of 0.4 mg N l^{-1} (Bonetto et al. 1991), which would result in the export of about 0.14 Tg DIN yr^{-1}. No data are available for TN concentrations in the Paraná River at the border of Brazil, but if we assume that the export of DIN in tropical rivers is roughly about two-thirds that of TN, according to the data presented in Lewis et al. (1999), then we can estimate that the export of N in the Parana River at the Brazilian border is approximately 0.21 Tg N yr^{-1}.

The Paraguay River drains about 140,000 km^2 of the great Pantanal wetland in Brazil, and has a wide surface of floodplains which are inundated during flood periods and undergo anoxia, promoting N losses by denitrification (Bonetto et al. 1991). Therefore, N limitation has been commonly observed in the Paraguay River (Bonetto et al. 1991; Villar et al. 1998), and the riverine N exports in Brazil are not likely to amount to more than about 0.03 Tg N yr^{-1} (Lewis et al. 1999).

The export of nitrogen in the remaining major watersheds in Brazil is not presented here because of unavailability of N concentration data for most rivers. Discharge data are widely available for Brazilian rivers as they are commonly monitored for their high potential for hydroelectric energy production in the country. However, water quality monitoring has been limited to a few rivers, mostly in the state of São Paulo, where water pollution has become a major issue.

Even without a complete estimate of the total amount of nitrogen that is exported from Brazil in all rivers both discharging into the Atlantic Ocean or flowing to neighboring countries, we speculate that the total riverine nitrogen export in Brazil is not likely to differ widely from the combined value for the Amazon, Paraná, Paraguay and Tocantins rivers (Table 3), because these rivers account for most of the water discharge in the country. Therefore, we estimate that the total riverine export of nitrogen in Brazil approximates 3.5 Tg N yr^{-1} (Table 3), which is 42% or more of the total N inputs in the country from anthropogenic sources (Table 1). The percent export of N in Brazil is higher than those observed for countries of the temperate zone probably because of the naturally high water flux rates observed in the Amazon River.

Nitrogen inputs vs. outputs in Brazil

Overall, we estimate that net inputs of anthropogenic N to Brazil in 2002 were 9.4 Tg, which is an increase of 4.4 Tg N since 1995 (Table 1). In contrast to what is observed for North America and globally, the main source of the new N inputs in Brazil is BNF in agriculture, which has more than doubled since 1995 (Table 1) especially because of the expansion of soybean cultivation (Figure 11). In 2004, the area planted with soybeans in Brazil was estimated at over 21 million ha (IBGE 2004), or about 2.7 million ha more than in 2002, that translates to an additional 0.46 Tg N in 2 years, assuming conservative rates of fixation.

We speculate that the importance of BNF in agriculture in Brazil is likely to increase in the future not only because of further expansion of soybean

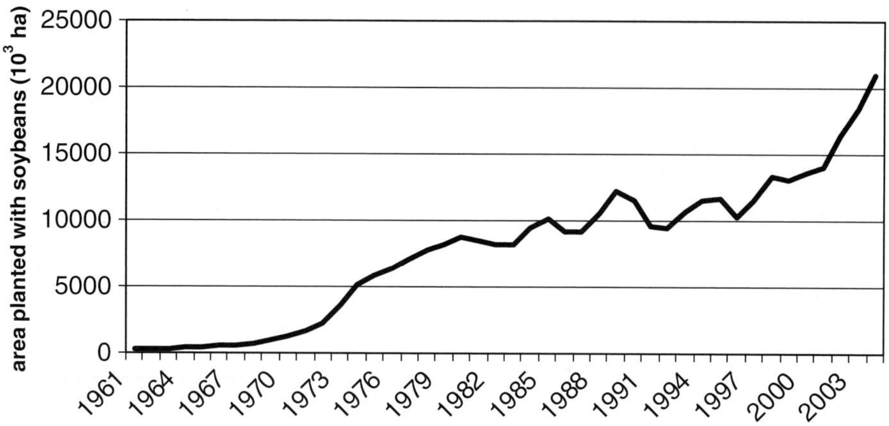

Figure 11. Area planted with soybeans in Brazil in the period between 1960 and 2003. Source: IBGE (Brazilian Institute of Geography and Statistics).

cultivation, but also because of the expansion of area planted with sugarcane for the production of ethanol, which Brazil plans to export in the next few years. Moreover, new technology and management strategies are now being developed in Brazil to improve rates of BNF in pastures (S. Urquiaga, personal communication 2004).

Besides BNF in agriculture, the use of nitrogen fertilizer in Brazil has more than doubled between 1995 and 2002 (Table 1), and is the second largest source of anthropogenic Nr in Brazil. Yet, the total amount of the fertilizer that is used presently in the country is about one-fifth of the amount used in North America, and is only a minor fraction of the global consumption (Table 1). The relatively low use of N fertilizer in Brazil is associated with high costs of the product in the country, which is mostly imported (~70%). However, as prices of fertilizer increase, roads improve, and agriculture intensifies in Brazil, the consumption of N fertilizer will tend to increase considerably, as has been the trend observed in the country over the past decade (Figure 3).

Energy production contributes a relative small amount of anthropogenic N in Brazil (Table 1) and has contributed little in the Nr budgets. However, because about 95% of the NO_x emissions in Brazil associated with energy production is related to the combustion of oil and derivatives for on-road vehicles, emissions are likely to increase with the number of vehicles increase (Figure 12).

Since the pre-industrial era, inputs of Nr from anthropogenic sources in Brazil have increased significantly but, according to our estimates for 2002, natural sources are still dominant, contributing approximately 11.4 Tg N yr^{-1} from both BNF in natural vegetation, and lightning (Table 1). The inputs from natural sources have been estimated using biological N fixation rates from the scientific literature for different types of natural vegetation in the tropics

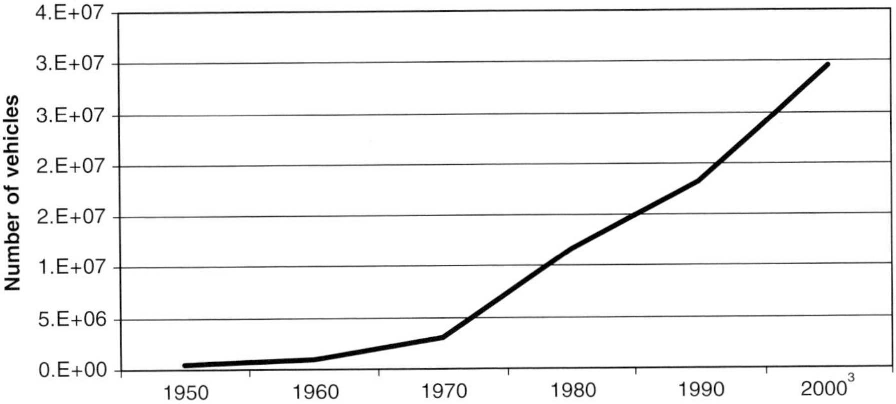

Figure 12. Number of on-road vehicles in Brazil between 1950 and 2000. Source: DENATRAN (Brazilian National Department of Transit).

(Cleveland et al. 1999) and model simulations of N deposition from lightning in South America in the early 1990s (Dentener and Crutzen 1994; Lelieveld and Dentener 2000). Therefore, these values are an approximation and should be interpreted with caution.

The values of BNF in the Amazon forest, for instance, were calculated based on the assumption that 70% of the basin soils are highly weathered *Ultisols* and *Oxisols* (Martinelli et al. 1999), and therefore should have BNF rates, based on 5% average cover of symbiotic N fixing vegetation (Cleveland et al. 1999) (Table 4). For the remaining 30% of the basin with richer soils, fixation rates were assumed to be higher and based on 15% cover of symbiotic N fixers (Cleveland et al., 1999). The other values used for estimation of BNF in different biomes of Brazil were multiplied by the estimated area of natural vegetation in each biome (Table 4).

Because much of the new anthropogenic N inputs in Brazil are associated with land use change and the expansion of agricultural area, there is a con-

Table 4. Natural N fixation rates in main Brazilian biomes based on values from Cleveland et al. (1999). Fixation rates used for 70% and 30% of the Amazon basin were 14.7 and 25.8, respectively.

Biome	N fixation rate
Amazonia (Rainforest)	14.7 or 25.4
Cerrado (Tropical savanna)	30.2
Caatinga (Xeromorphic forest)	9.4
Atlantic forest (Tropical deciduous forest)	21.6
Araucaria (Tropical evergreen forest)	15.0
Pantanal (Tropical non-forest floodplain)	28.5
Campos (Short grassland)	2.7

Values are in Kg ha^{-1} yr^{-1}.

comitant loss of natural vegetation in the country and consequent reduction of natural inputs of N from BNF. For instance, we estimated that under pristine conditions, BNF in natural vegetation contributed 17.1 Tg N yr^{-1} to the N budget in Brazil and that presently, because of a reduction in area, this contribution is approximately 11.4 Tg N yr^{-1}. Therefore, while net N inputs from anthropogenic sources increased about 9.4 Tg N yr^{-1} since the period when Brazil vegetation was pristine, there was a decrease of 5.7 Tg N yr^{-1} from natural biological fixation. Therefore, anthropogenic inputs of N in Brazil have actually increased the overall N budget by only about 3.7 Tg yr^{-1}. This is a relatively small number in comparison to the total inputs in the country, and is probably within the margin of error because the fixation rates can vary significantly depending on the assumed density of N fixers in the different biomes in Brazil (Cleveland et al. 1999) and on the fixation rates in soybeans (ranging from 70 to 250 kg ha^{-1} yr^{-1}) and pastures (averaging 15–30 kg ha^{-1} yr^{-1}). However, we can conclude that the N cycle in Brazil is increasingly controlled by anthropogenic activities and less controlled by natural processes. In different regions of Brazil, however, the importance of a natural versus an anthropogenically controlled N cycle becomes a much more prominent issue, especially because of the uneven distribution of the human population throughout the country and the intensity of agriculture in the different regions. Therefore, in order to assess this variability in Brazil, below we discuss some preliminary estimates of the major Nr sources in contrasting regions of Brazil, including the Amazon, the *Cerrado*, and the Southeast during the late 1990s and early 2000s.

Three contrasting regions in Brazil (Amazon, Cerrado and southeast)

The Amazon basin represents a relatively pristine region of Brazil and occupies close to 60% of the Brazilian territory, while the Brazilian *Cerrado* encompasses more than 20% of the country area and has vast areas of pastures and natural vegetation which are being converted to high-intensity agriculture. The Southeast region is the most developed part of Brazil and also one of the first areas to go through the process of land use changes, urbanization and industrialization.

The total area of the legal Amazon basin is over 6.3 million km^2, with about 5 million km^2 of continuous tropical forest extending to portions of eight countries, including 47.1% of the Brazilian territory. In Brazil, the Amazon basin extends for approximately 5 million ha, most of which is forested and tropical savanna landscapes, while land undergoing first stages of development (i.e., pastures and low intensity agriculture) are the second most dominant land use. The representation of more advanced stages of development in the basin, including high intensity agriculture and urbanization, is still relatively small but growing steadily in the southwestern region. Deforestation rates in the Brazilian Amazon averaged nearly 2 Mha yr^{-1} in the past 25 years

(INPE, 2000), and large-scale cattle ranches appear to be responsible for ~70% of all forest loss (Fearnside 2005). Fire is the most prevalent type of agricultural management in the region, especially for cattle ranching and slash-and-burn farming (Nepstad et al., 1999).

The *Cerrado* is an extensive phytogeographic zone located in the central portion of Brazil, covering over 200 Mha or 20% of the Brazilian territory (Ratter et al. 1997). This phytogeographic zone is located entirely within Brazil (Figure 13), extending from just south of the equator to south of the Tropic of Capricorn. Longitudinally, the *Cerrado* extends from east of the Tocantins River basin to the border of Brazil with Bolivia and Paraguay. The region includes a wide variety of vegetation types ranging from native grasslands to dense arboreal formations and dendritic forests that follow courses of water (Ratter et al. 1997). The three primary structural types of plant communities in the *Cerrado* include medium to tall woodlands with closed or semi-closed canopies (*cerradão*), savanna woodland or low trees and shrubs (*cerrado sensu stricto*), and open savanna with scattered trees or shrubs (*campo sujo*) (Pinto et al. 2002). One of the main characteristics of the *Cerrado* is that it is situated over a plateau and, therefore, it has vast areas (~127 Mha) of gentle topography suitable for mechanized agriculture. In the last 35 years, the *Cerrado* has lost over 40% of its native vegetation (Machado et al. 2004) due to the enormous advances in the agricultural frontier and the expansion of soybean plantations in the region (FAOSTAT 2004). Yet, the Brazilian Ministry of

Figure 13. Map of Brazil with the delineation of the major biomes in the country. Area shaded corresponds to the state of São Paulo. Source: WWF (http://www.wwf.org).

Agriculture estimates that Brazil's potential for agricultural expansion is about 90 million hectares (USDA 2003), and that most of the expansion is likely to occur in the *Cerrado* in the coming decades. These impacts are likely to have major ecological consequences for many of the major rivers with headwaters in the *Cerrado* region, such as the Paraná, São Francisco, Tocantins, and Araguaia.

Also, as in other tropical savannas, natural fires are a common feature of the *Cerrado*. Most savanna ecosystems are burned every 1–4 years during the dry season, with the highest frequency in the humid savannas (Nardoto and Bustamante 2003). Fires in the *Cerrado* have been occurring for thousands of years, but because of land use changes and agricultural management practices common in the region, they are becoming increasingly more frequent, and changing the grass/wood biomass ratio (Coutinho 1990).

The southeastern region of Brazil, here represented by the state of São Paulo because of data availability, has the most advanced stages of development in the Southern Hemisphere, with intensive agriculture, and high industry and population densities. The region also includes several dams constructed for hydroelectric power, which commonly characterizes development in the tropics (Downing et al. 1999). In terms of land use, the southeastern region is also the most diverse in terms of stages of development, especially because of the long-term history of deforestation, agriculture, and industrialization processes which started during the 17th century in São Paulo with deforestation followed by cultivation of coffee and sugar-cane (Martinelli et al. 2000).

Anthropogenic N sources in the Amazon, Cerrado and São Paulo

Our estimates show that BNF in agriculture is the largest source of anthropogenic N in all three regions (Table 5). In the Amazon, N inputs from BNF in agricultural systems are dominant mainly because of N fixation in actively grazed cattle pastures which cover an estimated area of 25 million ha (Bernoux et al. 2001; Müeller et al. 2004) and create approximately 0.75 Tg N yr^{-1} to the region. Soybean plantations, on the other hand, cover only about of 73,000 ha of the Amazon (Müeller and Bustamante 2002) and contribute to an estimated 0.01 Tg N yr^{-1} to the total budget. Inputs from BNF in actively grazed cattle pastures of the Amazon are also high because water stress is not as prevalent as

Table 5. Anthropogenic N sources in the Amazon, Cerrado and São Paulo State in Tg yr^{-1}.

	Amazon	Cerrado	S. Paulo
Lightning	0.26	0.12	0.01
BNF	7.5	1.74	0.05
Fertilizer	0.14	0.51	0.30
BNF-ag.	0.92	2.5	0.36
Atm. deposition	0.07	0.08	0.06

Table 6. Anthropogenic N sources in the Amazon, Cerrado and São Paulo State in kg km^{-2} yr^{-1}.

	Amazon	Cerrado	S. Paulo
Lightning	74	59	40
BNF	2142	853	201
Fertilizer	40	250	1210
BNF-ag.	263	1225	1452
Atm. deposition	20	39	246

in pastures of other major regions of Brazil such as the *Cerrado*. Therefore, fixation rates average about 30 kg ha^{-1} yr^{-1} (S. Urquiaga, personal communication, 2004), in contrast to 15 kg ha^{-1} yr^{-1} assumed for the other regions.

Since fixation rates in active cattle pastures of the Amazon are comparable to those of the rainforest (Table 4), the new inputs of N in the region through BNF in pasture land are not likely to affect N fluxes in the ecosystems on a large scale. However, as soybean plantations expand further into the Amazon, rates of N inputs from anthropogenic sources will far exceed natural inputs (Table 6) and potentially increase delivery rates of N to aquatic systems. This, however, will depend on whether or not N from BNF in soybeans accumulates in the soil over time. Soybeans are generally very efficient at utilizing available N in the soil before investing in N fixation (Peoples et al. 1995; Alves et al. 2003). Consequently, it is possible that soybean production in the Amazon will lead to the depletion of N from the soil rather than accumulation.

In the *Cerrado*, inputs of agricultural BNF are mostly from soybean cultivation. In 2002, the *Cerrado* had approximately 10 million ha of soybean fields, with fixation rates as high as 250 kg ha^{-1} yr^{-1} (Alves et al. 2003). Using a conservative rate of 170 kg ha^{-1} yr^{-1}, we estimated that soybeans added at least 1.7 Tg N to the *Cerrado* N budget in 2002. However, as most of this newly fixed N, and potentially some soil N, is exported from the region with harvested soybeans, the net input of N in the *Cerrado* is much lower than the gross input.

Fixation rates in *Cerrado* pastures ($\sim$15 kg ha^{-1} yr^{-1}, S. Urquiaga, personal communication, 2004) are lower than those of soybeans while the export of N from the region in cattle meat should be on the order of 9 kg ha^{-1} yr^{-1} in pastures grazed by 2–4 beef cattle ha^{-1} (Boddey et al. 2004). Therefore, well-managed pastures could potentially lead to N accumulation in the *Cerrado*, except that significant quantities of N can be lost from *Brachiaria* pastures in Brazil from animal excreta through volatilization, denitrification or leaching (Boddey et al. 2004). Losses from animal excreta increase with the number of grazing animal per hectare, and when these losses surpass the potential rates of BNF in pasture they can actually yield to a depletion of N from pastures in the *Cerrado*. In addition, given that the natural vegetation of the *Cerrado* can fix, on average, about 30 kg ha^{-1} yr^{-1} (Cleveland et al. 1999) (Table 4), the replacement of natural vegetation in the *Cerrado* with cattle ranching can lead to increasing N deficiency in the region's ecosystems.

In São Paulo state, where advanced stages of development such as urbanization, industrialization, and intensive agriculture are prevalent, estimated agricultural BNF is the largest input, but is not significantly different from inputs from fertilizer. According to our budget, these two sources combined contributed 0.66 Tg yr^{-1} or over 80% of all the inputs (Table 5), with about 0.08 Tg N yr^{-1} originating from biological fixation in 2.3 Mha of sugarcane plantations, 0.15 Tg N yr^{-1} from 10.3 Mha of cattle pastures, and 0.13 Tg N yr^{-1} from 0.74 Mha of soybeans. With a population of about 37 million people in 2000 (SIDRA/IBGE 2004) and an estimated consumption of N in food on the order of 0.15 Tg, assuming a per capita consumption of 4 kg N yr^{-1}, we predict that significant amounts of the N inputs from fertilizer and BNF in agriculture are not exported with foodstuffs from the state and are likely to be transferred to rivers and other aquatic systems with sewage discharges (Filoso et al. 2003). However, because the state of São Paulo exports much of its agricultural products to other parts of Brazil, the net input of N from anthropogenic sources in São Paulo should be considerably lower than gross inputs. Unfortunately, lack of data on imports and exports of agricultural products among the different states and territories in Brazil makes it difficult to estimate net N inputs more accurately for individual regions.

Among all of the anthropogenic N sources in the three comparison regions, atmospheric deposition was the least important (Table 5). Results from the model simulations for atmospheric N deposition in South America predicted that N inputs from the atmosphere can be significant in the overall budget of Brazil (5 to 12.5 kg ha^{-1} yr^{-1}), or in pristine regions such as the Amazon because of natural soil emissions and biomass burning. However, when emissions from fossil fuel are considered the only new source of N in the regions, the importance of atmospheric deposition inputs are reduced significantly, especially in the Amazon and *Cerrado*, where population densities average about 1 and 5 people km^{-2}, respectively (SIDRA/IBGE 2004). In São Paulo, where population density is about 150 people km^{-2}, the relative contribution of atmospheric deposition is considerably higher because fossil fuel emissions are substantial.

When comparing N inputs in the Amazon, *Cerrado* and São Paulo on a per area basis, we observe that the highest N input rates from anthropogenic sources occur in São Paulo (Table 6), probably because more than 90% of the territory in the state has been converted to agricultural or urban lands in the past few centuries. In the *Cerrado*, inputs of N from BNF in agricultural systems are about seven times higher than those in the state of São Paulo (Table 5), but rates on a per area basis are significantly lower because of the vast area of natural vegetation (~60%, IBGE 2004) that still covers the region. However, with the fast expansion of soybean cultivation in the *Cerrado*, and also with improving pasture management practices, rates of BNF in agriculture are likely to surpass those of the Southeast region of Brazil in the near future. Moreover, agriculture in the *Cerrado* is increasingly less associated with the monoculture of soybeans, and more with a diversity of crops including corn and cotton (Machado et al. 2004). Therefore, rates of N fertilizer use are also likely to grow following the trend observed for the country in general. Higher input rates of N fertilizer in the *Cerrado*, however, would have a much higher impact on the total N budget of Brazil because of the size of the region. Presently, total inputs of N from fertilizer use in the *Cerrado* are already higher than in São Paulo (Table 5), even though rates per area are lower. The same pattern is observed for inputs of N through BNF in agriculture.

Considering all the sources of N accounted for in the different regions of Brazil, the highest rates were estimated for natural BNF in the Amazon basin (Table 6), which is reflected in the total N budget for Brazil (Table 1) because of the vast area of the region. However, as land use change alters the N cycle in the Amazon, the importance of natural BNF in the overall budget will tend to decrease, while biological fixation of N is likely to remain the dominant mechanism for the creation of new N in Brazil. This is not just because of the potential expansion of soybean plantations and well managed pastures, but also because of the expansion of secondary forests in the Amazon. In tropical humid forests, BNF tends to be higher in young and secondary forest stands than in older stands in primary forests (Martinelli et al. 1999), indicating that the increasing area of secondary forests in the Amazon may be contributing to the increase of anthropogenic nitrogen inputs to the global budget (Galloway et al. 2004).

Conclusions

Although the alteration of the N cycle by human activities have been mainly associated with the temperate zone, especially Europe and the US, our evaluation of Nr inputs and outputs in Brazil presented in this paper suggests that the N cycle in developing regions of the tropics is increasingly less controlled by natural processes and more by human activities. Natural biological N fixation is still the main source of N in Brazil, but anthropogenic inputs are practically equivalent. In contrast to developed regions of the temperate zone, most of the newly created N originates from BNF in agricultural systems such as soybean crops and pastures, not from nitrogenous fertilizers or fossil fuel combustion. However, in regions of Brazil such as the southeast, which are undergoing more advanced stages of development, inputs of N from fertilizer application and combustion of fossil fuel are relatively high. In more pristine regions such as the Amazon, rates of N inputs from anthropogenic sources are still small, but the overall inputs are considerable because of the relatively large area of the region in Brazil (~40% of the overall territory). In the Brazilian *Cerrado*, which is the second largest biome of Brazil, the impact of anthropogenic activities on the N cycle is already visible on the N budget for Brazil and Latin America in general.

Except for N inputs from inorganic fertilizer, however, the estimates of N flows presented in this paper indicate general patterns and trends that are occurring to the N cycle in Brazil and regions, and should not be interpreted as absolute values. In many cases, such as the N inputs from BNF in soybeans, the real fluxes may be higher than the average estimates. In other cases, such as inputs from BNF in sugarcane, fluxes may be lower than estimated. Yet, the present study provides the first overview of the factors contributing to the changing N cycling in Brazil and other developing regions of the tropics, which should be helpful in outlining the priorities for improving our understanding of the effects of human activities on the N cycle in this part of the world. For instance, we now know that Nr inputs through BNF in cultivated pastures of the Amazon are potentially a major source of newly fixed N to the local and global budgets, so improved estimates that include field measurements of BNF rates and long-term monitoring of land use are necessary for improving the budgets. Challenges related to data availability and quality are not only unique to Brazil, and should be considered as a major priority for the advancement of our understanding of the impacts of human activities on the N cycle at the global scale.

References

Aber J.D., Magill A.H., McNutty S.G., Boone R.D., Nadelhoffer K.J., Downs M. and Hallett R.A. 1995. Forest biogeochemistry and primary production altered by nitrogen saturation. Water, Air Soil Pollut. 85: 1665–1670.

Aber J.D., Goodale C.L., Ollinger S.V., Smith M., Magill A.H., Martin M.E., Hallet R.A. and Stoddard J.L. 2003. Is nitrogen deposition altering the nitrogen status of Northeastern forests? BioScience 53: 375–389.

Alves B.J.R., Boddey R.M. and Urquiaga S. 2003. The success of BNF in soybean in Brazil. Plant Soil 252: 1–9.

Agriculture and Agrifood Canada. 2004. Opportunities in the food market in Brazil, [online] URL: http://www.atn-riae.agr.ca/latin/3795_e.htm.

ANA. Regiões Hidrográficas. Agência Nacional de Águas, [online] URL: http://www.ana.gov.br.

ANA. Water Resources Management in Brazil. Agência Nacional de Águas, [online] URL: http://www.hidroweb.ana.gov.br/.

ANDA. 2003. Estatísticas. Associação Nacional para Difusão de Adubos.[online] URL: http://www.anda.gov.br.

AQUASTAT. 2004. Food and Agriculture Organization of the United Nations, Information System on Water and Agriculture [online] URL: http://www.fao.org/.

Bernoux M. 2001. CO_2 emissions from liming of agricultural soils in Brazil. Global Biogeochem. Cycles 17: Art. No. 1049.

Boddey R.M., Macedo R., Tarre R.M., Ferreira E., de Oliveira O.C., Rezende C.P., Cantarutti R.B., Pereira J.M., Alves B.J.R. and Urquiaga S. 2004. Nitrogen cycling in Brachiaria pastures: the key to understanding the process of pasture decline. Agric. Ecosys. Environ. 103: 389–403.

Boddey R.M., Sá J.C.M., Alves B.J.R. and Urquiaga S. 1997. The contribution of biological N fixation for sustainable agricultural systems in the tropics. Soil Biol. Biochem. 29: 787–799.

Boddey R.M., Urquiaga S., Alves B.J.R. and Reis V. 2003. Endophytic nitrogen fixation in sugarcare: present knowledge and future applications. Plant Soil 252: 139–149.

Boddey R.M., Urquiaga S., Reis V.M. and Döbereiner J. 1991. Biological nitrogen fixation associated with sugarcane. Plant Soil 137: 111–117.

Boddey R.M. and Victoria R.L. 1986. Estimation of biological nitrogen-fixation associated with *Brachiaria* and *Paspalum* grasses using ^{15}N labeled organic matter and fertilizer. Plant Soil 90: 265–294.
Bonetto C., Zalocar Y., Planas D. and Pedrozo F. 1991. Responses of phytoplankton to experimental nutrient enrichment in the Paraguay, Bermejo and upper Parana rivers. Trop. Ecol. 32: 47–64.
Boyer E.W., Goodale C.L., Jaworski N.A. and Howarth R.W. 2002. Effects of anthropogenic nitrogen loading on riverine nitrogen export in the northeastern US. Biogeochemistry 57 & 58: 137–169.
Bustamante M.M.C., Nardoto G.B. and Martinelli L.A. 2004. Aspectos Comparativos Del Ciclaje De Nutrientes Entre Bosques Amazónicos De Terra-Firme Y Sabanas Tropicales (Cerrado Brasileiro). In: Hernán Marino Cabrera (ed.), Fisiología Ecológica En Plantas. Mecanismos y Respuestas a Estrés en los Ecosistemas, Ediciones Universitárias de Valparaiso, Pontificia Universidad Católica de Valparaíso, Valparaiso, Chile, pp. 189–205.
Cleveland C.C., Townsend A.R., Schimel D.S., Fisher H., Howarth R.W., Hedin L.O., Perakis S.S., Latty E.F., Von Fischer J.C., Elseroad A. and Wasson M.F. 1999. Global patterns of terrestrial biological nitrogen (N_2) fixation in natural ecosystems. Global Biogeochem. Cycles 13: 623–645.
Costa M.H., Botta A. and Cardille J.A. 2003. Effects of large-scale changes in land cover on the discharge of the Tocantins River, Southeastern Amazonia. J. Hydrol. 283: 206–217.
Dentener F.J. and Crutzen P.J. 1994. A three-dimentional model of the global ammonia cycle. J. Atmos. Chem. 19: 331–369.
Dobereiner J. 1997. Biological nitrogen fixation in the tropics: social and economic contributions. Soil Biol. Biochem. 29: 771–774.
Downing J.A., McClain M., Twilley R., Melack J.M., Elser J., Rabalais N., Lewis W.M., Turner R.E., Corredor J., Soto D., Yanez-Aranciba A., Kopaska J.A. and Howarth R.W. 1999. The impact of accelerating land-use change on the N cycle of tropical aquatic ecosystems: current conditions and projected changes. Biogeochemistry 46: 109–148.
EMBRAPA. 2002. Brazilian Agency of Agriculture and Pecuary.
FAOSTAT. 2004. Food and Agriculture Organization of the United Nations, Statistical Databases [online] URL: http://www.apps.fao.org/.
Fearnside P.M. 2005. Deforestation in Brazilian Amazonia: history, rates and consequences. Conservation Biology 19: 680–688.
Filoso S., Williams M.R. and Melack J.M. 1999. Composition and deposition of throughfall in a flooded forest archipelago (Negro River, Brazil). Biogeochemistry 45: 169–195.
Filoso S., Martinelli L.A., Williams M.R., Lara L.B., Krusche A., Ballester M.V., Victoria R.L. and Camargo P.B. 2003. Land use and nitrogen export in the Piracicaba River basin, Southeast Brazil. Biogeochemistry 65: 275–294.
Galloway J.N., Dentener F.J., Capone D.G., Boyer E.W., Howarth R.W., Seitzinger S.P., Asner G.P., Cleveland C., Green P., Holland E., Karl D.M., Michaels A.F., Porter J.H., Townsend A. and Vorösmarty C. 2004. Nitrogen cycles: past and future. Biogeochemistry 70: 153–226.
Galloway J.N. and Cowling E.B. 2002. Reactive nitrogen and the world: 200 years of change. Ambio 31: 64–71.
Galloway J.N., Schlesinger W.H., Levy H., Michaels A. II and Schnoor J.L. 1995. Nitrogen fixation: anthropogenic enhancement-environmental response. Global Biogeochem. Cycles 9: 235–252.
Howarth R.W., Boyer E.W., Pabich W.J. and Galloway J.N. 2002. Nitrogen use in the United States from 1961–2000 and potential future trends. Ambio 31: 88–96.
Howarth R.W., Anderson D., Cloern J., Elfring E., Hopkinson C., Lapointe B., Malone T., Marcus N., McGlathery K., Sharpley A. and Walker D. 2000. Nutrient pollution of coastal rivers, bays, and seas. Issues Ecol. 7: 1–15.
Howarth R.W., Billen G., Swaney D., Townsend A., Jarworski N., Lajtha K., Downing J.A., Elmgren R., Caraco N., Jordan T., Berendse F., Freney J., Kueyarov V., Murdoch P. and Zhu

Zhao-liang 1996. Riverine inputs of nitrogen to the North Atlantic Ocean: fluxes and human influences. Biogeochemistry 35: 75–139.

IBGE. 2004. Levantamento sistemático da produção agrícola. Instituto Brasileiro de Geografia e Estatística, [online] URL: http://www.ibge.gov.br/.

SIDRA/IBGE. 2004. Sistema IBGE de Recuperação Automática, Banco de Dados Agregados, Instituto Brasileiro de Geografia e Estatística, [online] URL: http://www.sidra.ibge.gov.br/bda/popul/.

INPE. 2004. Monitoramento da floresta Amazonica brasileira por satelite – Projeto PRODES. Instituto Nacional de Pesquisas Espaciais, [online] URL: http://www.obt.inpe.br/prodes/.

Keller M., Palace M., Asner G.P., Pereira R. and Silva J.N.M. 2004. Coarse woody debris in undisturbed and logged forests in the eastern Brazilian Amazon. Global Change Biol. 10: 784–795.

Laurance W.F., Albernaz A.K.M., Fearnside P.M., Vasconcelos H.L. and Ferreira L.V. 2004. Deforestation in Amazonia. Science 304(5674): 1109–1111.

Lara L.S., Artaxo P. and Martinelli L.A. 2001. Chemical composition of rainwater and anthropogenic influences in the Piracicaba River Basin, Southeast Brazil. Atmos. Environ. 35: 4937–4945.

Lelieveld J. and Dentener F. 2000. What controls tropospheric ozone? J. Geophys. Res. 105: 3531–3551.

Lesack L.F. and Melack J.M. 1991. The deposition, composition, and potential sources of major ionic solutes in rain of the central Amazon basin. Water Resourc. Res. 27: 2953–2977.

Lewis W.M., Melack J.M., McDowell W.H., McClain M. and Richey J. 1999. Nitrogen yields from undisturbed watersheds in the Americas. Biogeochemistry 46: 149–162.

Machado R., Ramos-Neto M, Pereira P.G., Caldas E., Gonçalves D., Santos N., Tabor K. and Steininger M. 2004. Estimativas de perda da área do Cerrado brasileiro. Relatório Técnico. Conservation International – Brazil, Brasília-DF., 23 pp.

Martinelli L.A., Piccolo M.C., Townsend A.R., Vitousek P.M., Cuevas E., Mcdowell W., Robertson G.P., Santos O.C. and Treseder K. 1999. Nitrogen stable isotopic composition of leaves and soil: tropical versus temperate forests. Biogeochemistry 46: 45–65.

Martinelli L.A., Almeida S., Brown I.F., Moreira M.Z., Victoria R.L., Filoso S., Ferreira C.A.C. and Thomas W.W. 2000. Variation in nutrient distribution in a humid tropical forest in Rondônia, Brazil. Biotropica 32: 597–613.

Marufu L., Dentener F., Lelieveld J. Andreae M.O. and Helas G. 2000. Photochemistry of the African troposphere: the influence of biomass burning emissions. J. Geophys. Res. 105: 14513–14530.

Matson P.A., McDowell W.H., Townsend A.R. and Vitousek P.M. 1999. The globalization of N deposition: Ecosystem consequences in tropical environments. Biogeochemistry 46: 67–83.

Matson P.A., Lohse K.A. and Hall S.J. 2002. The globalization of nitrogen deposition: consequences for terrestrial ecosystems. Ambio 31: 113–119.

Melillo J.M., Steudler P.A., Feigl B.J., Neill C., Garcia D., Piccolo M.C., Cerri C.C. and Tian H. 2001. Nitrous oxide emissions from forests and pastures of various ages in the Brazilian Amazon. J. Geophys. Res. Atmos. 106: 34179–34188.

Miranda C.H.B. and Boddey R.M. 1987. Estimation of biological nitrogen fixation with 11 ecotypes of *Panicum maximum* grown in nitrogen-15 labeled soil. Agron. J. 79: 558–563.

Mostasso L., Mostasso F.L., Dias B.G., Vargas M.T. and Hungria H. 2002. Selection of bean (*Phseoulus vulgaris L.*) rhizobial strans for the Brazilian Cerrados. Field Crops Res. 73: 121–132.

Müeller C.C. and Bustamante M. 2002. Análise da expansão da soja no Brasil. [online] URL: http://www.worldbank.org/rfpp/news/debates/mueller.pdf.

Müeller M.L., Guimarães M.F., Desjardins T. and Mitja D. 2004. The relationship between pasture degradation and soil properties in the Brazilian Amazon: a case study. Agric. Ecosys. Environ. 103: 279–288.

Nardoto G.B. and Bustamante M.C. 2003. Effects of fire on soil nitrogen dynamics and microbioal biomass in savannas of Central Brazil. Pesq. Agropec. Bras. 38: 955–962.

Neff J.C., Holland E.A., Dentener F.J. McDowell W.H. and Russell K.M. 2002. The origin, composition and rates of organic nitrogen deposition: a missing piece of the nitrogen cycle? Biogeochemistry 57/58: 99–136.

Neill C., Piccolo M.C., Cerri C.C., Steudler P.A., Melillo J.M. and Brito M. 1997. Net nitrogen mineralization and net nitrification rates in soils following deforestation for pasture across the southwestern Brazilian Amazon Basin landscape. Oecologia 110: 243–252.

Nepstad D.C., Verissimo A., Alencar A., Nobre C., Lima E., Lefebvre P., Schlesinger P., Potter C., Mouthinho P., Mendoza E., Cochrane M., and Brooks V. 1999. Nature 398: 505–508.

NRC 2000. Clean Coastal Waters: Understanding and Reducing the Effects of Nutrient Pollution. National Academy Press, Washington, DC.

Oliveira M.W., Trivelin P.C.O., Gava G.J.C. and Vitti A.C. 1999. Lixiviação de nitrogênio em solo cultivado com cana-de-açúcar: experimento em lisímetro. Stab. Álcool, Açúcar e Sub-produtos 18: 28–31.

Pedrozo F. and Bonetto C. 1989. Influence of river regulation on nitrogen and phosphorus mass transport in a large South American river. Regul. Rivers Res. Manage. 4: 59–70.

Peoples M., Gault R., Lean B., Sykes J. and Brockwell J. 1995. Nitrogen fixation by soybean in commercial irrigated crops in Central and Southern New South Wales. Soil Biol. Biochem. 27: 553–561.

Pinto A.S., Bustamante M.C., Kisselle K., Burke R., Zepp R., Viana L.T., Varella R.F. and Molina M. 2002. Soil emissions of N_2O, NO, and CO_2 in Brazilian savannas: effects of vegetation type, seasonality, and prescribed fires. J. Geophys. Res. 107: 8089doi: 10.1029./2001 JD000342.

POTAFOS-Brasil. 2004. Consumo de fertilizantes. Potash and Phosphate Institute of Brazil. [online] URL: http://www.potafos.org/ppiweb/brazil.nsf.

Rabalais N.N. 2002. Nitrogen in aquatic ecosystems. Ambio 31: 102–112.

Ratter J.A., Ribeiro J.F. and Bridgewater S. 1997. The Brazilian Cerrado vegetation and threats to its biodiversity. Ann. Bot. 80: 223–230.

Rodhe H., Dentener F. and Schulz M. 2002. The global distribution of acidifying wet deposition. Environ. Sci. Tech. 36: 4382–4388.

Seitzinger S.P., Kroeze C., Bouwman A.F., Caraco N., Dentener F. and Styles R.V. 2002. Global Patterns of dissolved and particulate nitrogen inputs to coastal systems: Recent conditions and future projections. Estuaries 25: 640–655.

Smil V. 1999. Nitrogen in crop production: an account of global flows. Global Biogeochem. Cycles 13: 647–662.

Smil V. 2001. Enriching the Earth. MIT Press, Cambridge.

Urquiaga S., Cruz K.H.S. and Boddey R.M. 1992. Contribution of nitrogen fixation to sugarcane: nitrogen-15 and nitrogen balance estimates. Soil Sci. Soc. Am. 22: 104–114.

USDA. 2003. Brazil: Future Agricultural Expansion Potential Underrated. United States Department of Agriculture, Production Estimates and Crop Assessment Division, Foreign Agricultural Service. [online] URL: http://www.fas.usda.gov/pecad/highlights/2003/01/Ag_expansion/index.htm.

Villar C.A., de Cabo L., Vaithiyanathan P. and Bonetto C. 1998. River-floodplain interactions: nutrient concentrations in the Lower Parana River. Archiv fur Hydrobioligie 142: 433–450.

Vitousek P.M. and Field C.B. 1999. Ecosystem constrains to symbiotic nitrogen fixers: a simple model and its implications. Biogeochemistry 46: 179–202.

Williams M.R. and Melack J.M. 1997. Solute export from forested and partially deforested catchments in the central Amazon. Biogeochemistry 38: 67–102.

Yoneyama T., Muraoka T., Kim T.H., Dacanay E.V. and Nakanishi Y. 1997. The natural 15N abundance of sugarcane and neighbouring plants in Brazil, the Phillippines and Myako, Japan. Plant Soil 189: 239–244.

Biogeochemistry (2006) 79: 91–108
DOI 10.1007/s10533-006-9004-z

Assessment of nitrogen flows into the Cuban landscape

JULIO A. BAISRE
Ministerio de la Industria Pesquera, Barlovento Santa Fe, Playa 19 100, La Habana, Cuba (e-mail: baisre@telemar.cu)

Key words: Atmospheric deposition, Fertilizer, Nitrogen, Nitrogen budgets

Abstract. The alteration of the nitrogen (N) cycle by human activities is widespread and has often resulted in increased flows of nitrogen to the marine environment. In this paper we have attempted to know the changes of N fluxes in Cuba by quantifying the N inputs to the landscape from (1) fertilizer applications, (2) atmospheric deposition, (3) biological nitrogen fixation and (4) net import of food and feeds. N-inputs to the country progressively increased until the end of the 20th century, reaching a peak during the 80s when low cost fertilizer imported from the former Soviet Union led to heavy rates of application. This rapid growth represented more than a 5-fold increase with respect to pristine values; higher than the two-fold global increase of anthropogenic N reported by Vitousek et al. (1997 Human alteration of the global nitrogen cycle: sources and consequences. Ecol. Appl. 7:737–750). Inorganic fertilizer was the largest single source of reactive N, followed by atmospheric deposition, biological fixation, and net imports of foods and feedstocks. Nitrogen inputs peaked in 1987 and data expressed on an area basis show that N flux to the Cuban landscape, in the 80s, was one of the highest reported in the literature. During the 90s, there was a dramatic drop in nitrogen inputs mainly associated to a decrease in the use of inorganic fertilizer. Other factors reducing nutrient inflows to Cuba, during the same period, were imports of foodstuff and livestock feeds, a decrease of nitrogen oxide emissions, and a decrease in the sugar cane crop area. Using an empirical relationship (Howarth et al. 1996 Regional nitrogen budgets and riverine N & O fluxes for the drainages to the North Atlantic Ocean: Natural and human influences. Biogeochemistry 35:75–139) we present a very preliminary estimate of N-inputs to coastal waters and discuss the consequences of these changes on the coastal zone.

Introduction

Human activities have more than doubled the inputs of reactive nitrogen (Nr) to the terrestrial landscape (Galloway et al. 1995; Howarth et al. 1996; Smil 1997; Vitousek et al. 1997). This increase was originally limited to developed countries, but is now being extended to developing countries (Matson et al. 1999). The accumulation of Nr in ecosystems is one of the most important research questions associated with the impact of humans on the nitrogen cycle (Galloway and Cowling, 2002). The influx of excess nitrogen has caused serious alterations to the natural nutrient cycle and disrupted terrestrial and aquatic ecosystems especially where intensive agriculture and high fossil fuel combustion coincide (WRI 2001).

One of the best-documented consequences of human alteration of the nitrogen cycle is the eutrophication of estuarine and coastal waters (Vitousek et al. 1997). While in moderation, nutrient inputs to oligotrophic coastal waters may increase fish production (Howarth et al. 2000) especially of pelagic species (Caddy 1993), a more intensive input of Nr has serious effects on the fisheries, biodiversity and ecosystem functioning (Boesch 2001).

Due to the location of Cuba in the oligotrophic Caribbean Sea, in absence of significant processes of coastal upwelling and because the very small tidal range; river discharge, delivering terrestrial material in particulate and dissolved form, is the most important source of nutrients supporting Cuban marine coastal fisheries (Baisre 1985). An analysis of the dynamics of its fisheries resources shows that, in 1995, approximately 38.9% were in the senescent phase (with declining catches), 48.7% were in the mature phase at a high exploitation level and only 12.4% were still in the developing phase with some possibility of increased landings. The decline is noteworthy in the cases of some of the most important estuarine species like shrimps, mullets and mangrove oyster. Baisre (2000) has hypothesized that a reduction of nutrients inputs to coastal waters is one of the factors that must be considered when analyzing the causes of this decline.

This paper represent a first step toward more specific studies about the ecological consequences of human alterations on the N-cycle in Cuba and the impact of anthropogenic nutrients in the coastal waters and marine fisheries. By considering the whole Cuban territory as a large watershed, I estimate N-inputs to the landscape from a combination of fertilizer use, atmospheric deposition, biological fixation and net imports of foods and feedstock. A preliminary assessment of the probable fluxes to coastal areas was also attempted by using a model developed for temperate regions.

Material and methods

Study area

The Cuban Archipelago, has a total surface area of 110860 km^2 which includes the Isle of Cuba (104945 km^2), the Isle of Youth (2200 km^2) and more than 1600 small and unpopulated islands and keys (3715 km^2). Approximately two-thirds of Cuba consists of plain or rolling lands with low elevation. The rest of the territory is formed by three groups of mountains where most of the natural forest occurs. The country lies within the northern tropics and has a sub tropical climate moderated by trade wind. Average daily temperatures range from 21 °C in winter to about 27 °C in summer. Annual average rainfall is 1300 mm with a marked difference between the rainy season (May to October) and the dry season (November to April).The combined area of the country is made up of agricultural lands (34.8%), non-cultivated lands (34.1%), and forest (21.5%), with a small fraction of urban land (6.6%) and freshwater aquatic areas (2.9%).

Many watershed runs like a spine along the length of the main island and as a consequence, most of the rivers are of limited length and reduced flow.

Furthermore, many parts of the country are underlined by limestone and some rivers may flow underground for at least parts of their lengths (Cubagua 2004).

In marine waters, Baisre (1985) described in sequence three fishery ecological complexes in terms of the decreasing influence of land. The three ecological complexes are: estuarine–littoral, seagrasses-coral reef and oceanic. The estuarine–littoral complex is particularly influenced by terrestrial fluxes and is typical of low coast, estuaries and lagoons dominated by terrigenous material and mangroves. Ecologically, they represent highly fluctuating habitats subject to great environmental changes, often accentuated by the unpredictable rainy season. There are elevated levels of primary production and organic detritus, which are the basis of relatively simple food webs. Estimated area of the estuarine–littoral complex is about 8500 km^2, approximately 16% of the total shelf area and it reaches its greatest extent on the southeast coast in the Gulfs of Ana Maria and Guacanayabo into which the largest river systems of the country drain (Baisre 1985).

Changes in the nitrogen cycle in Cuba was determined in this paper by quantifying net new anthropogenic Nr inputs to the country landscape and comparing to baseline inputs. New refers to Nr that is either newly fixed within, or transported into a region (Howarth et al. 2002).

Natural inputs

The estimation of natural biological N fixation in Cuban pristine areas could provide a baseline for further comparison with anthropogenic inputs in human dominated ecosystems. Available data suggest that approximately 60% of Cuba was covered with forest in pre-Columbian times (Smith 1954). This forest was largely deciduous broad-leaf while pine forests probably occupy the same 4% they do today and the tropical rain forest was limited to a small area in the northeastern mountains (Marrero 1950). Considering that 60% of the area of the country was covered by deciduous broad-leaf forest; 26% by tropical dry savanna, while the remaining area is occupied by different types of vegetation (xeromorphic shrubs, wetlands, mangroves, pinelands, (Smith 1954), I calculated N-fixation in Cuban pristine landscape using the area covered by the different types of vegetation and reference values from the literature.

Natural nitrogen fixation rates are difficult to estimate because of ecosystems heterogeneity. Cleveland et al. (1999) reviewed the available published estimates of both symbiotic and non symbiotic components of N fixation for each typed ecosystem, using the ecosystem classification of Schimel et al. (1996). For the dry tropical broad-leaved forest, the most common vegetation type in Cuba, Cleveland et al. (1999) reported a range of values from 9.4 to 34.0 kg N ha^{-1} yr^{-1} with an average of 21.7 kg N ha^{-1} yr^{-1}. Estimates of N-fixation in tropical savannas are higher because of the sparse presence of some legumes, and range from 16.3 to 44.0 kg N ha^{-1} yr^{-1} with a mean estimate of 30.2 kg N ha^{-1} yr^{-1} (Cleveland et al. 1999). In this paper, I used the lower and more conservative values for estimating inputs from N fixation in natural systems, because as pointed by Galloway et al. (2004), "there are several compelling reasons to believe that an estimate in lower portion of the range is more realistic than higher estimates".

Anthropogenic inputs

Besides determining annual N inputs from fertilizer application, atmospheric deposition, biological nitrogen fixation by agricultural crops, and net movement of food and animal feed stocks into or out the country, I also assessed the inter-annual variability of these inputs. Except for data on nitrate wet-deposition, from 1982 to 1994, the remaining data were available over a longer period of time, between 1961 and 2000.

Data on inorganic-N fertilizer was obtained from the International Industrial Fertilizer Association database (IFA 2004), while atmospheric deposition was estimated with data from a network of 5 monitoring stations operated by the Institute of Meteorology across the country (Figure 1). Atmospheric deposition (wet and dry) in Cuba is mainly associated with industrial and agricultural (biogenic) N emissions. The two major sources of nitrogen pollution to the air are fossil fuel combustion (e.g. vehicle and power plant emissions) and agriculture (e.g. fertilizer and manure emissions). Nitrate (NO_3^-) and ammonium (NH_4^+) are the dominant forms of inorganic N in atmospheric deposition (Centella et al. 2000). I considered only the emissions of oxidized nitrogen compounds to the atmosphere (NO_x), because the emission of ammonia and ammonium (NH_x) supposedly originates from the volatilization of fertilizer and animal wastes, and represents N recycling within the system (Howarth et al. 1996). According to Cuesta et al. (1998), atmospheric wet deposition in Cuba is about 60% and the dry one is 40% while the oxidized nitrogen compounds contribute 40% and the reduced ones 60%. Final values of total atmospheric deposition of NO_x were calculated using these figures. I have not considered the NH_x deposition from oceanic sources although we recognized that this is a potential source of N particularly because Cuba is a small and elongated island. Estimates of inputs of nitrogen in food and feedstocks were based on import and export data from the Food Balance Sheets prepared by FAO for different countries (FAOSTAT 2004), and the N content of the different food items obtained from the Food Composition Table published by the Research Institute for the Food Industry (IIIA 1985).

Data on N fixation refers to the sum of symbiotic N fixation by cultivation of legume crops, and non-symbiotic N fixation (Yan 2003). The cultivation of rice and other non-legume crops supply an additional source of biologically fixed N by microorganisms (Watanabe 1986). I estimated agricultural fixation rates of N in Cuba by multiplying the area of the different crops, and pasturelands by N fixation rates from the literature. The fixation rate used for edible beans was 40 kg N ha^{-1} yr^{-1}; which is the average for several types of

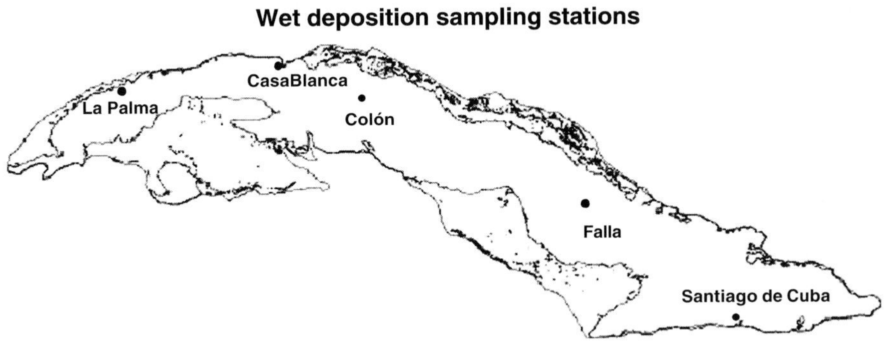

Figure 1. Map of Cuba showing the stations where wet atmospheric deposition have been monitored. This network is operated by the Institute of Meteorology.

beans calculated by Jordan and Weller (1996). Other fixation rates used were 30 kg ha^{-1} yr^{-1} for cyanobacteria associated with rice and 50 kg ha^{-1} yr^{-1} for endophytic diazotrophs associated with sugarcane (Smil 1999). For pasture, a fixation rate of 4 kg ha^{-1} yr^{-1} was used (Jordan and Weller 1996). Data on cultivated areas for all of these crops were also available at the FAO database (FAOSTAT 2004).

Estimates of Nr export to coastal waters were made using a simple linear regression model relating net anthropogenic Nr inputs to total Nr exports in rivers (Howarth et al. 1996). This model proved to be useful for estimating Nr fluxes in 16 major watersheds in the northeastern USA (Boyer et al. 2002) and it seems to be the best choice from several models used to estimate Nr fluxes from those watersheds (Alexander et al. 2002).

All fluxes were expressed as mass per unit area (Gg N km^{-2}) and as mass per unit area, per unit time as kg N km^{-2} yr^{-1} or kg N ha^{-1} yr^{-1}. (1 Gg = 1000 ton; 1 km^2 = 100 ha). When presenting the data on an area basis, I considered the combined area of the Isle of Cuba and Isle of Youth to be 107145 km^2.

Results

Natural inputs

Although electric discharges is one of the two natural sources of Nr, we assumed that this source do not have changed very much and have not been included in the N-budget. Natural biological fixation in Cuban pristine areas in pre Columbian times, accounts for 123 Gg N yr^{-1} and when expressed on an area basis this is equivalent to roughly 1119 kg N ha^{-1} yr^{-1}. This figure is about 6- fold the actual data for biological N fixation in primary and secondary forest which averaged 20 Gg N yr^{-1} from 1961 to 2000.

Anthropogenic inputs

Nr inputs from fertilizers

The most commonly used fertilizers in Cuba were urea (36.2%), ammonium nitrate (34.5%) and ammonium sulphate (24.3%). N-inorganic from fertilizer use from 1961 to 2000 averaged 180 Gg N yr^{-1}, although there were great differences between years (Figure 2). From 1961 to 1989, a progressive increase of nearly 9-fold occurred, from 41 Gg N yr^{-1}; to 367 Gg N yr^{-1}. Fertilizer use then declined drastically to 78 Gg N yr^{-1} (nearly 5-fold) by 2000.

Considering the entire area of the country, the yearly average of fertilizer application is 17 kg N ha^{-1} yr^{-1}, although during the period of intensive use of inorganic fertilizers in Cuban agriculture (1980–1989), the average inputs rose to 28 kg N ha^{-1} yr^{-1}. If only the area of agricultural land is considered, fertilizer application rate in Cuba increases to more than 90 ha^{-1} yr^{-1}, reaching a maximum of 94 ha^{-1} yr^{-1}.

Net atmospheric Nr inputs from fossil-fuel combustion

The principal sources of oxidized N in Cuba are the combustion of fossil fuel, motor vehicles and different industrial processes (Centella et al. 2000). Total atmospheric deposition of NO_x from 1982 to 1994 is presented in Figure 3, ranging from a minimum of 32 Gg N yr^{-1} in 1992 to a maximum of 171 Gg N yr^{-1} in 1987. With the decrease in the use of inorganic fertilizer, atmospheric deposition becomes the most important source of anthropogenic N in the last decade accounting for 34% of all the inputs in 1994.

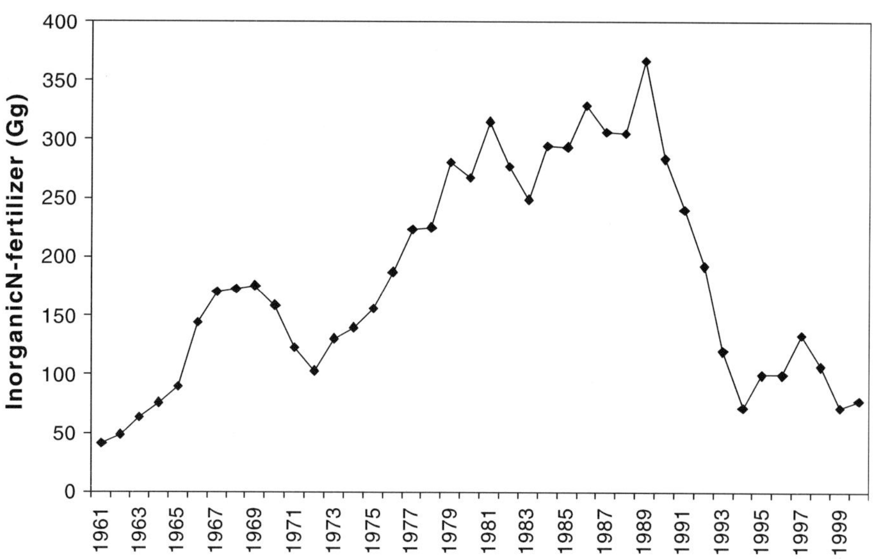

Figure 2. N inputs to Cuba from 1961–2000 from use of inorganic fertilizer (Data from the International Industrial Fertilizer Association).

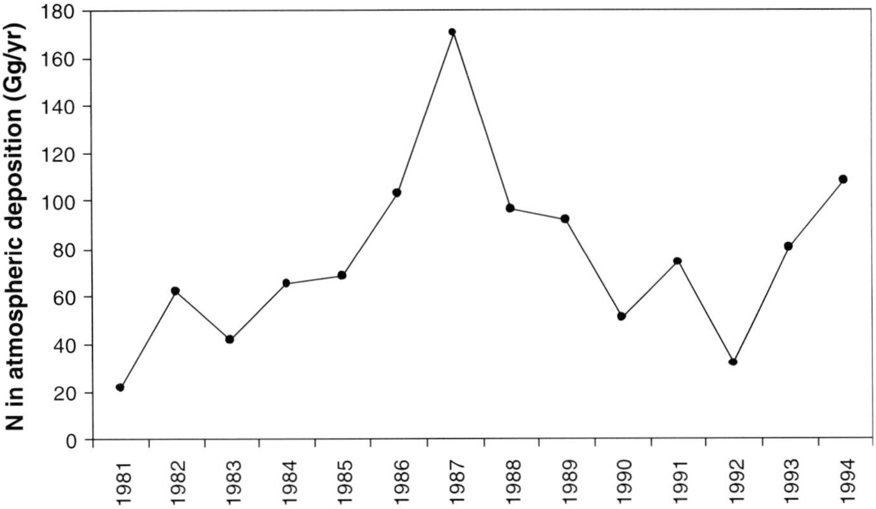

Figure 3. N inputs from total atmospheric deposition of NO_x. The calculations were based on annual average data from wet deposition and the estimates that wet deposition in Cuba represents about 60% while the oxidized nitrogen compounds contribute 40% (from Cuesta et al. 1998).

Nr inputs from N_2 fixation

Nr is also introduced to the Cuban landscape in significant quantities by biological fixation in agricultural systems (Figure 4). Available estimates indicated that biological N fixation in 1961–2000, excluding primary and

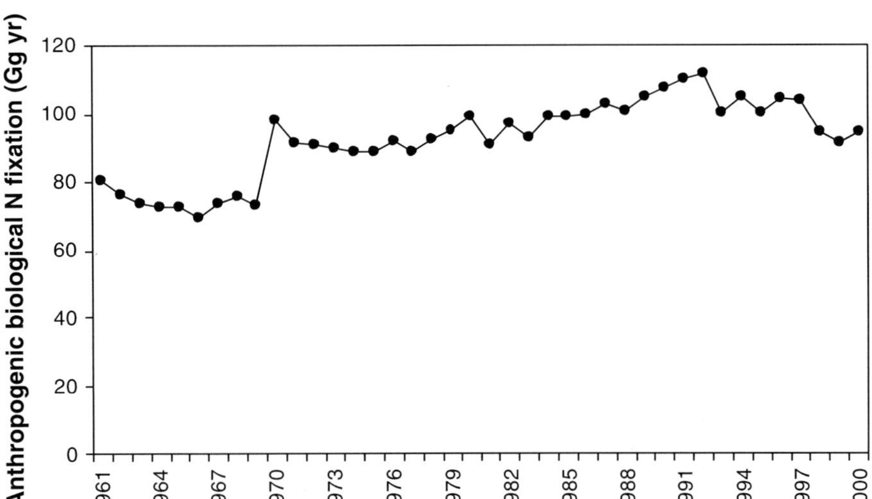

Figure 4. N-inputs from biological fixation in agricultural system from 1961 to 2000. Data on biological N fixation in primary and secondary forest were not included.

secondary forest, averaged approximately 94.5 Gg yr^{-1} and varied from about 72.5 Gg N yr^{-1} in 1965 to some 110 Gg N yr^{-1} in 1991, while it is the most stable of the four sources. Sugar cane crop (66.1%), and pasture (11.1%), were the most important sources of N-biological fixation.

Net Nr import in foods and feeds

The fourth source of nitrogen input in Cuba is from the net movement of food and feedstocks into and out the country. Although Cuba is typically a food exporter, most products, except for sea foods, are nitrogen poor (e.g. sugar and fruits). On the other hand, there are relatively large inputs of nitrogen from imported cereals, fish, chicken meat and milk. Most of the N in imported foods comes from cereals (73.9%), fish (9.7%) and beans (8.8%). The net import of nitrogen from 1961 to 2000 averaged 43.5 Gg N yr^{-1} and there was also a progressive increase on imported N from 1961 to 1980 (Figure 5), which reached more than 66 Gg N.

Summary of Nr inputs

Total N inputs into the country increased progressively from 136 Gg N in 1961 to 640 Gg N in 1987 (Figure 6). Excepting the biological N fixation which do not varies very much, there is a general trend of the other N inputs to increase rapidly until the end of the 1980s and then to decrease even more rapidly after the 1990s (Figure 7). From 1989 onwards, there is a significant decline of imported N in food and feedstock, reaching only 37.1 Gg N in 2000. From 1961 to 1965, biological fixation of N in agricultural systems was the largest source of Nr in Cuba, but since 1966 the use of inorganic fertilizer becomes more important. From 1994 onwards, NO_x deposition

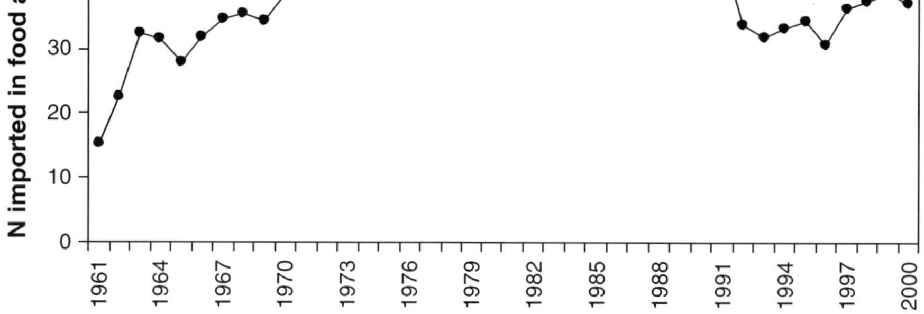

Figure 5. Net import of N to Cuba in foods and feedstocks from 1961–2000.

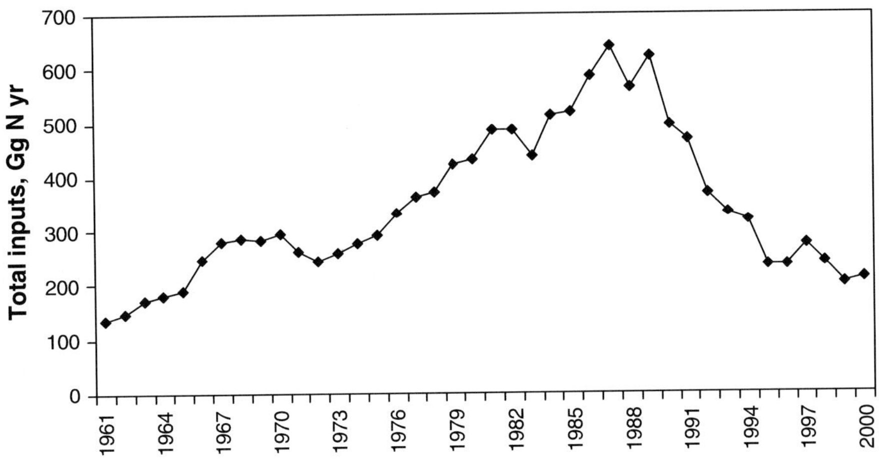

Figure 6. Total N-input to Cuba from 1961–2000. Data on atmospheric deposition was available only from 1983 to 1994 and refers only to deposition of NO_x.

becomes the most important anthropogenic source of Nr in the country budget.

The total inputs in 1987 practically represent more than a 5-fold increase with respect to the pristine values before the Spanish settlement (Figure 8).

In order to compare our data with those previously reported in the literature (Howarth et al. 1996; Yan et al. 2003), total Nr input is presented for three different years (Table 1), which are representative of three different periods of N inputs in the country. The first year (1961) represents an early stage of Cuban agricultural development and the first year when the statistical database

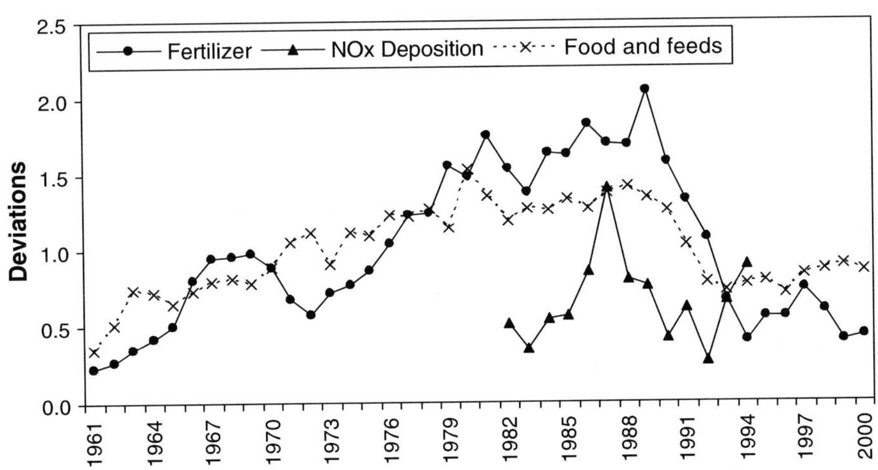

Figure 7. Trends of the inputs from fertilizer, atmospheric deposition and net imports of food and feeds based on standardized data expressed as deviations of the average value.

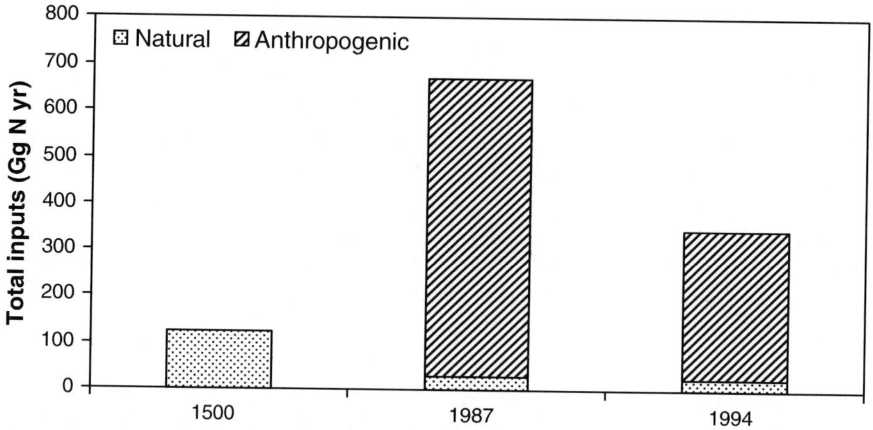

Figure 8. Comparison of N inputs to the Cuban terrestrial landscape from natural and anthropogenic sources, in selected years. 1500 (During the Spanish settlement), 1987 (the year of the highest N-input) and 1994 (representative of the actual conditions).

was more or less complete. The year of maximum input of Nr (1987) occurred after a relatively continuous development and was typical of a period of accelerated use of Nr. This peak was reached one a few years before the partial collapse of the Cuban economy due to the loss of imports from the Soviet Union, which rapidly led to a drastic reduction of imported fuel, fertilizer and food and feedstock.

Table 1. N-inputs to the Cuban landscape for three different years. Results from Cuba are compared with data from 10 temperate regions surrounding the North Atlantic (Howarth et al. 1996) and with data from the Chianjtse River watershed (Yan et al. 2003). Fluxes are in N km^{-2} yr^{-1}. Country area used in the calculation was 107.145 km^{-2}.

	NO_y deposition	Fertilizer	Fixation by crops	Net import in foods	Total
North Canada rivers	70	160	30	−50	210
St. Lawrence basin	610	330	260	−30	1170
NE coasts of US	1200	600	750	1000	3550
SE coasts of US	1020	1170	370	450	3010
Eastern Gulf of Mexico	760	1260	250	580	2850
Mississippi River basin	620	1840	1060	−1300	2220
Baltic Sea drainages	480	1730	30	20	2220
North Sea drainages	1090	5960	5	−5	7050
NW European coast	1090	2870	50	−320	3700
SW European coast	460	3370	15	−65	3780
Chiangtse River	(NA)	3510	1070	(NA)	4580
Cuba 1961	(NA)	383	860	143	1386
Cuba 1987	1707	2860	984	558	6109
Cuba 2000	(NA)	728	822	346	1896

(NA) Not available.

Finally, the year 2000 is representative of present conditions. This phase, began in 1990, and shows the results of a drastic decrease of Nr inputs due to a reduction in the use of inorganic fertilizers concomitantly with similar reductions in nitrogen oxides emission, sugar cane area and imported foods and feedstuff. Table 1 also shows that in the 80s, nitrogen input into the Cuban landscape was one of the highest in the literature, only below that of the North Sea.

It is also interesting to express Nr creation and use on a per capita basis to illustrate the average amount of Nr mobilized per person. North Americans for example, mobilize about 100 kg N $person^{-1}$ yr^{-1} (Howarth, et al. 2002). At the other extreme, people in Africa mobilize about an order of magnitude less, about 7 kg N $person^{-1}$ yr^{-1}. The world average is about 24 kg $person^{-1}$ yr^{-1} (Galloway and Cowling 2002). In 1987, when all N inputs peaked as 640 Gg, the Cuban population mobilized 62 kg $person^{-1}$ yr^{-1} while in 2000 this figure dropped dramatically to 19 kg $person^{-1}$ yr^{-1}.

Nr export to coastal waters

Average estimates of Nr export in rivers using the empirical model presented by Howarth et al. (1996) was 926 kg N km^{-2} yr^{-1} and values ranged from 379 kg N km^{-2} yr^{-1} in 1961 to 1461 kg N km^{-2} yr^{-1} in 1987.

Discussion

Human activity has roughly doubled the availability of nitrogen to terrestrial ecosystems of the planet (Vitousek et al. 1997). This nitrogen fixation increased by some two- to three- fold over the three decades between 1960 and 1990, and continues to grow (Galloway et al. 1995). The effects of N cycling in temperate and tropical ecosystem can be better understood if we can compare N dynamics in pristine or minimally disturbed ecosystems and highly modified ecosystems.

In the pristine biosphere, it is generally believed that N-biological fixation was the dominant source of newly fixed N to the landscape (Cleveland et al. 1999). Before the Spanish settlement in Cuba, we can assume that the small number of persons living in the region, their diet based on marine foods, and subsistence agriculture (Tabío and Rey 1985; Dacal Moure and Rivero de la Calle 1986), caused negligible human impacts on N-cycle. Our estimate of N-biological fixation represents then, a baseline for further assessment of the Nr mobilized by human activities.

It has been emphasized (Downing et al. 1999) that, with the progressive intensification of human development, vast areas of forest have been commonly cleared and replaced by agricultural crops or urban areas. The real extent of the destruction of Cuban forest during the Spanish rule is unknown but all the evidences suggest that this impact has been overemphasized. Although some of the so-called India's Chroniclers calculated the number of Indians who occupied Cuba to have been 200,000 or 300,000, this figures seems to overestimate the true population. Evidences supporting it come from the first census carried out in 1774 (ONE 2002), almost 300 years after the colonization. In this Census, the whole population of Cuba was calculated in only 171,620 inhabitants of all races. A rapid development of the population took place since that time and by 1899 the census reported that about 50% of the country was still forested even when the population reached 1.6 millions inhabitants. Then, although deforestation obviously took place since first Spanish settlement, do not reach disastrous proportions until the first half of the XX century. According to Smith (1954), the trend in the destruction and removal of Cuba's forest date from the treaty of 1903 between Cuba and the United States. This treaty gave sugar a place on the preferred list of imports into the USA. For that reason, land was rapidly cleaned for cane planting, and the forest were extensively cut and burned. In 1950, the forest covered only 18% of the landscape (Levi Marrero 1950; Smith 1954). Since then, anthropogenic N inputs have increased significantly until it reached one of the highest values reported in the literature in the 1980s. As mentioned earlier, the Cuban population in 1987 mobilized 62 kg N person^{-1} yr^{-1}, some 3-fold the global average. The increase of N availability between 1961 and the end of the 80s was mainly associated with the intensification of agriculture and livestock production, supplemented by an increase of the use of fossil fuel consumption and imports of feeds and foodstuff.

Nr inputs from fertilizers

The relative importance of different sources of nutrients varies greatly among different coastal regions of Cuba, depending on the characteristics of the drainage basin, their human populations, the intensity of agricultural activities, and the amount of atmospheric deposition. But overall, fertilizer applications have been the most important source of Nr inputs to the Cuban territory.

In order to sustain economic levels of production, fertilizer use on tropical lands may eventually exceed rates of application in temperate systems (Downing et al. 1999). Special problems arise with crops as sugar cane and rice, which receive large applications of N, but also lose large amount of N by denitrification and volatilization (Peoples et al. 1995). Sugar cane dominates agriculture in Cuba, accounting for approximately half of the cultivated area and has historically consumed most of the fertilizers (FAO 2003).

Agricultural N inputs by fertilizer application (per total area, including non agricultural lands) in the major drainage basin of the United States ranged from 0.9 kg N ha^{-1} yr^{-1} in the Great Basin, to 40 kg N ha^{-1} yr^{-1} in the Upper Mississippi region (Jordan and Weller 1996). When averaged over the entire area of the country, the rate of inorganic fertilizer use in the USA is approximately 13 kg N ha^{-1} yr^{-1} or some 2.2 fold greater than the global average (Howarth et al. 2002). In Cuba, considering the entire area of the country, the yearly average from 1961 to 2000 was 16.8 kg N ha^{-1} yr^{-1}, a 29% higher than the average in USA. Furthermore, in the period of intensive use of inorganic fertilizers in Cuban agriculture, from 1980 to 1989, the average was 28 kg N ha^{-1} yr^{-1}, or 2.1 fold greater than in the USA. Only in the Upper Mississippi region, where a high proportion of the land is devoted to intensive corn and soybean farming (Jordan and Weller 1996), rates of fertilizer use were higher than those in Cuba. Consequently, a large area of anoxia now exists offshore of the Mississippi estuary (Howarth et al. 2000).

Presently, crops and pasture lands cover approximately 53% of the landscape in Cuba, so inorganic fertilizer input to agricultural land was less concentrated than in the USA where this area covers only about 20% of the landscape (Howarth et al. 2002). Nevertheless, by considering only agricultural crops, the rate of application in Cuba, in 1981 and 1989, increased to more than 90 kg N ha^{-1} yr^{-1}, reaching the highest value in 1981 (94 kg N ha^{-1} yr^{-1}). This average is relatively similar to those rates seen in lands intensively farmed with corn in the American mid-west, where approximately 100 kg N ha^{-1} yr^{-1} are applied (Galloway et al. 2004). This figure is likely to be higher if we consider inputs to a particular river basin. This trend has been changing over the last years and nitrogen inputs are now quite reduced due to the drastic shortage in funds for purchase of inorganic fertilizer since 1990 (FAO 2003).

Nr inputs from atmospheric deposition

Total (wet and dry) atmospheric deposition of NO_x averaged 81 Gg N yr^{-1} from 1982 to 1994, with a peak 171 Gg N yr^{-1} in 1987. The figure for 1990 was 51 Gg N yr^{-1} which do not seem to be very high when we compare it with the 142 Gg N yr^{-1} released as NO_x emissions (Centella et al. 2000) in the same year.

Previous studies on atmospheric deposition in Cuba during a 6-year period (Cuesta Santos et al. 1998) reported total deposition values ranging from a minimum of 7.1 kg ha^{-1} yr^{-1} to a maximum of 33.2 kg ha^{-1} yr^{-1}, measured in an urban station. These authors also reported that the maximum values coincides with the growing of the Cuban economy up to 1989, as well as with the increasing generation of pollutants caused by the industry and transportation. They also stated that the higher values are similar to those reported in most of Europe. These values represent a total annual deposition of atmospheric N ranging from 76 to 355 Gg N yr, while the values for oxidized nitrogen compounds ranged from 30 to 1420 Gg N yr. Although this values seems to be relatively high, atmospheric deposition is a strong function of emissions. Once emitted to the atmosphere, nitrogen compounds, particularly NO_x, can travel great distances (~600 to 800 km or farther) from the point of emission to the point of deposition (Dennis 1997). Considering that Cuba is a long island, and also that northeast trade winds are predominant, it is also probable that some of the NO_x emitted from anthropogenic activities in the southeastern United States might be redeposited on the Cuba's land surface, increasing the estimated amount of Nr deposition.

Nr inputs to coastal waters

Once it has been introduced into a terrestrial system, nitrogen has three fates: (1) storage within the system, (2) transferred to another system (discharged to the water or emitted to the atmosphere), or (3) denitrified. Nitrogen inputs can be exported in many forms and through different pathways to the soil, atmosphere, groundwater and rivers. The extent to which agriculture and other human activities contributed to the nitrogen fluxes are very variable and depend on how intense the y are (Howarth et al. 2002). In the United States and Europe, discharges of nitrogen from rivers are highly correlated with increasing human generated nitrogen inputs into the watersheds, particularly from fertilizers and atmospheric deposition (Howarth et al. 1996; Jordan and Weller 1996).

Boyer et al. (2002) examined the relationships between N inputs and riverine N export for 16 catchments basin in the northeast USA using the empirical model from Howarth et al. (1996) for the large regions that drain to the North Atlantic Ocean. These basins encompass a range of climatic variability and are major drainages to the coast of the North Atlantic Ocean along a latitudinal profile from Maine to Virginia. In spite of the possible limitations of this model for tropical regions, we used it in order to have a first approximation of the magnitude of terrestrial N export to Cuban coastal waters. In any case, the drastic reduction experienced by the N inputs during the 1990s, must have also impacted significantly, the receiving marine waters.

One of the best documented and understood consequences of human alterations of the nitrogen cycle is the eutrophication of estuaries and coastal waters (Vitousek et al. 1997). The impact of anthropogenic N in coastal waters have been widely documented in the literature (Ryther and Dunstan 1971; Pearl 1985; Nixon 1995; Caraco and Cole 1999; Cloern 2001; Rabalais 2002); but the dramatic decline of nutrients that occurred in Cuba in the past decade, seems to follow the opposite trend of that seen for other regions of the world and therefore must be seriously considered in future ecological studies.

Some studies suggest (Caddy 2000) that drastic reductions in nutrient inputs to terrestrial landscapes might introduce a rapid change in coastal ecosystems. A good example comes from the dramatic decline of application of fertilizers by some of the countries of Eastern Europe. Discharges of phosphorus and subsequently nitrogen rapidly declined at the beginning of the 90s and by 1996; there was no hypoxic zone on the shelf of the Black Sea for the first time in 23 years (Boesch 2001). According to Kideys (2002), the decrease in nutrient inputs to the Black Sea were immediately reflected in measurements of nutrient from coastal waters.

It is recognized that there are three main categories of nutrient enrichment processes in the coastal zone: (1) coastal upwelling, (2) tidal mixing and (3) land-based runoff and major river outflow (Caddy and Bakun 1994). There is strong circumstantial evidence worldwide that nutrient-enriched riverine discharges enhance fishery production in adjacent shelves (Grimes 2001). It has been also hypothesized (Baisre 2000; Caddy per.com.), that there is a relationship between the drastic decrease of fertilizer applications in agricultural crops in the 1990s in Cuba, and a drop in coastal fisheries during the same period.

The reduction of the extent of brackish areas and wetlands by excessive freshwater extraction upstream will also have adverse impacts on marine species which are dependent on brackish habitat for part of their life history (Aleem 1972; Deegan et al. 1986; Caddy and Bakun 1995). A reduction of freshwater flow might concomitantly reduce the extent of the brackish water area and to provoke the salt-water invasions to the estuary and lower river system that may have severe impacts on estuarine dependent species (Deegan et al. 1986). The building of dams on rivers increases water retention and rates of degradation and sedimentation of particulate organic matter with the new impoundment and reservoirs become effective nutrient sinks (Stockner et al. 2000). In Cuba, the main rivers have been regulated and water volume in artificial reservoirs now represents 24.9% of the nation's internal renewable natural water resources (Cubagua 2004). The effects of river damming in Cuba and the concomitant reduction of nutrient inputs to the coastal zone are reflected in the dramatic decrease of the landings of the more typical estuarine species like mullets (Mugilidae), shrimps (Penaeidae), gerrids (Gerridae) and mangrove oyster (*Crassostrea rhizophorae*) (Figure 9).

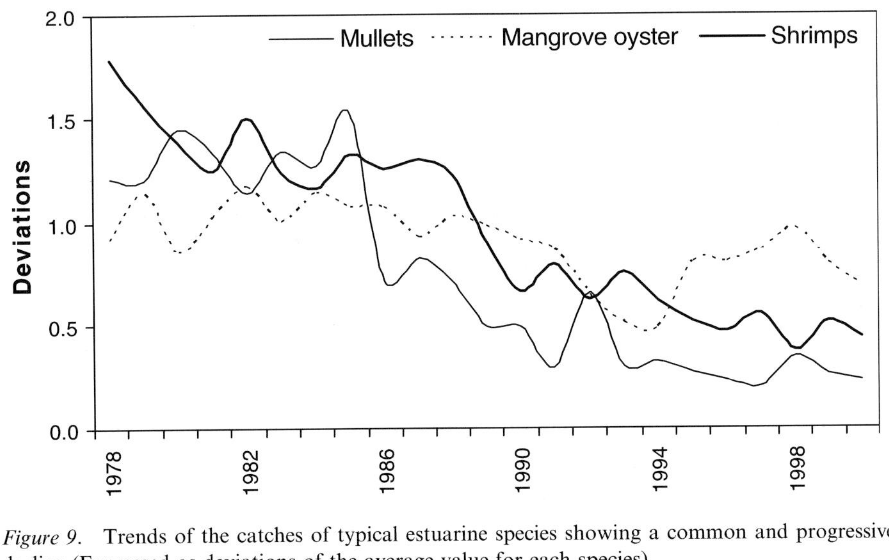

Figure 9. Trends of the catches of typical estuarine species showing a common and progressive decline (Expressed as deviations of the average value for each species).

Synchronous anthropogenic effects on marine coastal systems make it difficult to separate effects of fishing from terrestrial inputs, especially of those caused by nutrient runoff because as far as the different fisheries are close or beyond the top of the yield curve, the effects of environmental changes, natural or anthropogenic, are likely to predominate (Caddy 2000). In Cuba, it seems to be possible that river damming, acting synergistically with the drastic nutrient reduction previously discussed, provide a more comprehensive explanations for the decline of marine coastal fisheries (Baisre 2000) experienced since 1990.

Acknowledgments

Drs. Robert Howarth, Solange Filoso and Luiz Martinelli, not only stimulated my interest in this problem but also provided considerable help during the preparation of this paper as well as reviewing the first draft. I also thank to Dr. Carlos López who kindly provide the information on atmospheric deposition and to Dr. John F. Caddy who reviewed the manuscript and called my attention to the fishery implications of reduction in the fertilizer use in Cuba. I also thank to two anonymous reviewer whose comments improved very much the manuscript.

References

Aleem A.A. 1972. Effect of river outflow management on marine life. Mar. Biol. 15: 200–208.

Alexander R.A., Johnes P.J., Boyer E.W. and Smith R.A. 2002. A comparison of methods for estimating the riverine export of nitrogen from large watersheds. Biogeochemistry 57/58: 295–339.

Baisre J.A. 1985. Los Complejos Ecológicos de Pesca: Su importancia en la administración de las pesquerías cubanas. FAO Fish Rep. 278(Suppl.): 251–272.

Baisre J.A. 2000. Chronicle of Cuban Marine Fisheries 1935–1995: Trend analysis and fisheries potential. FAO Fish. Tech. Pap. 394, 26 pp.

Boesch D.F. 2001. Causes and consequences of nutrient overenrichment of coastal waters. In: Ragaini R. (ed.), International Seminar on Nuclear War and Planetary Emergencies. 25th Session, World Scientific publishing, Singapore, pp. 165–180.

Boyer E.W., Goodale C.L., Jaworski N.A. and Howarth R.W. 2002. Anthropogenic nitrogen sources and relationships to riverine nitrogen export in the northeastern U.S.A. Biogeochemistry 57/58: 137–169.

Caddy J.F. 1993. Towards a comparative evaluation of human impacts on fishery ecosystems of enclosed and semi-enclosed seas. Rev. Fish. Sci. 1(1): 57–95.

Caddy J.F. and Bakun A. 1994. A tentative classification of coastal marine ecosystems based on dominant processes of nutrient supply. Ocean Coastal Manag. 23: 201–211.

Caddy J.F. and Bakun A. 1995. Marine Catchment Basins and anthropogenic effects on coastal fishery ecosystems. FAO Fish. Rep. 349: 119–133.

Caddy J.F. 2000. Marine catchment basin effects versus impact of fisheries on semienclosed seas. ICES J. Mar. Sci. 57: 628–640.

Caraco N.F. and Cole J.J. 1999. Human impact of nitrate export: an analysis using major world rivers. Ambio 28(2): 167–170.

Centella A., Llanes J., López C. and Limia M. 2000. First National Report to the United Nation Convention on Climate Change (Spanish). Instituto de Meteorología, La Habana, Cuba.

Cleveland C.C., Townsend A.R., Schimel D.S., Fisher H., Howarth R.W., Hedin L.O., Perakis S.S., Latty E-.F-., Von Fischer J.C., Elseroad A. and Mason M.F. 1999. Global patterns of terrestrial biological nitrogen (N_2) fixation in natural ecosystems. Global Biochem. Cycles 13: 623–645.

Cloern J.E. 2001. Our evolving conceptual model of the coastal eutrophication problem. Mar. Ecol. Prog. Ser. 210: 223–253.

Cubagua 2004. Cuban data base on water resources. (http://www.hidro.cu).

Cuesta Santos O., Ortiz Bulto P.L. and González M.L. 1998. Deposition and atmospheric nitrogen concentrations trends in Cuba. Water, Air, Soil Pollution 106: 163–169.

Dacal Moure R. and Rivero de la Calle M. 1986. Arqueología aborigen de Cuba. Editorial Gente Nueva, La Habana, 174 pp.

Deegan L.A., Day J.W. Jr., Yañez-Arencibia A., Gosselink J.G., Soberón-Chávez G. and Sánchez-Gil P. 1986. Relationships among physical characteristics, vegetation, distributions, and fisheries yield in Gulf of Mexico estuaries. In: Wolfe D.A. (ed.), Estuarine Variability, Academic Press, NewYork, pp. 83–100.

Dennis R. 1997. Using the Regional Acid Deposition Model to determine the nitrogen deposition airshed of the Chesapeake Bay watershed. In: Baker J.E. (ed.), Atmospheric Deposition of Contaminants to the Great Lakes and Coastal Waters, SETAC Press, Pensacola, Florida, pp. 393–413.

Downing J.A., Mcclain M., Twilley R., Melack J.M., Elser J., Rabalais N.N., Lewis W.M. Jr., Turner R.E., Corredor J., Soto D., Yañes-Arencibia A., Kopaska J.A. and Howarth R.W. 1999. The impact of accelerating land-use change on the N-Cycle of tropical aquatic ecosystems: current conditions and projected changes. Biogeochemistry 46: 109–148.

FAO 2003. Fertilizer Used by Crop in Cuba. Food and Agriculture Organization of the United Nations, Rome, 28 pp.

FAOSTAT 2004. Food and Agriculture Organization of the United Nations FAOSTAT Agriculture Data. (http://www.apps.fao.org/).

Galloway J.N. and Cowling E.B. 2002. Reactive nitrogen and the world: 200 Years of change. Ambio 31(2): 64–71.

Galloway J.N., Dentener F.J., Capone D.G., Boyer E.W., Howarth R.W., Seitzinger S.P., Asner G.P., Cleveland C., Green P., Holland E., Karl D.M., Michaels A.F., Porter J.H., Townsed A. and Vorösmarty C. 2004. Nitrogen cycles: past, present and future. Biogeochemistry 70: 153–226.

Galloway J.N., Schlesinger W., Levy H. II, Michaels A. and Schnoor J.L. 1995. Nitrogen fixation: Atmospheric enhancement-environmental response. Global Biochem. cycles 9: 235–252.

Grimes C.B. 2001. Fishery production and the Mississippi river discharge. Fisheries, 26(8): 17–26.

Howarth R.W., Billen G., Swaney D., Townsend A., Jaworski N., Lajtha K., Downing J.A., Elmgen R., Caraco N., Jordan T., Berendse F., Freney J., Kudeyarov V., Murdoch P. and Zhu Zhao-Liang 1996. Regional nitrogen budgets and riverine N & O fluxes for the drainages to the North Atlantic Ocean: Natural and human influences. Biogeochemistry 35: 75–139.

Howarth R.W., Anderson D., Cloern J., Elfring C., Hopkinson C., Lapointe B., Malone T., Marcus N., McGlathery K., Sharpley A. and Walker D. 2000. Nutrient Pollution of coastal rivers, bays and seas. Issues Ecol. 7, 15 pp.

Howarth R.H., Boyer E., Pabich W. and Galloway J.N. 2002. Nitrogen use in the United States from 1961–2000, and estimates of potential future trends. Ambio 31: 88–96.

IIIA (1985). Tabla de Composición de Alimentos. Instituto de Investigaciones de la Industria Alimenticia e Instituto Nacional de Higiene de los Alimentos, 6 pp.

IFA (International Fertilizer Industry Association) 2004. Nitrogen, phosphate and potash statistics (http://www.fertilizer.org/ifa/statistics/IFADATA/dataline.asp).

Jordan T.E. and Weller D.E. 1996. Human contributions to terrestrial nitrogen flux. Bioscience 46: 655–664.

Kideys A.E. 2002. Fall and rise of the black sea ecosystem. Science 297: 1482–1484.

Marrero L. 1950. *Geografía de Cuba.* Talleres Tipográficos Alfa, La Habana, 736 pp.

Nixon S.W. 1995. Coastal marine eutrophication: a definition, social causes, and future concerns. Ophelia 41: 199–219.

Matson P.A., Mc Dowell W.H., Townsend A.R. and Vitousek P.M. 1999. The globalization of N deposition: ecosystem consequences in tropical environments. Biogeochemistry 46: 67–83.
ONE 2002. Oficina Nacional de Estadísticas. Anuario Estadístico de Cuba 2001, (Yearbook of Statistics 2001) (Spanish), 335 pp.
Pearl H. 1985. Enhancement of marine primary production by nitrogen-enriched acid rain. Nature 315: 747–749.
Peoples M.B., Herridgeand D.F. and Ladha J.K. 1995. Biological nitrogen fixation: an efficient source of nitrogen for sustainable agricultural production?. Plant and soil 174: 3–28.
Rabalais N.N. 2002. Nitrogen in aquatic ecosystems. Ambio 31: 102–112.
Ryther J. and Dunstan W. 1971. Nitrogen, phosphorus, and eutrophication in the coastal marine environment. Science 171: 1008–1012.
Schimel D.S., Braswell B.H., McKeown R., Ojima D.S., Parton W.J. and Pulliam W. 1996. Climate and nitrogen controls on the geography and timescales of terrestrial biochemical cycling. Global Biogeochem. Cycles 10: 677–692.
Smil V. 1997. Global population and the nitrogen cycle. Scientific American July: 76–81.
Smil V. 1999. Nitrogen in crop production: an account of global flows. Global Biochem. Cycles 13: 647–662.
Smith E.E. 1954. The Forest of Cuba. Maria Moors Cabot Foundation Publ. 2, 98 pp.
Stockner J.G., Rydin E. and Hyenstrand P. 2000. Cultural oligotrophication: causes and consequences for fisheries resources. Fisheries 25(5): 7–14.
Tabío E.E. and Rey E. 1985. *Prehistoria de Cuba.* Editorial de Ciencias Sociales, La Habana, 234 pp.
Vitousek P.M., Aber J.D., Howarth R.W., Likens G.E., Matson P.A., Schindler D.W., Schlesinger W.H. and Tilman D.G. 1997. Human alteration of the global nitrogen cycle: sources and consequences. Ecol. Appl. 7: 737–750.
Watanabe I. 1986. Nitrogen fixation by non-legumes in tropical agriculture with special reference to wetland rice. Plant Soil 90: 343–357.
Yan W., Zhang S., Sun P. and Seitzinger S.P. 2003. How do nitrogen inputs to the Changjiang basin impact the Changjiang river nitrate: A temporal analysis for 1968–1997. Global Biochem. Cycles 17(4): 1091–1100.
WRI 2001. Nutrient Overload Unbalancing the Global Nitrogen Cycle. Earth Trends 2001. World Research Institute, pp.1–3.

Biogeochemistry (2006) 79: 109–133
DOI 10.1007/s10533-006-9005-y

Urban influences on the nitrogen cycle in Puerto Rico

JORGE R. ORTIZ-ZAYAS[1,*], ELVIRA CUEVAS[2], OLGA L. MAYOL-BRACERO[1], LORETO DONOSO[3] IVONNE TREBS[4], DEBORA FIGUEROA-NIEVES[5] and WILLIAM H. MCDOWELL[5]
[1]*Institute for Tropical Ecosystem Studies, San Juan, Puerto Rico;* [2]*Department of Biology, University of Puerto Rico, 21910, San Juan, 00931-1910, Puerto Rico;* [3]*Instituto Venezolano de Investigaciones Científicas (IVIC), Caracas, Venezuela;* [4]*Max Planck Institute for Chemistry, Mainz, Germany;* [5]*University of New Hampshire, Durham, NH, USA; *Author for correspondence (e-mail: jrortiz@ites.upr.edu)*

Key words: Nitrogen, Peurto Rico, Urbanization, Land use change, Waste water disposal, Atmospheric nitrogen deposition, Caribbean, Water quality, Air quality

Abstract. Anthropogenic actions are altering fluxes of nitrogen (N) in the biosphere at unprecedented rates. Efforts to study these impacts have concentrated in the Northern hemisphere, where experimental data are available. In tropical developing countries, however, experimental studies are lacking. This paper summarizes available data and assesses the impacts of human activities on N fluxes in Puerto Rico, a densely populated Caribbean island that has experienced drastic landscape transformations over the last century associated with rapid socioeconomic changes. N yield calculations conducted in several watersheds of different anthropogenic influences revealed that disturbed watersheds export more N per unit area than undisturbed forested watersheds. Export of N from urban watersheds ranged from 4.8 kg ha^{-1} $year^{-1}$ in the Río Bayamón watershed to 32.9 kg ha^{-1} $year^{-1}$ in the highly urbanized Río Piedras watershed and 33.3 kg ha^{-1} $year^{-1}$ in the rural-agricultural Río Grande de Añasco watershed. Along with land use, mean annual runoff explained most of the variance in fluvial N yield. Wastewater generated in the San Juan Metropolitan Area receives primary treatment before it is discharged into the Atlantic Ocean. These discharges are N-rich and export large amounts of N to the ocean at a rate of about 140 kg ha^{-1} $year^{-1}$. Data on wet deposition of inorganic N ($NH_4^+ + NO_3^-$) suggest that rates of atmospheric N deposition are increasing in the pristine forests of Puerto Rico. Stationary and mobile sources of NO_x ($NO + NO_2$) and N_2O generated in the large urban centers may be responsible for this trend. Comprehensive measurements are required in Puerto Rico to quantitatively characterize the local N cycle. More research is required to assess rates of atmospheric N deposition, N fixation in natural and human-dominated landscapes, N-balance associated with food and feed trade, and denitrification.

Introduction

Human activities have caused unprecedented changes in the nitrogen (N) gaseous fluxes between the atmosphere and the Earth's ecosystems (Galloway et al. 1995; Holland et al. 1999; Matson et al. 1999; Socolow 1999). It is estimated that the application of fertilizers, the development of N-fixing crops (legumes and forages), and fossil fuel burning adds about 140 Tg of N per year to terrestrial environments (Vitousek et al. 1997). This rate matches the upper end of the natural N fixation rate associated with lightning and N-fixing algae and bacteria, estimated at 90–140 Tg per year prior to the Industrial Revolution (Galloway et al. 1996). In addition, human actions such as burning of forests, wood fuels, and grasslands, the draining of wetlands and the clearing of land for crops are depleting the N stored in soil organic matter and in tree biomass at a rate of about 60 Tg per year (Vitousek et al. 1997).

The impacts of these anthropogenic activities are evident in the atmospheric N budget and in the ecosystem functioning. For example, increased emissions of nitrous oxide (N_2O) are altering the heat balance of the Earth (Vitousek et al. 1997). This gas absorbs infrared radiation emitted by the Earth and is thought to contribute significantly to the greenhouse effect. In addition, N_2O contributes to the thinning of the stratospheric ozone layer that shields the Earth from damaging UV cosmic radiation. In urban areas, the production of nitric oxide (NO) from the combustion of fossil fuels controls ozone formation. High ambient ozone concentrations initiate and exacerbate respiratory illness (Townsend et al. 2003). In the atmosphere, NO is oxidized in two steps to form nitric acid (HNO_3), a compound that is responsible for increased rain acidity (e.g., Meixner 1994). Gaseous ammonia (NH_3), a byproduct of forest burning, of animal waste oxidation, and application of fertilizers, buffers the acidifying effect of NO (through its oxidation product HNO_3), and, therefore influences the chemistry of atmospheric aerosols, water vapor, and rain. When deposited to soil surfaces NH_3 can be considered as an acid, since it is rapidly nitrified in soils, a process that directly forms H^+ and NO_3^- (Galloway 1998).

Changes in the global N cycle create a cascade of effects with direct impacts on the global carbon cycle. The increase in wet and N dry deposition associated with increasing concentrations of atmospheric N species may promote plant growth and hence increase carbon sequestration by plants, particularly in N-limited ecosystems of the temperate and boreal regions (Vitousek et al. 1997). This increase in carbon storage may represent an important sink in the global carbon budget that counteracts the increase in atmospheric carbon dioxide associated with the combustion of fossil fuels. However, as these ecosystems become saturated with N, other elements such as phosphorus, calcium, and water may become limited thus reducing plant growth. This condition leads to high losses of N that have caused acidification of lakes and streams in some regions (Aber et al. 2003). Other ecosystem-level impacts associated with increased N inputs include changes in biodiversity, intensified trace gas (particularly NO and N_2O) exchange, increased cation leaching, and changes in estuarine trophic structure (Howarth et al. 1996; Aber et al. 1998; Matson et al. 1999).

The complex nature of interactions between humans and the global N cycle mandates the need for the determination of N fluxes at the regional and/or national level. Although research and environmental monitoring must continue in temperate regions where much of the anthropogenic N has been generated (Howarth et al. 1996), additional research is needed in developing tropical regions where particularly land-use change is expected to alter natural N fluxes (Matson et al. 1999; Trebs et al. 2006).

This paper evaluates the effects of human activities on the N flux in Puerto Rico, a densely populated tropical Caribbean island that has experienced drastic landscape transformations associated with rapid socioeconomic changes (Hunter and Arbona 1995; Thomlinson and Rivera 2000; López et al. 2001; Grau et al. 2004). Available hydrologic data are used to assess fluvial N fluxes from four watersheds in Puerto Rico with varying levels of human impacts and water yield. These are compared with N fluxes to the ocean from wastewater treatment plants located in urban centers. Collected data on rain chemistry were evaluated to estimate rates of N wet deposition. Fertilizer use in Puerto Rico was reviewed to assess its effect on local N budgets. Finally, future research needs are proposed to further improve our understanding of N dynamics in Puerto Rico and in the tropics.

Methods

The study site

With an area of 889,500 ha, Puerto Rico is the fourth largest island in the Caribbean (Figure 1). It is centered at 18°15′ N, 66°30′ W. The interior of the island has rugged topography with elevations reaching up to 1300 m above mean sea level. Geologic substrates include: alluvial, sedimentary, volcanic, limestone, and serpentine (Pico 1975). Average annual precipitation ranges from 2100 mm in the humid uplands, 1600 mm in the humid coastal plains, 1150 mm in the semiarid mountains and valleys, to 900 mm in the semiarid coastal plains. The average annual temperature is 24 °C in the humid uplands, 25 °C in the humid coastal plains, 26 °C in the semiarid mountains and valleys, and 26 °C in the semiarid coastal plains. On average, the island receives about 1753 mm of rain per year, of which 1057 mm are evaporated and transpired by vegetation, 622 mm are discharged to the ocean through rivers and underground aquifers, with the remaining stored in rivers, reservoirs, and cycled internally (DNER 2005). Seventeen major rivers drain the island, with the largest ones found on the Northern slopes, which are wetter than the Southern slopes.

During the 20th century, Puerto Rico experienced dramatic socioeconomic development of unprecedented proportions for many tropical countries. From its discovery in 1493 until 1899, when Puerto Rico became a United States possession, its population increased from about 30,000 to 953,243 inhabitants at an average rate of 2274 inhabitants year^{-1} (Picó 1975). Over the 20th century, the population increase was much faster, at a rate of 28,271 inhabitants year^{-1}. With a population of 428 inhabitants km^{-2} (Puerto Rico Planning Board 1995), Puerto Rico is one of the densest territories of the world. Major urban settlements are concentrated on the coastal plains in the cities of the San Juan Metropolitan Area (SJMA; e.g., Trujillo Alto, Bayamón, Carolina),

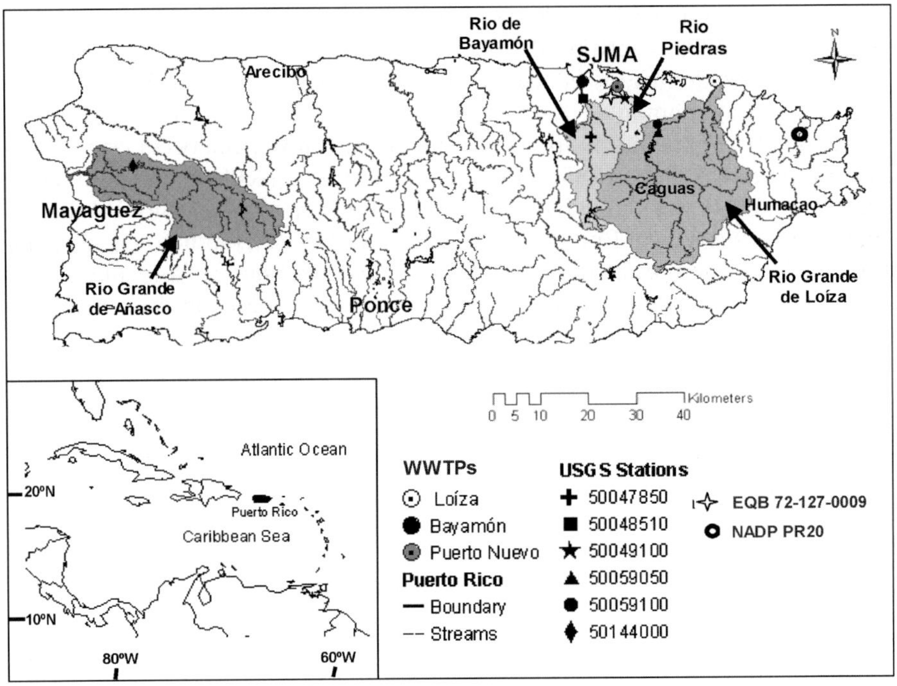

Figure 1. Location of wastewater treatment facilities, hydrologic USGS stations, and atmospheric deposition measurement stations in Puerto Rico referenced in this study.

Ponce, Mayaguez, Arecibo, and Humacao and inland cities such as Caguas (Figure 1). Rural areas, however, are heavily populated. Thus Puerto Rico can be considered to be an urban island. Only 35,000 ha or 4% of its territory are protected lands.

These demographic patterns have dramatically altered the Puerto Rican landscape. Puerto Rico was mostly forested in the 16th century. However, by 1828 only 61% forest cover remained due to timber cutting and development of non-forested land uses. By 1899, forest cover was reduced to 20% and coffee was the principal crop. By 1931, only 9% of forest remained while the production of coffee was declining. Boosted by an expansion of the sugar cane industry, deforestation continued. By late 1940, only 6% of the land area remained forested. In 1948, a new economic model based on manufacturing (Operation Bootstrap) was initiated and the sugar cane production began to decrease (Pico 1975). In 1952, Puerto Rico became a commonwealth associated with the United States. This political status allowed for the free trade of products and unrestricted immigration between the two countries. Between 1950 and 1980, Puerto Rico experienced rapid economic growth with $>$10-fold increase in its gross domestic product, labor wages, and personal annual income (Pico 1975; Dietz 1986). With industrialization, forest cover began to increase due to natural succession in abandoned agricultural lands, reaching

34% by 1985 (Birdsey and Weaver 1987) and 42% by 1992 (Helmer et al. 2002). The per capita GDP in Puerto Rico is one of the highest of Latin American and Caribbean ($16,800 in 2003 in purchasing power parity; http://www.cia.gov).

The socioeconomic transformations that led to reforestation in Puerto Rico also resulted in an increase in energy consumption. Because Puerto Rico is dependent almost exclusively on fossil fuels, per capita carbon emissions are among the highest in the Neotropics (about 2.5 Mg C person^{-1} year^{-1}; Grau et al. 2004). The effects of these emissions on the atmospheric deposition of N are still uncertain; however, there is consensus that N emissions associated with fuel combustion in Puerto Rico are increasing (DNER 1999).

Fluvial export of nitrogen

Flow-weighted means of the total N concentration were calculated for four watersheds in Puerto Rico based on data collected at US Geological Survey (USGS) streamflow and water quality stations (Table 1). Total N concentration and instantaneous discharge data are available quarterly at each water quality station. At streamflow stations, mean daily discharge data are also available.

Río Grande de Loiza watershed

This is the largest watershed in Puerto Rico. The Río Grande de Loiza drains into the Atlantic Ocean near the municipalities of Loiza and Carolina. This watershed supplies the San Juan Metropolitan area with an average of 3.7×10^5 m^3 day^{-1} withdrawn from the Carraizo Reservoir. The reservoir was built in 1953, and its current capacity is estimated at 18.1×10^6 m^3; Ortiz-Zayas et al. 2004). Estimates of mean annual runoff below the reservoir are impacted by these extractions (Table 1). It is a semi-rural basin with about 13% of its area urbanized (Gould et al. 2005). According to Osterkamp (2001), about 21% of the watershed is forested and about 58% of this basin is dominated by agriculture and pasture. Two major wastewater treatment plants service most of the population in the watershed: the Caguas Regional Wastewater Treatment Plant and the Carolina Regional Wastewater Treatment Plant. The Caguas plant discharges secondary treated effluent back into the Río Grande de Loíza while Carolina discharges primary treated effluent to the Atlantic Ocean via a submerged outfall.

N export from the Río Grande de Loíza was calculated from data on total N collected at the Río Grande de Loíza below Trujillo Alto (station 50059100; Table 1) and discharge data collected at the Río Grande de Loíza below the dam site (station 50059050) as an approximation of that in station 50059100. Station 50059050 is located 4 km upstream from station 50059100 and there are no major tributaries discharging between the stations. To evaluate the relationship between discharge and N concentration at the 50059100 station, a linear regression was developed between the logarithm of the concentration of

Table 1. Summary of information on the water-quality stations in Puerto Rico analyzed. Based on data from Díaz et al. (2004).

Station name and number	Period of record		Elevation, m amsl	Drainage area, km^2	Runoff, mm $year^{-1}$	% Urban land in watershed[a]
	Start	End				
Río Grande de Loíza below dam site, 50059050	Dec-1986	Sep-2002	10	541.0	427[b]	13
Río Grande de Loíza below Trujillo Alto, 50059100	Jan-1981	Aug-2001	8	552.0	–	13
Río de Bayamón near Bayamón, 50047850	Oct-1988	May-2002	30	108.3	302	22
Río de Bayamón at flood channel, 50048510	Dec-1973	May-2002	5	186.2	–	22
Río Piedras at Hato Rey, 50049100	Dec-1973	Apr-2002	5	39.4	1208	63
Río Grande de Añasco nr San Sebastían, 50144000	Jan-1974	Sep-2002	31.6	244.2	1202	5

[a]Land use data based on Gould et al. (2005) and watershed boundaries of the USGS (2002).
[b]Water yield at this site is affected by water withdraws for water supply at the Carraizo Reservoir.

total N (TN, mg l^{-1}) and the logarithm of instantaneous discharge (Q, ft^3 s^{-1}) at the time of sampling for the period of record. The equation obtained, Log TN = 0.2069*log Q − 0.1784 ($r^2 = 0.19$, p-value < 0.0001), was used to estimate TN concentration from a mean daily discharge series at the 50059050 station. An average annual total N load (kg $year^{-1}$) was calculated by summing the mean daily N concentration for each year over the period of record. The N yield per unit area of watershed (kg ha^{-1} $year^{-1}$) was calculated as the mean of the annual concentration series divided by the watershed area (cf. McDowell and Asbury 1994; Swistock et al. 1997).

Río de Bayamón watershed

The Río de Bayamón watershed drains the Western part of the SJMA (Figure 1). This basin is highly influenced by human activities with about 22% of its area urbanized. Secondary-growth forests and fallow lands predominate in the upper watershed. The lower 6 km of the river are impacted by a straight earthen channel with raised levees for flood control. Cidra Reservoir, with a volume of 6.6×10^6 m^3 (Ortiz-Zayas et al. 2004) is located in the upper part of the watershed and provides water supply to the Guaynabo area (Ramos-Ginés 1997). The Bayamón Regional Wastewater Treatment Plant services the inhabitants of the watershed and nearby communities (Figure 1). This plant provides primary treatment to wastewater before it is discharged to the Atlantic Ocean through a submerged outfall.

The USGS operates a water-quality station located 5.1 km above the river mouth (50048510; Table 1) and a streamflow gauging station (50047850) located upstream (Figure 1). The lack of a significant relationship between total N concentration and instantaneous discharge at the time of water sampling suggests that total N in the lower Río de Bayamón is not sensitive to flow. Therefore, the concentration of total N from 1974 to 2002 was averaged (1.88 mg N/L; SD = 1.16). Mean daily flow at station 50048510 was obtained from a relationship developed between the instantaneous flow at station 50048510 and mean daily flow at station 50047850. A flow-weighted mean concentration of total N was calculated for each year as the product of the mean daily flow at station 50048510 and the mean concentration of total N. An average annual total N load was computed from the annual series. The N yield per unit area of the watershed was calculated as described above.

Río Piedras watershed

This watershed, located in the center of the SJMA (Figure 1), is the most urbanized of the four watersheds studied with urban land covering 63% of the watershed. While some rural sections in the upper watershed are not connected to the public sewer system, the lower section is connected to the Puerto Nuevo Wastewater Treatment Plant. This plant provides primary treatment to wastewater, which is then discharged to the ocean via the Bayamón submerged outfall.

Since 1973, the USGS has monitored water quality at a station located in Hato Rey (Table 1). The USGS has also monitored streamflow at this station since 1988. There was no relationship between the concentration of total N and discharge indicating that total N in Río Piedras is not sensitive to streamflow. Therefore, the mean concentration of total N was calculated for the period of record (2.87 mg N/L; SD = 1.97). The average annual total N load and the N yield per unit area were calculated as described above.

Río Grande de Añasco watershed

This is a rural watershed located in Western Puerto Rico (Figure 1). Urban land cover is minimal (Table 1) and agricultural activity is sparse at mid and high elevations. Three reservoirs, located in the upper part of the watershed (Guayo, Prieto, and Yahuecas) transfer part of runoff generated in this watershed to an outside watershed for hydroelectric generation, irrigation, and water supply (Ortiz-Zayas et al. 2004).

The USGS operates a combined water quality and streamflow station in this watershed (50144000; Table 1). Water quality records are available since 1974 while streamflow data are available since 1963. Total N concentration and discharge were statistically related (Log TN = 0.5047*log Q − 1.1170; $r^2 = 0.43$, p-value < 0.0001), at this station, and thus the mean daily discharge series for the period of record was used to estimate the mean daily total N concentration. A flow-weighted mean concentration of total N was calculated for each year as the product of the mean daily flow at station 50014400 and the mean daily concentration of total N. An average annual total N load was computed from the annual series from which a nitrogen yield per unit area was calculated.

Nitrogen export via wastewater

Wastewater generated in the SJMA is treated in three regional wastewater treatment plants (WWTPs): Carolina, Puerto Nuevo, and Bayamón (Figure 1; Table 2). The three plants serve about 798,000 inhabitants and are operated by the Puerto Rico Aqueduct and Sewer Authority (PRASA). The WWTPs offer primary treatment to wastewater with up to 80% removal of suspended solids and up to 50% removal of organic matter (Table 2). The three plants discharge

Table 2. Design parameters of the wastewater treatment system for the SJMA. Data from PRASA (2005).

WWTP	Population served	Flow (mgd)	BOD_5		Total suspended solids	
			% Removal	Effluent quality (mg l^{-1})	% Removal	Effluent quality (mg l^{-1})
Bayamón	224,155	40	50	130	80	75
Puerto Nuevo	395,041	72	38	130	78	75
Carolina	178,901	45	–	130	57	70
Total	798,097	157				

their treated effluents to the Atlantic Ocean through two ocean outfalls: the Bayamón/Puerto Nuevo combined outfall and the Carolina outfall.

The WWTPs are not designed to remove N from wastewater. An unknown amount of organic N associated with the nitrogenous oxygen demand might be incidentally removed with primary treatment. However, given the low removal efficiency of organic matter, this amount would be small. Therefore, effluents discharged to the ocean are N-rich with a designed effluent concentration of inorganic N (nitrate, nitrite, and ammonium) of 54 mg l^{-1} (PRASA 2005).

The mean concentration of total N in wastewater and the total load of N discharged to the ocean from each of the three WWTPs were calculated based on historic flow and effluent chemistry data reported by PRASA in their discharge monitoring reports to EPA. PRASA does not have precise data on areas serviced by its WWTPs. Therefore, in order to estimate the N yield per unit area, the serviced area was estimated based on the product of the population served and the population density of each serviced area using data from the Puerto Rico Planning Board (2005). N yield per unit area associated with each of PRASAs sewer systems was estimated from the load and estimated serviced area of each WWTP.

Atmospheric N deposition

Data about total (wet + dry) atmospheric N deposition based on field measurements are lacking in Puerto Rico. There is only one field station on the island where rain water has been collected continuously since 1985. The field site (PR-20) is operated within the framework of the National Atmospheric Deposition Program (NADP) and is located in the Luquillo Experimental Forest in Northeastern Puerto Rico (Figure 1). Annual wet deposition data for inorganic N ($NH_4^+ + NO_3^-$) are available for the NADP site from 1985 to 2003 (http://www.nadp.sws.uiuc.edu). Rain water was also collected at two locations in the Luquillo Mountains (El Verde field station and Bisley Experimental Watersheds). Data from the initial years of operation of the NADP PR-20 station, as well as bulk precipitation chemistry in the Luquillo Mountains, were summarized by McDowell et al. (1990).

The most important contributors to N dry deposition are the atmospheric trace gases NH_3, HNO_3, HNO_2 (nitrous acid) as well as NO_x ($NO + NO_2$). Up to date, however, no measurements to quantify the surface-atmosphere exchange of these compounds have been carried out in Puerto Rico. The Puerto Rico Environmental Quality Board (EQB) has been monitoring NO_2 at a station located in the SJMA (EQB-72-127-0009; Figure 1) from May 2000 to June 2004. However, because micrometeorological quantities were not measured at this urban site and NO_2 measurements have been affected by strong local pollution, the data were not used to estimate surface-atmosphere exchange of this compound. In order to obtain a preliminary estimate of the expected total N (dry + wet) deposition in Puerto Rico our data for N wet deposition were related to results from Howarth et al. (1996), Galy-Lacaux et al. (2003), and Trebs et al. (2006).

Fertilizer use

Data on fertilizer use in Puerto Rico were compiled and published by the Association of American Plant Food Control Officials (AAPFCO 2002). These data, available from 1994 to 2002, were used to calculate N inputs to Puerto Rico from fertilizers. These data as well as cropland data (USDA 2004) were analyzed to compute N application indexes for Puerto Rico. N-fertilizer use in Puerto Rico was compared to other Latin American and Caribbean countries based on similar indexes developed by Martinelli et al. (this issue). An indirect parameter to evaluate the efficiency of the N-fertilizer use is the ratio of total cereal production to N fertilizer consumption for all crops, which in turn is a crude proxy of the N use efficiency (NUE) (Cassman et al. 2002).

Results

Fluvial fluxes of nitrogen

Land use in the studied watersheds ranged from highly urbanized in the Río Piedras (63% urban) to rural areas in the Río Grande de Añasco (5% urban) with some agriculture mainly of bananas, coffee, citrus, plantains, and yams (Sotomayor-Ramirez et al. 2004). These watersheds and others with published N export data were grouped into three land use categories: urban, agricultural-rural, and forested (Table 3).

In general, urban and agricultural-rural watersheds had similar and highly variable N yields (Table 3). Export of N from urban watersheds ranged from 4.8 kg N ha^{-1} $year^{-1}$ in the Río Bayamón watershed to 32.9 kg N ha^{-1} $year^{-1}$ in the highly urbanized Río Piedras watershed. Published N yields for urban sewered and unsewered watersheds in Puerto Rico (Ramos-Ginés 1997) were within this range. N export rates in agricultural-rural watersheds ranged from 6.9 kg N ha^{-1} $year^{-1}$ in one of the tributary watersheds to Lago de Cidra to 33.3 kg N ha^{-1} $year^{-1}$ in the Río Grande de Añasco watershed, a value similar to that computed for the urban Rio Piedras watershed. Both urban and agricultural-rural watersheds had higher N yields than forested watersheds, which ranged from 2.7 to 9.8 kg N ha^{-1} $year^{-1}$ (Table 3).

Among all the watersheds, the Río Grande de Añasco and the Río Piedras watersheds had the highest runoff (>1200 mm $year^{-1}$; Table 1). To evaluate the relationship between runoff and N yield, linear regression analysis was developed for each land use category. No significant relationship was observed between N yield and mean annual runoff for the urban and rural-agricultural

Table 3. N yield of the studied watersheds and comparison with published estimates for other watersheds in Puerto Rico.

Land use	Basin	N yield, kg N ha^{-1} $year^{-1}$	References
Urban	Río Grande de Loiza	16.4 (3.8)	This study
	Río de Bayamon	4.8 (0.7)	This study
	Río Piedras	32.9 (3.0)	This study
	Lago de Cidra Site 4 (sewered)	6.6	Ramos-Ginés (1997)
	Lago de Cidra Site 5 (unsewered)	17.1	Ramos-Ginés (1997)
Agricultural-Rural	Río Grande de Añasco	33.3 (6.4)	This study
	Lago de Cidra Site 2	6.9	Ramos-Ginés (1997)
	Lago de Cidra Site 3	8.6	Ramos-Ginés (1997)
Forested	Río Icacos	9.8	McDowell and Asbury (1994)
	Quebrada Toronja	4.4	McDowell and Asbury (1994)
	Quebrada Sonadora	5.9	McDowell and Asbury (1994)
	Lago de Cidra Site 1	2.7	Ramos-Ginés (1997)

Values in parenthesis represent the standard error of the mean.

watershed groups. A significant relationship, however, was found when the watersheds from these two land use categories were combined. Similarly, the N yield from forested watersheds was significantly related to mean annual runoff (Figure 2). This analysis revealed that N yield in watersheds in Puerto Rico is related to runoff and that under similar runoff conditions, urban and agricultural-rural watersheds export more N per unit area than forested watersheds.

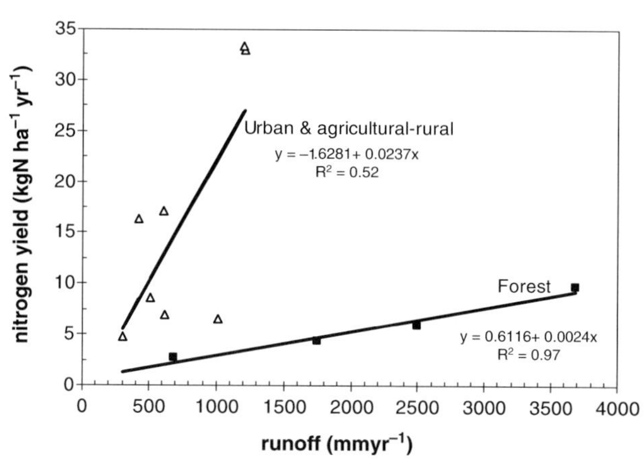

Figure 2. Relationship between runoff and N yield for two land use categories in Puerto Rico. The regression lines are significant at the $p = 0.05$ level.

Nitrogen export via wastewater

The average concentrations of total N in treated wastewater ranged from 18.0 mg l^{-1} at the Puerto Nuevo WWTP to 54 mg l^{-1} at the Carolina WWTP. Average effluent flow ranged from 25.8 million gallons per day (mgd) at Carolina to 55.6 mgd at Puerto Nuevo (Table 4). Per capita N load varied between facilities from 3.5 at Puerto Nuevo to 11.8 kg N $person^{-1}$ $year^{-1}$ at Bayamón. The average of the three facilities (9.6 kg N $person^{-1}$ $year^{-1}$) is higher than that reported by Meybeck et al. (1989) (3.3 kg N $person^{-1}$ $year^{-1}$) as a global average for sewered populations. When evaluated on a per unit area basis, N export through wastewater (mean = 139.6 kg N ha^{-1} $year^{-1}$; Table 4) is four times higher than the maximum rate of fluvial export reported for an urban-agricultural rural area (33.3 kg N ha^{-1} $year^{-1}$; Table 3) and as much as 52 times that of some forested basins in Puerto Rico (Table 3).

Atmospheric nitrogen deposition

In Puerto Rico, annual N_2O emissions have increased 11% per year from 1990 to 1994 from 395 tons $year^{-1}$ (full molecular basis) to 569 tons $year^{-1}$, while NO_x emissions have increased 2.4% per year over this period from 97,307 to 106,544 tons $year^{-1}$ (DNER 1999). Under a 'business as usual' scenario, N_2O and NO_x emissions are expected to continue to increase in the future at 7.8 and 1.1% per year, respectively, from the 1994 level (Figure 3). Fuel combustion, either from stationary or mobile sources, is the primary source of these emissions. Burning of agricultural waste and fertilizer use are considered to be negligible sources of NO_x and N_2O. No emission estimates exist for NH_3.

The NADP PR-20 Luquillo Experimental Forest station is located upwind from the SJMA and represents the rain chemistry associated with oceanic moisture carried by the Easterly Trade Winds. Concentrations of inorganic N (NO_3^- andNH_4^+) in precipitation have been increasing at this site over the last

Table 4. Export of N from primary WWTP in the SJMA. TN is total nitrogen.

WWTP	Average effluent conditions		Total N load, (kg $year^{-1}$)	Per capita N load, kg N $person^{-1}$ $year^{-1}$	Serviced area (ha)	N yield, kg N $ha^{-1}year^{-1}$
	Flow (mgd)	TN (mg l^{-1})				
Bayamón[a]	39.6	48.6	2,650,802	11.8	11,429	231.9
Puerto Nuevo[b]	55.9	18.0	1,387,403	3.5	11,259	123.2
Carolina[c]	25.7	54.0	1,910,437	10.7	19,956	95.7
All	121.0	40.2	5,948,642	9.6	42,644	139.5

[a]Data based on discharge monitoring reports from March 2002 to March 2004.
[b]Data based on discharge monitoring reports from February 1997 to March 2004.
[c]Data based on discharge monitoring reports from January 2000 to March 2004.

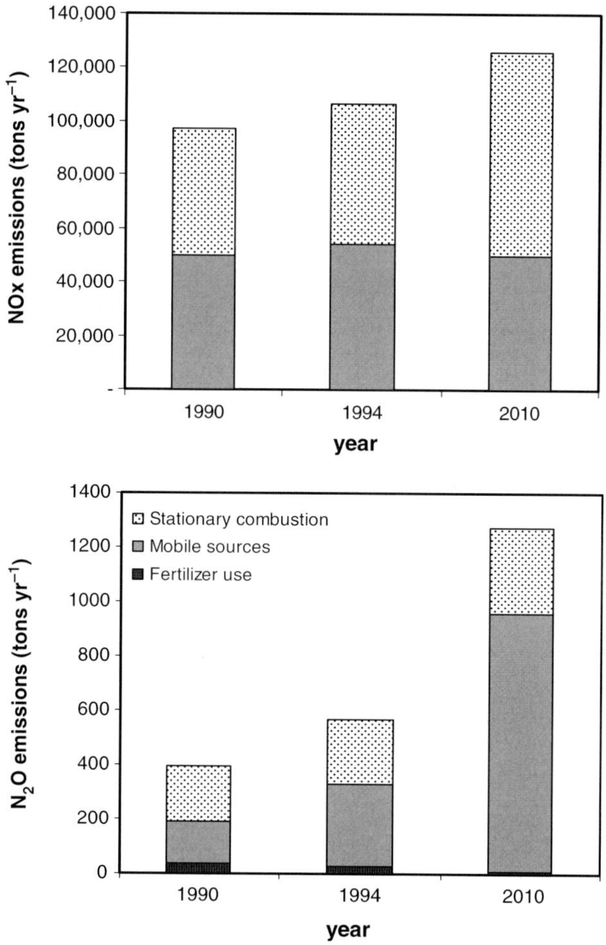

Figure 3. Change in NO_x and N_2O emissions by source from 1990 to 1994 and forecasts based on a 'business as usual' scenario. Data published by the Energy Affairs Administration of the Puerto Rico DNER. Emission rates are expressed on a full molecular basis.

18 years at a rate of 0.08 kg N ha^{-1} year^{-1} (Figure 4). Over the last five years, the wet input of N to the site was on average 2.7 kg N ha^{-1} year^{-1}. Howarth et al. (1996) estimated NO_y (ensemble of NO_x and its reservoirs, such as HNO_3 and N_2O_5) and NH_x ($NH_3 + NH_4^+$) deposition for the Caribbean by applying estimates of N wet + dry deposition to the North Atlantic region modeled by Propero et al. (1996). They estimated that, in the Caribbean, modern NO_y is deposited at a rate of 2.1 kg N ha^{-1} year^{-1} (wet + dry). Pre-industrial rates were estimated to 0.7 kg N ha^{-1} year^{-1} with the difference of 1.4 kg N ha^{-1} year^{-1}, obviously attributed to anthropogenic activities. NH_x deposition (wet + dry) was estimated to 1.82 (modern), 0.83 (pre-industrial), with a difference of

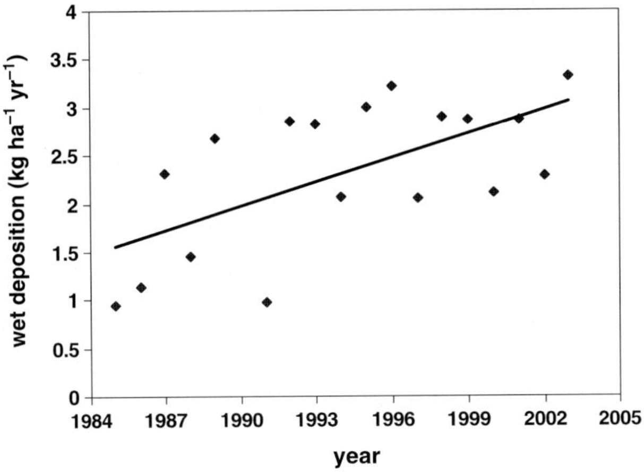

Figure 4. Wet deposition of inorganic N (NH_4^+ andNO_3^-) at the NADP PR-20 station located at El Verde. The trend line is significant at the $p < 0.01$ level.

0.99 kg N ha^{-1} $year^{-1}$ related to anthropogenic activities. The measured N wet deposition rates in the Luquillo Experimental Forest station (1999–2003 average = 2.7 kg N ha^{-1} $year^{-1}$) are lower than the total modern estimate of 3.92 kg N ha^{-1} $year^{-1}$ (for NO_y and NH_x; dry + wet deposition) by Howarth et al. (1996). The difference is due to the fact that N dry deposition in Puerto Rico is not included because of the lack of available data.

To date, there are only few studies that give estimates of N dry deposition in tropical environments based on field measurements. Galy-Lacaux et al. (2003) have estimated N wet + dry deposition to different ecosystems in West-central and South Africa (DEBIT study). The relative contributions of dry N deposition to the total N deposition rates in Africa given by Galy-Lacaux et al. (2003) are 65% for dry savanna, 73% for wet savanna, 75% for forest, 59% for rural dry Savanna and 37% for industrial area. Trebs et al. (2006) have estimated wet + dry N deposition for a pasture site in the state of Rondônia (Brazil), which is located in the Southwestern part of the Amazon Basin. Trebs et al. (2006) found that during the late dry season (biomass burning) in September 2002, about 46% of the total N deposition was attributed to dry deposition and the contribution of dry deposition to the total N deposition dropped to only 31 and 22% during October (transition period) and November (onset of the wet season, clean conditions), respectively. Thus, on average N dry deposition may contribute approximately 30% to the total N deposition at the disturbed Amazonian site.

Puerto Rico's tropical climate is strongly affected by the Northeast trade winds with significant maritime influence from the Atlantic Ocean, whereas evergreen forest regions in West-central Africa are largely influenced by continental flows associated with the Northeast or Southeast trade winds. While in Puerto Rico atmospheric chemistry is a complex mixture of marine aerosols, fuel combustion and others, in Rondônia, the main sources of atmospheric N deposition are biomass burning and cattle ranching. Thus, neither the African nor the Amazonian site may completely represent atmospheric N deposition patterns in Puerto Rico. However, considering that the contribution of N dry deposition to the total N deposition was about 70% for the African ecosystems determined during the DEBIT study we would obtain a dry N deposition rate of 6.3 kg N ha^{-1} $year^{-1}$ for the pristine Puerto Rican forest site. On the other hand, taking into account the average contribution of 30% obtained by Trebs et al. (2006) for the Amazonian pasture site, N dry deposition would be 1.16 kg N ha^{-1} $year^{-1}$, which is much lower than the African estimate. Hence, N dry deposition to the Puerto Rican forest site may be expected to range from 1.16 to 6.3 kg N ha^{-1} $year^{-1}$. The resulting total N deposition would vary between 3.86 and 9.0 kg N ha^{-1} $year^{-1}$. Nevertheless, it should be noted that (i) the wet deposition estimate for the pristine forest does not include the potentially important organic N fraction (cf. Cornell et al. 2003), (ii) N deposition rates are probably much higher in regions of the island that are directly affected by urban pollution from the SJMA, and (iii) intensive field measurements are needed to reliably quantify Puerto Rican N dry + wet deposition rates.

Other nitrogen fluxes

In Puerto Rico, there is evidence that links modern agriculture with increased nitrogen loading to ground water (Conde-Costas and Gómez-Gómez 1998). However, contrary to the increasing trend in the use of N fertilizers worldwide (Galloway and Cowling 2002; Howarth et al. 2002), the use of N fertilizer had decreased in Puerto Rico from a maximum use of 10,093 tons N $year^{-1}$ in 1996 to a minimum of 5067 tons N $year^{-1}$ in 2002 (Figure 5), a reduction of 50%. Over this period, cropland area in Puerto Rico decreased by 16% from 329,969 in 1993 to 271,440 ha in 2002 (USDA 2004). This decrease was counteracted by a 20% increase in food imports from 1995 to 2004 from \$1705.1 M to \$2052.2 M (Puerto Rico Planning Board 2005b). Table 5 shows that N applications in Puerto Rico are low compared to the Latin American-Caribbean region and the World. This reflects that Puerto Rico is a low-intensity agriculture country, dependent almost exclusively on food imports. It is expected that significant amounts of N enter Puerto Rico through food imports; however, the exact N fluxes associated with this activity are unknown.

Discussion

Nitrogen fluxes in tropical regions

Howarth et al. (1996) documented that N fluxes per unit area to the North Atlantic Ocean were the highest in the highly disturbed watersheds of Northern

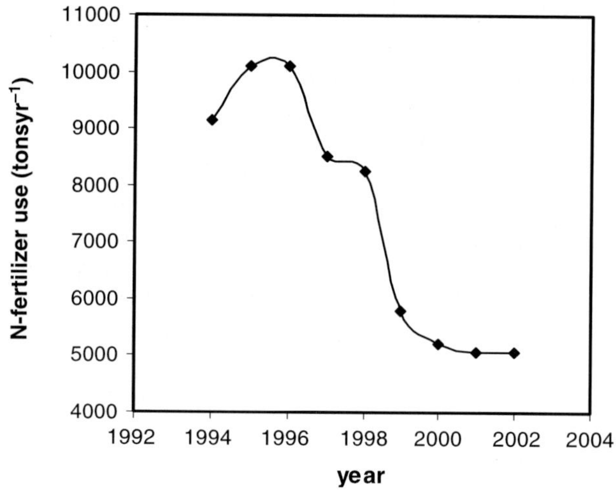

Figure 5. Annual N fertilizer use in Puerto Rico. Data from AAPFCO (2002).

Table 5. Comparison of N fertilizer use in Puerto Rico during 2002 and regional estimates for Latin American/Caribbean region and the World based on Martinelli et al. (this issue).

	Puerto Rico	LA-Ca	World
Fertilizer per unit area, kg N/ha (2002)	16.9	34	60
Per capita fertilizer consumption, kg N/person	1.2	9	14

Europe and Northeastern United States (14.5 and 10.7 kg N ha^{-1} $year^{-1}$, respectively) Atmospheric deposition of N and the application of fertilizers have been considered the two major sources of N in these regions. In temperate regions, N fixation becomes important only in the intensively cultivated Mississippi Valley, where N-fixing crops are abundant.

In the American tropics, Lewis et al. (1999) summarized N yields from 17 minimally disturbed watersheds in the Americas. They reported that N fluxes averaged 5.08 kg N ha^{-1} $year^{-1}$ (range = 0.73–9.98) and that are strongly correlated to mean annual runoff. When compared to Northern temperate rivers (Howarth et al. 1996), these findings suggest that tropical undisturbed watersheds have lower N yields. Apparently, in these tropical watersheds, lower population densities, lower fertilizer applications, extensive mature forests, and lower atmospheric N deposition kept these fluxes at a minimum.

Future land and demographic changes in the tropics associated with globalization, however, may change the role of tropical watersheds in the global N budget. This is because nearly 40% of global N fertilizer applications are already taking place in the tropics and subtropics, a fraction that is expected to increase to 75% by 2020 (Matthews 1994; Matson et al. 1999). Moreover, it is estimated that about 75% of the global fuel-related N emissions will occur in the tropics and subtropics (Galloway et al. 1994). If development and industrialization occurs in tropical developing countries as it has in the temperate Northern developed countries, it is expected that atmospheric N deposition will increase rapidly, such that tropical N-deposition may move to levels higher than those of present industrialized temperate countries (Downing et al. 1999).

In fact, some tropical regions are already experiencing rapid landscape transformations that match or exceed the N export from temperate industrialized areas. For example, the Piracicaba River is a heavily developed watershed located in Southeastern Brazil with a mean annual runoff of 403 mm. Fluvial N export in this watershed averaged 17.5 kg N ha^{-1} $year^{-1}$ (Filoso et al. 2003). Remarkably, fluvial N export in the Piracicaba River basin is similar to the basins reported in this study of similar runoff (Cidra and Río Grande de Loiza). The fluvial N export rates reported for the Piracicaba River, Río Piedras, and the Río Grande de Añasco basins are higher than those reported for the most impacted industrialized zones in Northern Europe (Howarth et al. 1996). This suggests that the role of disturbed tropical watersheds in the global N flux to oceans may be more important than previously thought.

Tropical regions are considered important global carbon sinks (Schlesinger 1991). Because tropical forests are not generally N-limited (Martinelli et al. 1999) it is expected that under a future scenario of increased N deposition, forest primary productivity in tropical regions may decrease due to increased soil acidity and loss of soil fertility due to depletion of base cations (Matson et al. 1999). Thus, increases in N fluxes in tropical regions can have significant effects on the global carbon balance with negative consequences for tropical societies, particularly on islands due to the effects of increased N fluxes on N-limited coastal waters (Corredor et al. 1999a, b).

Challenges for nitrogen control in developing tropical islands

Land use change and water pollution control

While industrialization may have reached a steady state in most Northern developed countries, developing tropical regions will likely experience rapid land and demographic transformations as their economies change from agrarian to industrial (e.g., Grau et al. 2003). A demographic trend that favors natural reforestation of abandoned croplands due to human migration from rural areas to cities has been documented in Puerto Rico and the Dominican Republic (Zweifler et al. 1994; Aide and Grau 2004). In these islands, inhabitants from rural mountainous areas that depended on agriculture for subsistence have migrated to urban centers out-competed by the highly efficient modern farmers typically concentrated in the lowlands and attracted by the wider job opportunities of coastal cities. This has resulted in the natural reforestation of abandoned agricultural highlands.

N fluxes in these countries are likely to change with reforestation favoring N sequestration and urbanization increasing N losses. Pre-development N export in tropical rivers averages 5 kg N ha^{-1} $year^{-1}$ (Lewis et al. 1999; McDowell 2002). For Puerto Rico, N export from forested watersheds ranges from 3 to 10 kg ha^{-1} $year^{-1}$ (Table 3). With urban growth, there is an increase in wastewater generation and a need for proper treatment and disposal. Without adequate wastewater treatment, pollution of inland and coastal waters occurs. In 2000, the World Health Organization (WHO) reported that about 77% of the population in Latin American and the Caribbean did not have access to adequate water sanitation systems (Gleick et al. 2002). The WHO defines adequate sanitation broadly as the hygienic separation of human feces from human contact. In coastal urban centers, such as those in Puerto Rico, this implies primary treatment of wastewater, which does not remove N. This result in an increase in the load of N to the ocean of about two orders of magnitude higher than the load associated with urban and forested watersheds (Table 3; Figure 2).

Although the current waste treatment practices generally meet regulatory standards, they alter the nutrient balance and biogeochemical processes of Caribbean coastal ecosystems (Corredor and Capone 1985; Corredor et al. 1992, 1999a, b; Morel and Corredor 1993; Corredor and Morel 1994; Mosquera et al. 1998). In tropical developing islands, such as Puerto Rico, improvements of the wastewater treatment system for nutrient removal, particularly N, are necessary to protect coastal ecosystems. This can be accomplished with tertiary treatment technology or less conventional procedures such as terrestrial application of wastewater or constructed wetlands.

Currently, the WWTPs in the SJMA are granted waivers that allow for disposal of sewage receiving primary treatment only into receiving waters. The waiver was granted by the U.S. Environmental Protection Agency following guidelines established under Section 301(h) of the Clean Water Act. This waiver allows eligible WWTPs that met environmentally stringent criteria to receive a modified National Pollutant Discharge Elimination System permit waiving the secondary treatment requirements for the conventional pollutants biochemical oxygen demand (BOD_5), suspended solids (SS), and pH. Under the waiver, EPA requires PRASA (among other requirements) to protect and propagate a balanced indigenous population of shellfish, fish, and wildlife in the receiving waters, to meet water quality standards set by EPA and by the Puerto Rico EQB, to establish a monitoring program to assess impacts. In 2000, the EPA and the Puerto Rico Commonwealth reached an agreement by which PRASA will upgrade its primary WWTPs to provide secondary treatment by year 2020. The capital investments associated with these improvements are large and thus represent a challenge to Puerto Rico. However, the health benefits associated with improved sanitation and the economic and ecological values of clean beaches should warrant these investments.

Control of atmospheric emissions

Balancing N fluxes for pollution control through reforestation and improved sanitation undoubtedly represent a challenge in developing tropical islands.

However, controlling N emissions will be equally important if forests are to continue sequestering N. The increased wet deposition of inorganic N observed in the Luquillo Experimental Forest in Puerto Rico (Figure 4) and the predicted inverse relationship between primary productivity in N-limited tropical forests and N deposition (Matson et al. 1999) suggests that N emissions to the atmosphere may alter the ability of tropical forests to retain N.

In 1999, the Commonwealth Government of Puerto Rico set a strategic plan to reduce by year 2010 the emissions of greenhouse gases by 10% of the 1990 emission level of 30.9 million tons of CO_2 equivalents; a net reduction of 10.4 million tons of CO_2 equivalents. The plan calls for improved efficiency in the energy sector with implementation of alternative energy sources (other than crude oil) such as Ocean Thermal Energy Conversion (OTEC), solar, wind, and natural gas. Also, the plan includes changes in transportation to favor the use of cars with increased fuel efficiency, reduction in traveled miles, and the use of mass transportation such as urban trains. Finally, reforestation and recycling of household wastes will be promoted under the plan. These plans represent an important effort to control emissions of greenhouse gases with benefits at the local and global scales. The observed increase in N wet deposition in the Luquillo Experimental Forest (Figure 5) suggests that local or regional emissions may already affect relatively pristine tropical forests.

Conclusions and recommendations

Quantifying N dynamics on the island of Puerto Rico is essential to protecting economically and ecologically valuable coastal resources that are threatened by increased N inputs. Our analysis indicates that there are serious gaps in understanding of N inputs, transformations (denitrification) and temporal trends that limit the ability of managers to deal with issues of coastal eutrophication. The most important gaps are: (a) those associated with atmospheric N deposition, (b) N fixation in natural and human-dominated landscapes, and (c) denitrification in both terrestrial and aquatic portions of the landscape. Each of these is a quantitatively significant process that affects N delivery to the coast, but both the spatial variability and trends over time of these processes are largely unknown.

A more comprehensive quantification of atmospheric N deposition for the island is central to a better understanding of Puerto Rico's N budget. The presence of only one wet deposition measurement station at a relatively pristine site and the absence of systematic dry deposition measurements provide only limited quantitative information about one of the major N inputs to the island. Given the significant increase of N wet deposition during the last 18 years ($\sim$1.44 kg N ha^{-1} $year^{-1}$) at a station that is relatively unaffected by pollution from the SJMA, it is imperative that atmospheric deposition (wet + dry) is monitored in parts of Puerto Rico that are receptors of urban pollution. N deposition rates in these areas are expected to be comparable to moderately polluted urban regions in the temperate latitudes and changes over time are probably much greater than those measured in the Luquillo Mountains of Northeastern Puerto Rico. Additional information on N inputs to forested lands due to N fixation would also be useful. Past estimates of N fixation range from 8 to 16 kg ha^{-1} $year^{-1}$ (McDowell and Asbury 1994), and rates in suburban and lowland forests are unknown. Conversion of agricultural lands to forest and urban development may have offsetting effects on landscape-scale N inputs due to N fixation, with higher rates in secondary forests offsetting the lower rates in urban areas.

Denitrification is a particularly important process for environmental management because it removes N from soils, streams and rivers and returns it the atmosphere as N_2. Moreover, denitrification is a process that either consumes or produces N_2O and NO, which are both environmentally relevant trace gases. The rates of denitrification in terrestrial and aquatic ecosystems are poorly constrained for the island of Puerto Rico, but are thought to be quantitatively significant in forests (Erickson et al. 2001; Mosier et al. 1998; McSwiney et al. 2001), at the land–water interface (McDowell et al. 1996; McDowell 2001; Chestnut and McDowell 2000), and perhaps in the channels of streams and rivers (Peterson et al. 2001). The warm, wet conditions in much of Puerto Rico, together with the high rates of primary productivity, make the potential for denitrification particularly large in the riparian zone of streams and rivers (McDowell 2003). A particularly important area of research will be to assess landscape-scale rates of denitrification under two contrasting scenarios – on-site disposal of human waste (septic systems) vs. centralized sewage treatment. With on-site disposal, there may be greater potential for denitrification, but there is less control over the quality and effectiveness of the treatment systems and hence greater potential for spatial variability in rates of N delivery to surface waters. With centralized sewage treatment plants, riparian denitrification no longer occurs, and instead N is introduced directly into surface waters, where denitrification may also be occurring. Assessing the environmental costs and benefits of both treatment approaches will be critical to protecting water quality as decisions are made about whether outmoded on-site systems are refurbished, or replaced with centralized sewage treatment plants.

In conclusion, Puerto Rico exemplifies how increased population, coupled with changes in land use and increased industrial development has affected the N dynamics in a tropical Caribbean island. In the future, the development of a comprehensive monitoring network that assesses inputs and outputs of N is needed in Puerto Rico. At the whole-island scale, measures of N imported in food and feed are required. For other fluxes, a watershed approach should be taken in which a few representative sites with varying land uses and some historical data on water quality are studied long-term to determine how trends of N export in rivers relates to changes in inputs and land use in the study watersheds. A particularly important aspect of watershed N dynamics would be to document the magnitude of denitrification in terrestrial and aquatic environments. Understanding the controls on denitrification might provide extremely useful tools for reducing non-point sources of N to coastal waters. Such information will allow a predictive understanding of the N budget under scenarios of climate and land use changes under tropical island conditions.

Acknowledgments

The authors are grateful to L Martinelli and B Howarth of the Interamerican Nitrogen Network for establishing a very productive forum for discussions and for providing helpful guidance and motivation. Funding for this project was provided by the CREST Center for Applied Tropical Ecology and Conservation of the University of Puerto Rico. D Sotomayor (UPR-Mayaguez) and an anonymous reviewer provided valuable comments that improved the article. The US Geological Survey provided access to water quality and streamflow data. J Coliño of the Puerto Rico Aqueduct and Sewer Authority and M del Valle (CSA Group) provided operational data for the wastewater treatment plants. J Bauza (CSA Group) shared valuable information on coastal N dynamics. A Morales (Puerto Rico Environmental Quality Board) assisted us with air chemistry data. This research is associated to the NSF-funded Luquillo LTER Program.

References

AAPFCO. 2002. Commercial Fertilizers 2002. Association of American Plant Food Control Officials. http://www.aapfco.org. Columbia, MO.

Aber J., McDowell W., Nadelhoffer K., Magill A., Bernston G., Kamakea M., McNulty S., Currie W., Rustad L. and Fernandez I. 1998. Nitrogen saturation in temperate forest ecosystems: hypotheses revisited. Bioscience 48: 921–934.

Aber J.D., Goodale C.L., Ollinger S.V., Smith M.L., Magill A.H., Martin M.E., Hallett R.A. and Stoddard J.L. 2003. Is nitrogen deposition altering the nitrogen status of northeastern forests? Bioscience 53: 375–389.

Aide T.M. and Grau H.R. 2004. Globalization, migration, and Latin American ecosystems. Science 305: 1915–1916.

Birdsey R.A. and Weaver P.L. 1987. Forest area trends in Puerto Rico. USDA Forest Service, Southern Experiment Station SO-331. 5 pp.

Cassman K.G., Doberman A. and Walters D.T. 2002. Agroecosystems, nitrogen-use, efficiency, and nitrogen management. Ambio 31: 132–140.

Chestnut T.J. and McDowell W.H. 2000. C and N dynamics in the riparian and hyporheic zones of a tropical stream, Luquillo Mountains, Puerto Rico. J. North Am. Benthol. Soc. 19: 199–214.

Conde-Costas C. and Gómez-Gómez F. 1998. Nitrate contamination of the upper aquifer in the Manatí-Vega Baja Area, Puerto Rico. Third International Symposium on Tropical Hydrology. San Juan, Puerto Rico, July 12–16, 1998. American Water Resources Association.

Cornell S.E., Jickells T.D., Cape J.N., Rowland A.P. and Duce R.A. 2003. Organic nitrogen deposition on land and coastal environments: a review of methods and data. Atmos. Environ. 37: 2173–2191.

Corredor J.E. and Capone D.G. 1985. Studies on nitrogen diagenesis in coral reef sands. In: Gomez E. (ed.), Proceedings 5th International Coral Reef Symposium, Vol. 3, pp. 395–399.

Corredor J.E., Morell J.M. and Diaz M.R. 1992. Environmental degradation, nitrogen dynamics and proliferation of the filamentous Cyanophyta *Microcoleus lyngbyaceus* in nearshore Caribbean waters. In Mediterraneo e Caraibi due mari in pericolo? Sversamenti accidentali di idrocarburi ed emergenze causate dalle alghe. Atti del cenvegno internazionale. Cuaderni delle ICRAM/IFREMER: 91–100.

Corredor J.E. and Morel J.M. 1994. Nitrate depuration of secondary sewage effluents in mangrove sediments. Estuaries 17: 295–300.

Corredor J.E., Morell J.M. and Bauza J. 1999a. Atmospheric nitrous oxide fluxes from mangrove sediments. Marine Pollut. Bull. 38: 473–478.

Corredor J.E., Howarth R.W., Twilley R.R. and Morell J.M. 1999b. Nitrogen cycling and anthropogenic impact in the tropical interamerican seas. Biogeochemistry 46: 163–178.

Dietz J.L. 1986. Economic history of Puerto Rico. Institutional change and Political and development. Princeton University, NJ.

Downing J.A., McClain M., Twilley R., Melack J.M., Elser J., Rabalis N.N., Lewis W.M. Jr., Turner R.E., Corredor J., Soto D., Yanez-Arancibia A., Kopaska J.A. and Howarth R.W. 1999. The impact of accelerating land-use change on the N-Cycle of tropical aquatic ecosystems: current conditions and projected changes. Biogeochemistry 46: 109–148.

Díaz P.L., Aquino Z., Figueroa-Alamo C., García, R. and Sánchez A.V. 2004. Water Resources Data: Puerto Rico and the U.S. Virgin Islands, Water Year 2002. US Geological Survey Water-Data Report PR-02-1. 672 pp.

DNER. 1999. Puerto Rico State Action Plan to reduce greenhouse gas emissions. Final Report, Department of Natural and Environmental Resources, Energy Affairs Administration, Government of Puerto Rico, 456 pp.

DNER. 2005. Plan Integral de Aguas de Puerto Rico. Oficina del Plan Integral de Aguas. Departamento de Recursos Naturales y Ambientales. San Juan, P.R. http://www.plandeaguaspr.com.

Erickson H., Keller M. and Davidson E.A. 2001. Nitrogen oxide fluxes and nitrogen cycling during postagricultural succession and forest fertilization in the humid tropics. Ecosystems 4: 67–84.

Filoso S., Martinelli L.A., Williams M.R., Lara L.B.L.S., Krusche A, Ballester M.V., Victoria R.L. and Camargo P.B. 2003. Land use and nitrogen export in the Piracidaba River Basin, Southeast Brazil. Biogeochemistry 65(3): 275–294.

Galloway J.N., Howarth R.W., Michaels A.F., Nixon S.W., Propero J.M. and Dentener F.J. 1996. Nitrogen and phosphorus budgets of the North Atlantic Ocean and its watershed. Biogeochemistry 35: 3–25.

Galloway J.N., Levy H. II and Kasibhatla P.S. 1994. Year 2020: Consequences of population growth and development on the deposition of oxidized nitrogen. Ambio 23: 120–123.

Galloway J.N., Schlesinger W.H., Levy H. II, Michaels A. and chnoor J.L. 1995. Nitrogen fixation: anthropogenic enhancement-environmental response. Global Biogeochem. Cycles 9: 235–252.

Galloway J.N. 1998. The global nitrogen cycle: changes and consequences. Environ. Pollut. 102: 15–24.

Galloway J.N. and Cowling E.B. 2002. Reactive nitrogen and the world: consequences for terrestrial ecosystems. Ambio 31: 64–71.

Galy-Lacaux C., Carmichaël G.R., Song C.H., Lacaux J.P. and Galwa J. 2003. Dry and wet atmospheric nitrogen deposition in Africa. IGAC Newslett. Article 27: 6–11.

Gleick P., Burns W.C.G., Chalecki E.L., Cohen M., Kao-Cushing K., Mann A.S., Reyes R., Wolf G.H. ?nd Wong A.K. (2002). The World's Water: The Biennal Report on Freshwater Resources 2002–2003. Island Press, Washington, 335 pp.

Grau H.R., Aide T.M., Zimmerman J.K. and Thomlinson J.B. 2004. Trends and scenarios of the carbon budget in postagricultural Puerto Rico (1936–2060). Global Change Biol. 10: 1–17.

Grau H.R., Aide T.M., Zimmerman J.K., Thomlinson J.B., Helmer E. and Zou X. 2003. The ecological consequences of socioeconomic and land-use changes in postagriculture Puerto Rico. Bioscience 53: 1159–1168.

Gould W., Martinuzzi S., Fevold B. and Ramos O. 2005. Urban and Developed Landcover of Puerto Rico, Vieques, and Culebra, 2000–2003. Map 2005-1. USDA Forest Service, International Institute of Tropical Forestry, Río Piedras, Puerto Rico.

Helmer E.H., Ramos O., López T. del M., Quiñones M. and Díaz W. 2002. Mapping the forest type and land cover of Puerto Rico, a component of the Caribbean Biodiversity Hotspot. Caribbean J. Sci. 38(3–4): 165–183.

Holland E.A., Dentener F.J., Braswell B.H. and Sulzman J.M. 1999. Contemporary and pre-industrial global reactive nitrogen budgets. Biogeochemistry 46: 7–43.

Howarth R.W., Billen G., Swaney D., Townsend A., Jaworski N., Lajtha K., Downing J.A., Elmgren R., Caraco N., Jordan T., Berendse F., Freney J., Kudeyarov V., Murdoch P. and Zhao-Liang Z. 1996. Regional nitrogen budgets and riverine N and P fluxes for the drainages to the North Atlantic Ocean: natural and human influences. Biogeochemistry 35: 75–139.

Howarth R.W., Boyer E.W., Pabich W.J. and Galloway J.V. 2002. Nitrogen use in the United States from 1961–2000 and potential future trends. Ambio 31: 88–96.

Hunter J.M. and Arbona S.I. 1995. Paradise lost: an introduction to the geography of water pollution in Puerto Rico. Soc. Sci. Med. 40: 1331–1355.

Lewis W.M. Jr., Melack J.M., McDowell W.H., McClain M.M. and Richey J.E. 1999. Nitrogen yields from undisturbed watersheds in the Americas. Biogeochemistry 46: 149–162.

López T. del M., Aide T.M. and Thomlinson J.R. 2001. Urban expansion and the loss of prime agricultural land in Puerto Rico. Ambio 30: 49–54.

Martinelli L.A., Howarth R.W., Cuevas E., Filoso S., Austin A., Donoso L., Huzsar V., Keeney D., Lara L.L., Llerena C., McIssac G., Medina E., Ortiz-Zayas, J.R., Scavia D., Schindler D.W., Soto D. and Townsend A. (this issue). The origin and fate of reactive nitrogen in Latin America and the Caribbean. Biogeochemistry.

Martinelli L.A., Piccolo M.C., Twonsend A.R., Vitousek P.M., Cuevas E., McDowell W.M., Robertson G.P., Santos O.C. and Treseder K. 1999. Nitrogen stable isotopic composition of leaves and soil: tropical versus temperate forests. Biogechemistry 46: 45–65.

Matson P.A., McDowell W.H., Townsend A.R. and Vitousek P.M. 1999. The globalization of N deposition: ecosystem consequences in tropical environments. Biogeochemistry 46: 67–83.

Matthews E. 1994. Nitrogenous fertilizers: global distribution of consumption and associated emissions of nitrous oxide and ammonia. Global Biogeochem. Cycles 8: 411–439.

McDowell W.H. 2001. Hurricanes, people, and riparian zones: controls on nutrient losses from forested Caribbean watersheds. For. Ecol. Manage. 154: 443–451.

McDowell W.H. 2002. Nutrient export from tropical rain forests. In: Ambasht R.S. and Ambasht N. K. (eds), Modern Trends in Applied Terrestrial Ecology. Kluwer Academic Press, pp. 149–163.

McDowell W.H. 2003. Dissolved organic matter in soils-future directions and unanswered questions. Geoderma 113: 179–186.

McDowell W.H., Ginés-Sanchez C., Asbury C.E. and Ramos-Pérez C.R. 1990. Influence of seasalt aerosols and long range transport on precipitation chemistry at El Verde, Puerto Rico. Atmos. Environ. 24: 2813–2821.

McDowell W.H. and Asbury C.E. 1994. Export of carbon, nitrogen, and major ions from three tropical montane watersheds. Limnol. Oceanogr. 39: 111–125.

McDowell W.H., McSwiney C.P. and Bowden W.B. 1996. Effects of hurricane disturbance on groundwater chemistry and riparian function in a tropical rain forest. Biotropica 28: 577–584.

McSwiney C.P., McDowell W.H. and Keller M. 2001. Distribution of nitrous oxide and regulators of its production across a tropical rainforest catena in the Luquillo Experimental Forest, Puerto Rico. Biogeochemistry 56: 265–286.

Meixner F.X. 1994. Surface exchange of odd nitrogen oxides. Nova Acta Leopoldina NF 70(288): 299–348.

Meybeck M., Chapman D.V. and Helmer R. 1989. Global Freshwater Quality: A First Assessment. World Health Organization/ UNEP Basil Blackwell, Inc., Cambridge, MA.

Mosier A.R., Delgado J.A. and Keller M. 1998. Methane and nitrous oxide fluxes in an acid oxisol in western Puerto Rico: Impact of tillage, liming, and fertilization. J. Soil Biol. Biochem. 30: 2087–2098.

Morel J.M. and Corredor J.E. 1993. Sediment nitrogen trapping in a mangrove lagoon. Esturine, Coastal Shelf Sci. 37: 203–212.

Mosquera A.I., Corredor J.E. and Morel J.M. 1998. Exponential collapse of benthic depurative capacity in a eutrophic tropical marine ecosystem: a threshold response to organic nitrogen loading. Chem. Ecol. 14: 341–335.

Ortiz-Zayas J.R., Quiñones F., Palacios S., Vélez A. and Más H. (2004). Características y condición de los embalses principales en Puerto Rico. Oficina del Plan Integral de Aguas, Departamento de Recursos Naturales y Ambientales, San Juan, P.R., 179 pp.

Osterkamp W.R. 2001. Earth-Surface Processes, Materials Use, and Urban Development – A Case Study of the SJMA, Northeastern Puerto Rico. U.S. Geological Survey Water Resources Division, Tucson, AZ, pp. 1–23.

Peterson B.J., Wollheim W., Mulholland P.J., Webster J.R., Meyer J.L., Tank J.L., Grimm N.B., Bowden W.B., Vallet H.M., Hershey A.E., McDowell W.H., Dodds W.K., Hamilton S.K., Gregory S.V. and D'Angelo D.J. 2001. Stream processes alter the amount and form of nitrogen exported from small watersheds. Science 292: 86–90.

Pico R. 1975. Nueva geografía de Puerto Rico. Editorial Universitaria Universidad de Puerto Rico, San Juan PR., 460 pp.

PRASA. 2005. Report on PRASA's WWTPs in operation in the SJMA. Puerto Rico Aqueduct and Sewer Authority. San Juan, P.R.

Propero J.M., Barrett K., Church T., Dentener F., Duce R.A., Galloway J.N., Levy H. II and Moody J. 1996. Atmospheric deposition of nutrients to the North Atlantic Basin. Biogeochemistry 35: 27–73.

Puerto Rico Planning Board 1995. Proyecciones de población: Puerto Rico 1990–2025. Area de Planificación Económica y Social, Oficina del Gobernador, San Juan, P.R.

Puerto Rico Planning Board. 2005. Population projections based in the 2000 Census. http://www.censo.gobierno.pr/Proyecciones_Poblacion.htm.

Puerto Rico Planning Board 2005b. Imports of Recorded Merchandise by North American Industrial Classification System (NAICS). Economic and Social Planning Program, San Juan, P.R.

Ramos-Ginés O. 1997. Water balance and quantification of total phosphorus and total nitrogen loads entering and leaving the Lago de Cidra, Central Puerto Rico. US Geological Survey, Water-Resources Investigations Report 96-4222. San Juan, Puerto Rico. 28 pp.

Schlesinger W.H. 1991. Biogeochemistry: An Analysis of Global Change. Academic Press, Inc., New York, 443 pp.

Socolow R.H. 1999. Nitrogen management and the future of food: lessons learned from management of energy and carbon. Proc. Nat. Acad. Sci. 96: 6001–6008.

Sotomayor-Ramírez D., Martínez G.A., Pérez-Alegría L.R. 2004. Nutrient Discharges from Mayaguez Bay Watershed: Final Progress Report. University of Puerto Rico, College of Agricultural Sciences, Mayaguez, P.R., 56 pp.

Swistock B.R., Edwards P.J., Wood F. and Dewalle D.R. 1997. Comparison of methods for calculating annual solute exports from six forested Appalachian watersheds. Hydrol. Process. 11: 655–669.

Thomlinson J.R. and Rivera L.Y. 2000. Suburban growth in Luquillo, Puerto Rico: some consequences of development on natural and semi-natural systems. Landscape Urban Plan. 49: 15–23.

Townsend A.R., Howarth R.W., Bazzaz F.A., Booth M.S., Cleveland C.C., Collinge S.K., Dobson A.P., Epstein P.R., Holland E.A., Keeney D.R., Malling M.A., Rogers C.A., Wayne P. and Wolfe A.H. 2003. Human health effects of a changing global nitrogen cycle. Fron. Ecol. Environ. 1: 240–246.

Trebs I., Lara L.S., Zeri L.M., Gatti L.V., Artaxo P., Dlugi R., Slanina J., Andreae M.O. and Meixner F.X. 2006. Dry and wet deposition of atmospheric inorganic nitrogen compounds in a tropical environment (Rondônia, Brazil). Atmos. Chem. Phys. 6: 447–469.

USDA 2004. Puerto Rico 2002 Census of Agriculture. National Agricultural Statistics Service. Jeffersonville, IN.

USGS 2002. Provisional Hydrologic Unit Boundaries for Puerto Rico – Watershed and Subwatershed Level. GIS Dataset. USGS Caribbean District, San Juan, Puerto Rico.

Vitousek P.M., Aber J., Howarth R.W., Likens G.E., Matson P.A., Schindler D.W., Schlesinger W.H. and Tilman G.D. 1997. Human alteration of the global nitrogen cycle: causes and consequences. Ecol. Appl. 7: 737–750.

Zweifler M., Gold M.A. and Thomas R.N. 1994. Land use evolution in hill regions of the Dominican Republic. Profess. Geographer 46: 39–53.

Biogeochemistry (2006) 79: 135–161
DOI 10.1007/s10533-006-9009-7

Nitrogen and phosphorus budgets for a tropical watershed impacted by agricultural land use: Guayas, Ecuador

MERCY J. BORBOR-CORDOVA[1,*], ELIZABETH W. BOYER[2]
WILLIAM H. MCDOWELL[3] and CHARLES A. HALL[1]
[1]*College of Environmental Science and Forestry, State University of New York, Syracuse, NY, USA;* [2]*Department of Environmental Science, Policy, and Management, University of California, Berkeley, CA, USA;* [3]*Department of Natural Resources, University of New Hampshire, Durham, NH, USA;* **Author for correspondence (e-mail: mjborbor@mailbox.syr.edu; phone: +1-593-4-2800013; fax: +1-593-4-2800013)*

Key words: Agricultural watershed, Anthropogenic, Ecuador, Fertilizers, Food production, Land use, Nitrogen, Phosphorus

Abstract. Large-scale changes in land use are occurring in many tropical regions, with significant impacts on nitrogen and phosphorus biogeochemistry. In this study we examine the relationships between land use, anthropogenic nutrient inputs, and riverine nutrient exports in a major agricultural watershed of the Pacific coast of South America, the Guayas River basin of Ecuador. We present comprehensive nutrient budgets for nitrogen (N) and phosphorous (P) for the Guayas River basin and 10 sub-watersheds. We quantify the four major anthropogenic nutrient fluxes into and out of the region: N and P fertilizer application, N fixation by leguminous crops, net import/export of N and P in agricultural products (food and feed), and atmospheric deposition. We also estimate inputs of N from biological N fixation in forests and of P from weathering sources in soils and bedrock. The sum of these sources represents net inputs of N and P to each watershed region. Overall, synthetic fertilizers are the largest input to the Guayas Basin for N (53%) and P (57%), and the largest outputs are N and P in crops. Losses of N and P in river export account for 14–38% of total N and P inputs, and there is significant accumulation of N and P, or unmeasured forms of N and P export, in most of the sub-basins. Nutrient balances are indicative of the sustainability of land use practices in a region, where a negative balance of N or P indicates nutrient depletion and subsequent loss of soil fertility, yield, and economic viability. Although the nutrient balance of the entire Guayas Basin is positive, there are negative or near zero balances in two sub-watersheds with extensive banana, coffee and permanent crops. In these basins, degradation of soil quality may be occurring due to these net nutrient losses. Our data show that nutrients are leaving the basin primarily as export crops, with riverine losses of nutrients smaller than crop exports. Nonetheless, there is a direct relationship between nutrient inputs and river outputs, suggesting that agricultural management practices in the basin may have direct impacts on N and P delivery to the highly productive Guayas estuary.

Introduction

Large-scale land-use changes are occurring in the tropical Americas that have important implications for the future of freshwater and coastal marine ecosystems and their management (Yanez-Arancibia and Lara-Dominguez 1998). However, relatively little is known about the potential impact of land use alteration on the large-scale biogeochemistry of tropical aquatic ecosystems

(Vitousek et al. 1997; Downing et al. 1999). Rapid population growth and the rising international demand for tropical crops has resulted in the conversion of vast land areas to intensive agricultural production in Ecuador, as is typical for many other developing tropical regions of the world (Houghton 1994; Skole et al. 1994). Agricultural production was dominated by cacao in the 1920–1930s, and from the 1950s to the present bananas have been the most important agricultural export product of Ecuador. Recently the land devoted to these export crops has increased; between 1980 and 2000 harvest area increased by about 140,000 ha for banana and 160,000 ha for cocoa (FAO 2002). Other export products, such as oil during the 1970s and shrimp farming in the 1980s, have exacerbated deforestation of the Amazonian region and the mangrove forest in coastal areas. The economic gain from exporting these products to meet international demands has been the driver for the deforestation and land use change in Ecuador and also in the Guayas Basin.

Extensive conversions in land use typically result in dramatic changes to regional hydrology (Bruijnzeel 1996; Cronan et al. 1999), alterations of biogeochemical cycles (Boyer and Howarth 2002), losses of nutrients such as nitrogen and phosphorus (Caraco 1993; Howarth et al. 1996), and soil degradation (Cole et al. 1993; Ojima et al. 1994). Previous studies in temperate and tropical areas have found strong linkages between large-scale changes in land use and associated nutrient fluxes in land and waters (Van den Bosch et al. 1988; Galloway et al. 1995; Howarth et al. 1996; Valiela and Bowen 2002; Boyer et al. 2002; Boyer and Howarth 2002; Filoso et al. 2003; Van Breemen et al. 2002).

Our work focuses on the Guayas Basin in Ecuador, which is the largest tropical agricultural watershed and estuarine system on the Pacific coast of South America. The Guayas Basin was chosen for study because it is experiencing severe environmental consequences associated with nutrient loading. This important estuary and coastal zone is experiencing major problems such as eutrophication, sedimentation, and pollution driven by urbanization, agriculture, aquaculture, and deforestation (Lacerda et al. 2002).

We examine the relationship between land use, anthropogenic nutrient inputs, and surface water nutrient exports in the Guayas watershed. We first develop comprehensive nutrient budgets for nitrogen (N) and phosphorous (P) for the overall Guayas river basin, 10 of its sub-watersheds, and subsequent nutrient loadings to the Gulf of Guayaquil. This budgeting method is useful as it provides a uniform way to quantify inputs (sources) and outputs (fate) in the region (Howarth et al. 1996; Van Breemen et al. 2002). Our nested-basin approach allows us to explore transfers and flows of nutrients within the Guayas region over a range of scales from cropland, watershed, and regional (Smaling and Oenema 1997). Next, we interpret the nitrogen and phosphorus budgets as an indicator of sustainable agricultural practices with respect to soil fertility (Pierzynski 1997). Further, we explore the relationship between nutrient inputs and subsequent nutrient loadings to the coastal zone, where water quality problems associated with eutrophication are a major concern.

Study area: the Guayas Basin

The Guayas Basin lies along the Pacific, draining a land area of 32,024 km^2 into the Gulf of Guayaquil (Figure 1a, 1b). A diversity of ecosystems is characteristic of this region: estuarine mangrove areas, dry and humid forests, Andean paramo, and agricultural land. Economic activities in this basin have changed the land use and water quality in the estuary enormously over the past 20 years (Southgate and Whitaker 1994; CAAM 1996; Twilley et al. 1998). Three main anthropogenic activities have impacted the Guayas Basin landscape: urban-industrial development, monoculture agriculture (banana, palm, soy, sugar cane, rice, and maize), and more recently shrimp pond aquaculture at the outlet of the Guayas Basin (Southgate and Whitaker 1994; Falconi-Benites 2000). The Guayas region is economically important in Ecuador because it produces 68% of the national crops, 73% of corn, 88% of bananas, 89.6% of export shrimp, 39% of cattle, and 50% of industry and manufacturing (Comisión Asesora del Medio Ambiente, CAAM 1996).

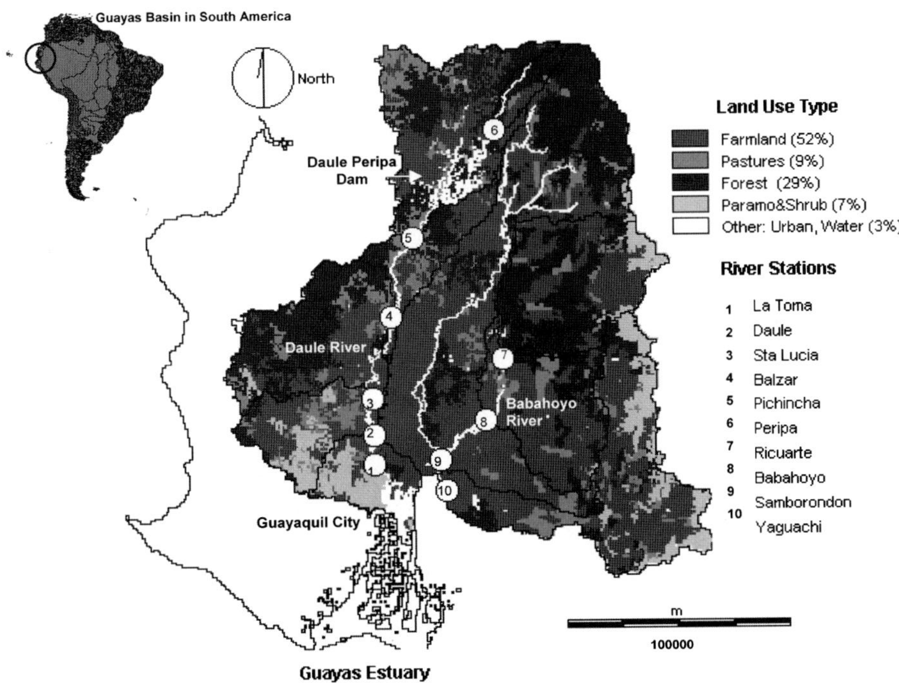

Figure 1. Watersheds and river systems in the Guayas Basin draining to the Guayas Estuary and the Gulf of Guayaquil in Ecuador.

Climate

The Guayas Basin is a humid tropical system, characterized by high solar radiation, temperatures, rainfall, and humidity. The annual mean temperature of the Guayas Basin is 24.8 °C, with lowest mean monthly temperatures in September (23 °C) and highest mean monthly temperature in April (35 °C) at Guayaquil. In the mountains toward the Andes, minimum temperatures can drop to 0–10 °C (Comision Asesora del Medio Ambiente, CAAM 1996). The basin is characterized by a rainy season from December to May each year. Mean annual precipitation is 1462 mm and ranges from 1060 to 2316 mm, with variations related to West-East orographic factors. In dry years precipitation can drop to 400 mm and it can increase to 4000 in wet years such as during El Nino events. March is the wettest month, with a mean precipitation of 365 mm; August is the driest month with a mean of 9.8 mm. Seventy eight percent of total precipitation occurs between January and April (CAAM 1996).

Watershed

We used elevation data from the USGS international elevation derivative database at 90 m resolution (USGS 2003) to delineate the boundaries of sub-watersheds within the Guayas River basin. We chose 10 sub-watersheds (Table 1) located along two main rivers, the Daule, and Babahoyo (Figure 1). The Daule and Babahoyo rivers merge downstream to form the larger Guayas River and subsequently the Guayas Estuary and Gulf of Guayaquil.

Table 1. Characteristics of the study sub-basins within the Guayas River basin.

ID	Name	Basin	Area (km^2)	%	Pop. density (hab km^{-2})	Elev. (m)	% Coverage				
							Forest	Crop	Pasture	Water	Other[a]
1	Guayaquil	Daule	1360	4.2	350	83	6.4	38.1	1.7	5.6	48.2
2	Daule	Daule	1795	5.6	53	198	7.9	51.3	17.8	0.4	22.6
3	Sta Lucia	Daule	3404	10.6	55	209	55.8	31.5	10.3	0.3	1.3
4	Balzar	Daule	1345	4.2	81	123	30.4	44.5	23.9	1.2	0.0
5	Pichincha	Daule	3768	11.8	30	216	40.6	35.1	16.2	6.8	0.9
6	Peripa	Daule	506	1.6	100	274	70.0	28.7	1.2	0.2	0.0
7	Ricaurte	Babahoyo	3567	11.1	49	1131	48.4	41.6	2.3	0.0	7.6
8	Babahoyo	Babahoyo	2894	9.0	53	551	24.1	68.7	6.1	0.4	0.7
9	Samborondon	Babahoyo	8767	27.4	89	305	31.5	62.3	4.4	0.2	0.9
10	Yaguachi	Babahoyo	4621	14.4	86	2152	11.2	56.5	7.7	0.1	23.9
	Total Guayas		32,026	100.0	90		29.0	52.0	9.0	2.0	8.0

When sub-basins are nested along the river network (downstream of each other), only the drainage area below any upstream sampling points is included in the sub-basin.

[a]Other is largely urban in watershed 1, shrubs in basin 2 and barren land (including paramo) in basin 10.

The Guayas Basin can be classified into four zones: the estuary of the Guayas (0–7 m a.s.l.), the lowlands (7–50 m a.s.l.), the uplands (> 200 m a.s.l.), and the highlands and paramos (> 1500 m a.s.l.). The Guayas estuary is a highly productive ecosystem both biologically and economically, and is a significant resource for shrimp aquaculture, which in 1998 generated 15% of the value of all Ecuadorian exports (Banco Central del Ecuador 1998). Mangrove forest ecosystems surround the estuary and are characterized by water-saturated soils (entisols and aquents) with poor drainage.

The lowlands of the Guayas Basin are primarily found downstream of the Babahoyo River, in the Lower Basin of the Guayas (Figure 1). This region is flat flood plain that is cross-cut by many rivers. Alluvial soils of volcanic origin are common in this region, and they are typically sandy clay soils with variable texture that are usually well drained. In these fertile river valleys the inhabitants have developed intensive agriculture focusing on banana, sugar cane and rice.

The Upper Basin of the Guayas is the area located upstream of the Daule River, and contains most of the uplands in the basin. The main characteristic of this region is its irregular topography that favors forest plantations and conservation of native forest; soils are entisoils and alfisoils. An important landmark of this region is the major Daule-Peripa Dam located in the Pichincha watershed (Figure 1). This reservoir has a capacity of 5.4 km^3, a flooded area of 27,000 ha, and a total average water discharge of about 71 $m^3 s^{-1}$ (CAAM 1996; CEDEGE 2001). Purposes of this dam include irrigation, water diversion, drinking water supply, salinity control, as well as a hydropower generation capacity of 213 MW yr^{-1} (CEDEGE 2001).

The highlands and paramos are located in the foothills of the Andes Mountains. Soils in this region have low fertility, and support low human density. Only a small portion of the basin (21%) is located in the highlands.

Land use

Biophysical factors such as soil and climate, as well as socio-economic variables such as availability of labor, local need for food crops and animal products, drive changes in land use (Southgate et al. 1991). We used a high resolution base map to describe land use in the Guayas Basin that was first developed by the Comisión de Estudios para la Cuenca del Guayas (CEDEGE 2000). This original vector map was processed using a geographic information system and classified into the 16 major land use and agricultural crop types of the region (Table 2). The original land use map had an aggregated classification of 'permanent' and 'annual' crops that was maintained in this analysis. Permanent crops consist of a combination of banana, sugar cane, fruit trees, plantain, African palm, cacao, and coffee; annual crops include a combination of maize, rice, soybeans, and vegetables (INIAP 2000). We assigned fertilizer inputs to these two land use categories (permanent and annual crops) in each sub-watershed of the Guayas by first estimating the area fraction of each

Table 2. Agricultural land use and fertilizer application rates within the Guayas Basin.

Land use	Area (km^2)	% of Guayas	N fertilizer (kg ha^{-1} yr^{-1})	P fertilizer (kg ha^{-1} yr^{-1})
Permanent crops and other	2783	9	99	13
Annual crops and other	8801	28	58	6
Cocoa and coffee	381	1	90	8
Banana	1221	4	250	20
Maize	331	1	46	5
Sugar cane	481	2	150	20
Rice	2222	7	68	0
Pastures, native & cultivated	2805	9	0	0
Paramo & pasture	1325	4	–	–
Forests (native & cultivated)	9206	29	0	0
Shrubs & barren lands	1169	4	–	–
Shrimp ponds	19	0	200	70
Urban	223	1	–	–
Mangroves	8	0	–	–
Water	487	2	–	–
Soybeans	495	2	50	10

individual crop (e.g. cacao, maize, etc.) from interviews with farmers and local agricultural agents, and multiplying this area by typical fertilizer application rates for each individual crop in this region of Ecuador (Table 2). Large quantities of inorganic fertilizer such as urea, ammonium sulfate, super phosphate and diammonium phosphate are applied as routine agricultural practice in the Guayas Basin (INIAP Manual Técnico No 26). Irrigation in the Guayas Basin is uncommon due to economic constraints, and most farmers must rely on the seasonality of rain for their agricultural practices.

The pastures and forest include both cultivated and natural varieties. The most common forest species in the basin are Teak (*Tectonis grandis*), Cedro (*Cederela odorata*), Pachaco (*Schizolobium parahybium*), Guadua Cane (*Guadua angustifolia*), and Balsa (*Ochrona kagopus*). The native forest includes species such as Samanes (*Albicia samanea*), Algarrobo (*Prosophis juliensis*), Guabo (*Inga coruscans*), and Leucaena (*Leucaena leucacephala*) that are leguminous and may represent between 5 to 10% of the forest biomass. Local management practices mix native forest species with pastures such as saboya (*Panicum maximun*), elefante (*Perisetum purpurem*), or Janeiro (*Eriochloa polystachya H.B.K.*) in order to improve soil fertility (Borbor V. J. 1960; Victor A. Borbor and Carlos Martinez, personal communication).

Population

We calculated population density using census data from the Instituto de Estadisticas y Censos del Ecuador (INEC 2001) geo-referenced to the 'parroquias' digital map (Almanaque Electronico Ecuatoriano (AEE) 1998), which is the basic administrative unit of Ecuador. In the year 2000 the population density ranged from 10 to 650 inhabitants ha^{-1}, an increase over 1990 values (4–472 inhabitants ha^{-1}). The most urbanized city is Guayaquil (650 people ha^{-1}) at the outlet of the Guayas Basin (Figure 1). Its population has grown from 1.5 million to 2.5 million over the past 10 years, and it is the largest city in Ecuador. The annual population growth for Guayaquil city has averaged 1.7%, compared to an average of 0.8% annually in rural areas (ECLAC 1999). These data indicate an intense migration to the urban area and even a reduction in population in some areas of the countryside. Five million people (40% of the national population) live in the Guayas Basin, which contains 27 of the 50 most populated cities of Ecuador (INEC 2002).

Methods

In overview, we developed nutrient budgets by combining spatial land use, demographic, and economic data obtained from census or remotely-sensed approaches with data derived from field and literature studies. The major fluxes of N and P within each of the 10 sub-watersheds of the Guayas Basin were calculated. Sources of nutrient inputs include atmospheric deposition, fertilizer application, biological fixation by agricultural and forest/leguminous crops, and net import or export in food and feed.

Nutrient inputs to the Guayas Watershed

Inputs from fertilizer use

Fertilizer use was calculated from knowledge of nitrogenous and phosphate fertilizer application rates based on data collected by INIAP (Instituto de Investigaciones Agropecuarias 2000) and from interviews that we conducted with farmers in the region during the 2002 sampling season. The quantities of N and P fertilizer entering the watershed were estimated by multiplying the rate of fertilizer application (kg ha^{-1} yr^{-1}) by land area harvested for each crop, subsequently adding them to obtain a weighted fertilizer input for each watershed (Table 2). The highest rates are for banana, coffee, sugar cane, and shrimp; all but sugar cane are export crops.

Inputs from biological fixation

Fixation in cultivated crop lands and in other vegetated lands that host symbiotic, N-fixing bacteria is a significant source of new nitrogen to each watershed region (Cleveland et al. 1999) Biological N fixation in agricultural lands (leguminous and pastures) and forests was calculated using N fixation rates reported in the literature. The most important rhizobium/legume crops in Ecuador are beans (*Phaseolus vulgaris*), soybeans (*Glycine max*), and peas (*Pisum sativum*). These crops normally fix 40–168 kg ha^{-1} yr^{-1} (Freire 1982;

FAO 1999; De Koning 1999). We chose values near the low end of this range because the P-availability in the soils of Ecuador is generally low, therefore limiting N-fixation (Smaling et al. 1993). Fixation rates used in this study are listed in Table 3.

Inputs from atmospheric deposition
The spatial distribution of nutrient inputs from wet deposition was estimated based on published correlations of deposition with rainfall in the region. Following the methods of Stoorvogel (1993) and De Koning (1999), nutrients in wet deposition can be approximated by $N = 0.14\sqrt{r_n}$ and $P = 0.023\sqrt{r_n}$, where r_n is rainfall (mm yr^{-1}), and N and P are the nutrient inputs (kg ha^{-1} yr^{-1}). Digital rainfall maps were obtained from the Almanaque Electronico Ecuatoriano (AEE 1998).

Inputs in food and feed transfers
Transfers of nutrients in agricultural products account for a significant re-distribution of N and P among the Guayas watersheds and to other regions, highlighting the importance of agricultural trade. The flows of nutrients from primary production (crops) to secondary production (animals) and then to households both inside and outside the region are driven by factors such as socio-economic status and agricultural practices. Animal waste and human waste are not considered as new inputs, as they represent recycling within a region; these terms are included in the estimate of N and P as part of the net food and feed (Boyer et al. 2002). We estimated the *net import or export of N and P in food (for humans) and feed (for animals)* in each watershed as a mass balance difference between food produced and consumed within the watershed (Boyer et al. 2002). This can be further disaggregated into the components associated with human food and animal feedstocks: Net food import = human consumption – crop production for humans – animal production for humans; and similarly Net feed import = animal consumption – crop production for animals. Details of the method are summarized below:

- We calculated rates of *human consumption* based on dietary intake of N and P per year, multiplying the intake per capita per year by the population density of each watershed. We assumed that Ecuadorians consume 4 kg N and 0.66 kg P per year (OPS 1994).

Table 3. Nitrogen fixation rates in cultivated croplands and in forests used to determine rates of N fixation in sub-basins.

Crop-legume type or forest	N fixation (kg N ha^{-1} yr^{-1})	Source
Bean, soybean	25	Freire (1982)
Grassland (cultivated and native)	10	De Koning (1999)
Wetland rice	15	De Koning (1999)
Forest (cultivated and native)	25	Freire (1982)

- We estimated *crop production for human food* from agricultural data on the nutrient content of food products grown in the watersheds, as detailed in Table 4.
- We calculated rates of *animal consumption* based on dietary needs for their intake of N and P consistent with agricultural management practices, based on the rates indicated for each type of livestock in Table 5. Livestock data were obtained from Ecuador's Agricultural Census (INEC 2001).
- We estimated *crop production for animal feed* from data on animal consumption in the region, which include maize, soybeans, sub-standard banana fruits, and annual crop residues. The rest is provided by pastures. The percentages of these crops used for animal consumption are 90, 90, 10, and 20, respectively. The total pasture production is for animal consumption (V. Borbor, personal communication).
- We quantified *animal production for human foods* (i.e. meat, milk, and eggs) as the difference between animal feed consumption (intake) and animal excretion (manure waste production). Nutrient demands for animals that are not met by locally grown produce are assumed supplemented by feed imports.
- There is little information about the rate of *excretion by animals* in Ecuador, thus we used rates from other regions (Table 5). We took the data for demand or excretion (in units of kg $head^{-1}$ yr^{-1}) and weighted them by the number of animal heads per watershed, based on animal inventory data for the region from the agricultural census. P demand in kg P $head^{-1}$ yr^{-1} is calculated as: P in dry matter/P excretion * 100; while P excretion in kg P $head^{-1}$ yr^{-1} is calculated as: amount manure * % solid manure * % dry matter * P content (Table 5). (Sibessen and Rutger Metzger 1995; Van Horn 1998).

Nutrient exports in the rivers of the Guayas Basin

We estimated annual exports of nutrients from each sub-watershed as the product of seasonal flow and nutrient chemistry for the wet and dry season at each station. Nutrients were sampled monthly during the wet and dry season at 10 stations along the Daule (stations 1–6) and Babahoyo Rivers (stations 7–10), the major sub-watersheds in the Guayas (Figure 1). We sampled three times during two wet seasons (February–May, in both 2002 and 2003) and twice in one dry season (August–September 2003). A total of 240 samples were collected during the fieldwork. For nested watersheds, we report the total nutrient export for the accumulated sub-watersheds located upstream of our sampling stations. Practical considerations (safety and access) dictated that some natural watershed units could not be sampled precisely at their mouths. In these cases, we have assumed that the chemistry at the point sampled is representative of chemistry at the watershed mouth.

At each station, we took three surface samples in a transect across the river for analysis of total dissolved nitrogen (TDN) and total phosphorus (TP). During all the fieldwork we filtered (Whatman GF/F;0.7 μm nominal pore size)

Table 4. Nutrient content of crops, crop residues, and pastures.

Land use type	Crop yield[a] (kg ha^{-1} yr^{-1})	Residue[a, d] (%) Wet weight basis	Moisture[c] (%)	Crop nutrient content[b] (% dry matter)		Crop residue[b] (g kg^{-1} dry matter)	
				N	P	N	P
Permanent crops	15,000	10	30	14.8	1.8	19.9	1.5
Temporary crops	12,600	10	90	15.8	1.8	27.7	2.5
Cocoa, coffee	1800	15	70	24.9	1.5	28.0	1.7
Banana	28,200	15	74	6.4	0.8	23.7	2.0
Maize	3636	15	52	23.0	5.0	16.7	3.3
Sugar cane	80,000	10	50	11.3	1.5	15.5	1.4
Rice	4200	10	50	14.1	3.8	5.5	1.3
Pasture (cultivated, native)	35,000	0	50	16.8	2.4	0.0	0.0
Forest (cultivated, native)	113,097	–	–	9.2	2.0	–	–
Shrimp from ponds[e]	3500	–	70	30.0	7.3	–	–
Soybeans	2727	10	70	41.4	3.2	43.6	3.6

Residue refers to crop residue as % of yield left on field following harvest. Moisture refers to crop moisture content as harvested.
[a]Instituto Nacional de Investigaciones Agropecuarias del Ecuador (INIAP) (2000).
[b]Van den Bosch et al. (1988).
[c]U.S. Department of Agriculture, National Nutrient Database for Standard Reference (2003).
[d]V. Borbor, Pers. Comm.
[e]Wet weight basis.

Table 5. Consumption (demand) and waste production (excretion) of nutrients by livestock.

Livestock	N demand (kg N head^{-1} yr^{-1})	N excretion (kg N head^{-1} yr^{-1})	P content mg (kg^{-1} dry matter)	Weight live (kg head^{-1})	% carcass fraction	P demand (kg P head^{-1} yr^{-1})	P excretion (kg P head^{-1} yr^{-1})	Source
Beef cattle	66.75	58.51	7.1	240	47	0.08	0.062	ARC (1984), Sibbesen (1989)
Dairy cattle	156	121	7.1	220	47	73.4	57.2	ARC (1984), Sibbesen (1989)
Chicken								ARC (1984), Sibbesen (1989)
Layers	0.84	0.55	5.8	2	80	0.93	0.54	ARC (1984), Sibbesen (1989)
Broilers	0.13	0.07	5.8	2	80	0.93	0.54	ARC (1984), Sibbesen (1989)
Pigs	8.51	5.84	5	65	70	22.7	17.2	Gunther (1972), Sibbesen (1989)
Sheep/goats	5.97	5	5	15	47	3.53	2.72	Gunther (1972), Sibbesen (1989)
Horses	44.8	40	7	180	47	59.2	45.0	Sibbesen (1989)

All data for N from Van Horn (1998), as cited in Boyer (2002). The P content of livestock commodities was calculated by multiplying the P content of live dry matter times the mean weight, then dividing by the carcass or dressing fraction which is the percentage of live weight that is considered edible. P excretion is calculated as a fraction of P demand. References as given in Tiessen et al. (1994).

and froze the 240 TDN samples immediately after collection in Ecuador and they were kept frozen until they were analyzed at the Water Quality Analysis Laboratory of the University of New Hampshire using high temperature combustion (Merriam et al. 1996). About 40 samples were not filtered and were kept frozen until analysis for TP in the Chemistry Laboratory of the Instituto Oceanografico de la Armada del Ecuador using persulfate digestion followed by ammonium molybdate analysis of phosphate. For 26 samples collected during the wet season 2003, we examined the magnitude of particulate N losses in this basin relative to TDN losses. Particulate N was measured directly on glass fiber filters (Whatman GF/F) with N analysis of the whole filter using a Perkin Elmer 2400 CHN analyzer.

We estimated average annual mass loadings of TDN and TP using measured concentrations and estimated streamflow at the 10 sampling stations. We obtained daily streamflow data for the Pichincha and Ricaurte (upstream) and Daule and Babahoyo (downstream) stations for the years 2002–2003. We used historical monthly stream flow data collected since 1980 to calculate average monthly and annual streamflow at these stations. These data were provided by the Instituto de Hidrologia y Meteorologia (INAMHI) and the Comision de Estudios para la Cuenca del Guayas (CEDEGE). We used these data to estimate flow in each of our sampling stations using a linear regression of streamflow as a function of the sub-watershed accumulated area. We used a simple volume-weighted mean concentration for calculating mean annual fluxes for 2002 and 2003 and subsequently averaged the 2 years (Webb et al. 2000). Annual Load $=\Sigma[c_i \cdot q_i]/\Sigma q_i) \cdot q_{\mathrm{annual}}$; where c_i = concentration for samples collected; q_i = flow at the time of collection; and q_{annual} = annual flow).

Results

Our watershed-specific N and P budgets allow us to evaluate the impacts of various land uses and agricultural practices on nutrient retention and losses.

Nitrogen budget

At the level of the Guayas Basin, fertilizer applications are the dominant source of N inputs (57% of total N inputs), and food exported out of the basin is the largest loss (30.1 kg N ha^{-1} yr^{-1}, which represents 43% of total N inputs) (Table 6). A relatively small fraction of total basin inputs (10 kg N ha^{-1} yr^{-1}; 14%) is delivered to the Guayas estuary by the Guayas River. At the sub-watershed level a positive N balance was found in 9 of the 10 sub-watersheds (Table 6). The specific fluxes of nitrogen are described as follows.

Atmospheric deposition

Nitrogen deposition varied from 3 to 7 kg ha^{-1} yr^{-1} in a gradient across the watershed in accordance with rainfall patterns. We estimated a minimum value

Table 6. Annual nitrogen budgets for the Guayas Basin and its sub-watersheds.

Watershed	ID	Fertilizer	Food[a]	Feed[b]	N fix. (crops)	N fix. (forest)	Atm. Dep.	Total input[c]	Net input[d]	River export[e]	N balance
Guayaquil	1	24.6	1.9	11.8	4.3	2.7	4.0	49.3	49.3	3.7	45.6
Daule	2	32.0	−16.2	−4.2	5.2	1.7	3.9	42.8	22.4	4.6	17.9
Sta Lucia	3	19.4	−13.2	6.3	2.9	9.6	4.7	42.8	29.6	3.8	25.8
Balzar	4	37.1	−38.7	−15.4	2.5	7.2	5.9	52.7	−1.4	3.1	−4.5
Pichincha	5	35.1	−30.1	6.9	2.6	9.6	6.3	62.5	30.9	3.0	27.9
Peripa	6	27.1	−46.0	9.2	2.2	8.3	7.4	54.2	8.2	7.4	0.8
Ricaurte	7	37.2	−31.5	57.6	2.8	9.1	5.9	112.6	81.2	12.6	68.5
Babahoyo	8	57.5	−48.4	36.9	6.9	6.1	5.8	113.2	64.8	18.6	46.2
Samborondon	9	54.7	−33.0	13.1	11.1	4.0	5.6	88.5	55.0	11.0	44.0
Yaguachi	10	41.5	−43.6	50.0	3.3	3.4	3.6	101.9	58.2	2.3	55.9
Overall Basin		36.8	−30.1	17.2	4.4	6.2	5.3	69.9	39.8	10.1	29.7

Fertilizer is total fertilizer application; food is net N input to basin in human food; feed is net N input in animal feed; N fix. (crops) is average in cultivated croplands; N fix. (forest) refers to N fixation by both managed and unmanaged forests; Atm. Dep. is atmospheric deposition. River export is hydrologic loss of N as total dissolved N (TDN) contributed by each sub-basin. When sub-basins are nested, flux from upstream basins is subtracted from flux at the downstream site to obtain the flux contributed by the sub-watershed under consideration. N balance is net inputs less river export, and is a measure of N accumulation in the sub-watershed. Negative values indicate N depletion is occurring in the sub-watershed. All values kg ha^{-1} yr^{-1}.

[a]Negative values for food signify a net export of N in food from the sub-watershed.

[b]Negative values for feed signify that a surplus of feed is left as organic N input within the watershed.

[c]Total N input = summation of positive inputs only.

[d]Net N input = fertilizer + food + feed + N fixation crops + N fixation forest + N atmospheric deposition.

[e]River export was measured at the outlet of each sub-watershed and at the outlet of the Guayas Basin.

of 3.6 kg N ha^{-1} yr^{-1} in the Yaguachi watershed (South-East) and a maximum value of 7.4 kg N ha^{-1} yr^{-1} at the Peripa watershed (North). Overall, this estimated atmospheric nitrogen input represented 8% of the total nitrogen input to the entire Guayas Basin.

Fertilizer use

Annual fertilizer N inputs ranged from a mean of 19.4 kg ha^{-1} yr^{-1} in a forested watershed (Sta Lucia) to over 57.5 kg ha^{-1} yr^{-1} in the highly agricultural Babahoyo watershed (Table 6). Watersheds with permanent export crops had the highest rates of fertilizer input per area. The Babahoyo and Samborondon watersheds received high inputs of fertilizer N, due to the high application rate to crops such as banana and sugar cane (150–250 kg N ha^{-1} yr^{-1}) and coffee and cocoa (90 kg N ha^{-1} yr^{-1}). The rate of N inputs to the individual sub-watersheds was related to the percentage of cropped lands in each sub-watershed ($r^2 = 0.41$, $p < 0.05$; Figure 2a).

Net food imports

The crop biomass produced for human consumption exceeded local consumption in all but the urban watershed (Guayaquil), which imports 1.9 kg N ha^{-1} yr^{-1} as food. The remaining study watersheds exported food products, resulting in a net N export that ranged from 13.2 to 48 kg N ha^{-1} yr^{-1} (Table 6).

Net feed imports

Livestock density varied greatly across the region and livestock were significant in watershed N budgets. Most of the sub-watersheds imported additional feed beyond that which was grown in the sub-watershed, and this import was up to 57 kg N ha^{-1} yr^{-1} (Ricaurte sub-watershed). Only two sub-watersheds produced surplus feed, primarily maize, which accounted for up to −15 kg N ha^{-1} yr^{-1} (Table 6). Overall, feed N imported into the Guayas Basin represented 25% of total N inputs.

Nitrogen fixation by plants

Nitrogen fixation by leguminous crops and agricultural land (pastures, wetland rice, and soybeans) ranged from an average of 2.2 kg N ha^{-1} yr^{-1} in Peripa (location above the dam reservoir) to 11.1 kg N ha^{-1} yr^{-1} in Samborondon where the dominant land use is wetland rice (32%), which is associated with nitrogen-fixing blue green algae. Forest fixation added a small amount of N input to the budget. The Daule watershed had the lowest value of N fixed by forested land at 1.7 kg N ha^{-1} yr^{-1}, in which forest is only 7% of the total area, and the maximum occurred in the Sta Lucia and Pichincha watersheds with a mean of 9.6 kg N ha^{-1} yr^{-1} (Table 6) with a dominant land use of forest (40–50%). In the overall watershed, N fixation by agricultural land and forest contributed 6 and 9% of the total N input, respectively.

Overall nitrogen inputs

The magnitude and relative importance of the N inputs varied widely by sub-watersheds along the Daule and Babahoyo Rivers (Table 6). N fertilizer input and net food imports/exports were the largest inputs and outputs across the sub-watersheds. Total inputs of N to each sub-watershed ranged from 42.8 kg N ha^{-1} yr^{-1} in the sub-watersheds of Sta Lucia and Daule to 113 kg N ha^{-1} yr^{-1} in the Babahoyo, where cultivation of banana (a permanent export crop) is the dominant land uses. Land use was a key driver of the N sources at the overall watershed scale of the Guayas Basin. Fertilizer was the largest input of N to the watershed (53% of total inputs), N fixation accounted for 15% of total inputs, and atmospheric deposition 8%. Crop production for animal feed was highly variable from one watershed to another; at the entire basin scale local food production was not enough to meet the needs of the human and animal populations, and thus we assumed there is a net import of feed from other regions (25% of total inputs to the Guayas basin).

Nitrogen fate

Combining the estimated net N inputs (including net food and feed) and the TDN riverine export we found a net negative N balance (a net loss of N from the basin) in the Balzar sub-watershed (−4.5 kg N ha^{-1} yr^{-1}; Table 6), but in all other sub-watersheds N inputs exceeded estimated and measured outputs. N exports in rivers as TDN ranged from 2.3 to 19 kg N ha^{-1} yr^{-1} representing a variable fraction of the total N input to each sub-watershed (4–29%). Much of the N entering individual sub-basins was exported as food or feed for use in other basins. Riverine export of TDN was higher in the Babahoyo River region (16–30% N export of N inputs; watersheds Ricaurte, Babahoyo and Samborondon) compared to the Daule River region (7–20.4% N export of net N inputs; from Peripa to Guayaquil watersheds). This variation is due to differences in the amount of cropland and forest in those regions, and the effect of the Daule Peripa Dam upstream on the Daule River. The dam not only regulates the streamflow but also appears to serve as a nutrient sink. Nitrogen flux above the dam was large (7.4 kg ha^{-1} yr^{-1} at the Peripa study site), but substantially lower at the first station below the dam, Pichincha (3.0 kg ha^{-1} yr^{-1}; Table 6) suggesting intense nutrient removal at the dam.

Phosphorus budget

At the basin level, P was lost from these watersheds primarily through the export and trade of agricultural commodities to other regions (38%), and we calculated 2.4 kg P ha^{-1} yr^{-1} exported to the Guayas estuary, which represents 38% of the P inputs to the basin (Table 7). A positive P balance was found in all of the agricultural watersheds, suggesting that phosphorus is retained and immobilized in the soils. The specific fluxes of phosphorus are described as follows.

Table 7. Annual phosphorus budgets for the Guayas Basin and its sub-watersheds.

Watershed	ID	Fertilizer	Food[a]	Feed	Weathering	Atm. Dep.	Total input[b]	Net input[c]	River export[d]	P balance
Guayaquil	1	1.45	0.95	1.19	0.10	0.67	4.36	4.36	0.93	3.43
Daule	2	2.81	−1.33	0.97	0.10	0.65	4.53	3.20	1.12	2.04
Sta Lucia	3	1.86	−1.51	1.58	0.10	0.77	4.31	2.79	1.10	1.69
Balzar	4	4.30	−3.10	0.42	0.10	0.95	5.35	2.65	0.56	2.10
Pichincha	5	3.89	−2.91	1.95	0.10	1.05	6.99	4.08	0.60	3.48
Peripa	6	3.64	−3.99	2.23	0.10	1.22	7.19	3.19	0.38	2.48
Ricaurte	7	3.87	−2.88	3.03	0.20	0.96	8.05	5.18	0.68	4.50
Babahoyo	8	5.39	−3.73	2.17	0.20	0.95	8.71	4.98	0.99	3.98
Samborondon	9	4.88	−2.55	0.75	0.20	0.96	6.04	4.23	1.42	2.81
Yaguachi	10	4.38	−3.17	2.69	0.20	0.59	7.87	4.70	0.59	4.32
Overall Basin		3.65	−2.43	1.70	0.14	0.88	6.36	3.93	2.4	1.53

Fertilizer is total fertilizer application; food is net P input to basin in human food; feed is net P input in animal feed; weathering is estimated background weathering rates; Atm. Dep. is atmospheric deposition. River export is hydrologic loss of N as total phosphorus contributed by each sub-basin. When sub-basins are nested, flux from upstream basins is subtracted from flux at the downstream site to obtain the flux contributed by the sub-watershed under consideration. P balance is net inputs less river export, and is a measure of P accumulation in the sub-watershed. All values kg ha^{-1} yr^{-1}.

[a] Negative values signify a net export of P in food from the sub-watershed.

[b] Total P input = summation of positive inputs only

[c] Net P input = fertilizer + food + feed + weathering + atmospheric deposition.

[d] River export was measured at the outlet of each sub-watershed and at the outlet of the Guayas Basin.

Atmospheric deposition

Inputs of P via atmospheric deposition ranged from an estimated 0.7 kg ha^{-1} yr^{-1} in dryer areas to 1.2 kg ha^{-1} yr^{-1} in the wetter mountainous areas (Table 7). At the basin level atmospheric deposition represents 14% of total P inputs.

Fertilizer use

In the phosphorus budget of each sub-watershed, fertilizer is typically the main input, ranging from 1.4 to 5.4 kg P ha^{-1} yr^{-1} (Table 7). The land use of Samborondon and Babahoyo, which are the sub-watersheds with the highest average P inputs from fertilizer, is dominated by permanent crops such as bananas and sugar cane, which use about 13–20 kg P ha^{-1} yr^{-1} as mineral fertilizer. Overall, 57% of the total input to the Guayas Basin comes from inorganic fertilizer.

Net imports in food and feed

The high population density in the city of Guayaquil and the intensive shrimp aquaculture surrounding the Guayas estuary are the drivers of the high TP loadings in the Guayaquil sub-watershed. Human consumption that ends as sewage waste in the highly urbanized Guayaquil demands an import of food of 0.95 kg P ha^{-1} yr^{-1} (Table 7). A net export of P as food is estimated in the rest of the watersheds, with a maximum value of 4 kg P ha^{-1} yr^{-1} in the Peripa sub-watershed, in which the predominant land use is permanent export crops (banana, African palm, coffee). In all the sub-watersheds the demand of P inputs for animal feed surpassed the production of P in crops and pastures used for animal consumption, thus imports of P in mineral fertilizers supplied this deficit. Inputs in feed ranged from 0.42 kg P ha^{-1} yr^{-1} in the Balzar watershed with 24% pastures to 3.03 kg P ha^{-1} yr^{-1} in Ricaurte with less than 3% pasture.

Weathering

P input by background weathering was estimated from the lowest phosphorus concentrations in upstream reaches of the basin, which is equivalent to 0.1–0.2 kg ha^{-1} yr^{-1} (Table 7).

Phosphorus inputs

At the sub-watershed scale, Babahoyo received the highest P inputs of the Guayas Basin with 8.7 kg ha^{-1} yr^{-1}; the minimum was 4.3 kg ha^{-1} yr^{-1} in the forested Sta Lucia watershed. The net P balance varied from 1.69 to 4.5 kg P ha^{-1} yr^{-1} (Table 7). The overall average for the Guayas Basin was 6.4 kg P ha^{-1} yr^{-1}. At the scale of the overall basin, 57% of the total P inputs were from the use of inorganic fertilizers, while 14% was from rainfall, and 2% from natural rock weathering. At the same time, 27% of the P inputs were estimated to be imported as feed for animals.

Fate of P inputs

Our nutrient budgets suggest a net export of P in crop production. Phosphorus exported to the outlet of the Guayas Basin accounted for 38% of the total P inputs to the landscape. The Daule River area carried lower TP loads (0.56–1.12 kg P ha^{-1} yr^{-1}) than the Babahoyo River (0.6–1.42 kg P ha^{-1} yr^{-1}) on average. We estimate that riverine export of TP increased downstream along the Daule and Babahoyo drainage network. The river reach with the largest P flux carried 1.42 kg P ha^{-1} yr^{-1} from the Samborondon watershed, which is subject to very intensive agriculture and livestock production.

Land use, N and P Inputs, and nutrient export in rivers

Nutrient inputs were directly related to agricultural land use in our sub-basins. We found a direct relationship between percent of cropland and N input ($R^2 = 0.41$), as well as percentage of cropland and TDN export (Figure 2b). High N inputs to the Ricuarte sub-watershed, related to livestock production and not to agricultural land, lowered the correlation coefficients relating N export to croplands and inputs. The Babahoyo, Ricaurte, and Samborondon sub-watersheds located along the Babahoyo River export greater amounts of N than those on the Daule river side of the Guayas Basin (Figure 2b).This is related to the intense agricultural management of the export crops of the Babahoyo river area versus the traditional annual crop management developed in the Daule River.

We also found that TDN exports were significantly correlated ($R^2 = 0.49$) with total N inputs (Figure 3), but not with Net N inputs ($R^2 = 0.32$; $p > 0.05$), because N is leaving the basin as export crops. The Net N input does not include losses from denitrification or volatilization, which we address in the discussion section.

Inputs of P were significantly correlated with the animal consumption represented by net feed inputs ($R^2 = 0.56$, $p < 0.05$) and were not correlated with agricultural land use. The main source of P is as fertilizer that is mainly applied to the permanent export crops. Phosphorus exported to the rivers was not correlated with total inputs, as much of the P inputs left the Guayas Basin as export crops (47% of total P input). Riverine P exports were directly correlated with river discharge at the sampling sites ($R^2 = 0.70$, $p < 0.05$; Figure 4) suggesting that phosphorus export is driven by runoff and erosion processes. The largest values of P export to rivers were found in the sub-watersheds with high livestock production (Ricaurte, Samborondon) and the Guayaquil urban area (Figure 4).

Discussion

Uncertainties in N losses

Our estimate of riverine export of N from these basins as TDN is an underestimate because we have not included particulate losses of N. Particulate N is

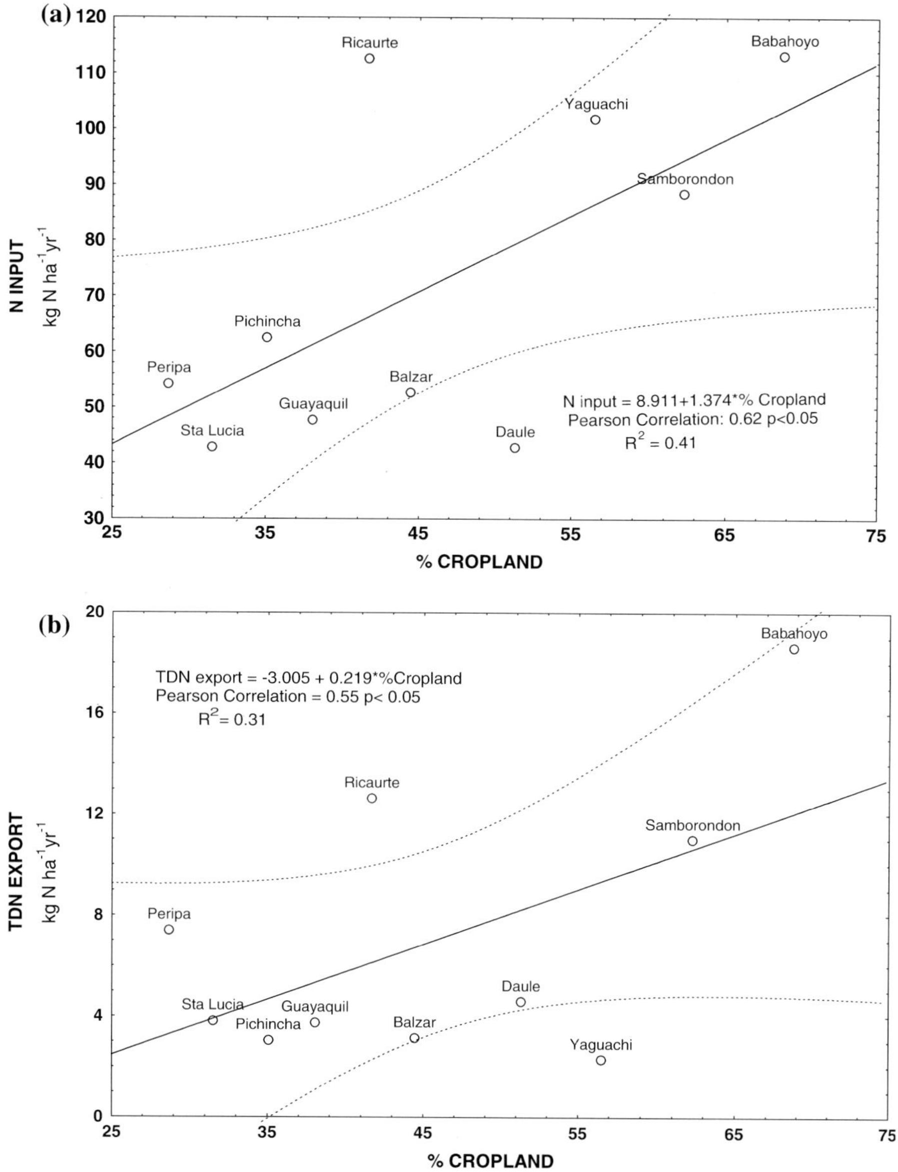

Figure 2. Nitrogen input (upper panel) and total dissolved nitrogen (TDN) export (lower panel) as a function of percentage of basin in cropland.

typically 30% of total N export in relatively undisturbed tropical rivers (Lewis et al. 1999). Comparison of TDN to particulate N for a subset of 26 samples from our study sites suggests that particulate N losses are significant in these basins, as particulate N was 72% of TDN concentrations. Because particulate

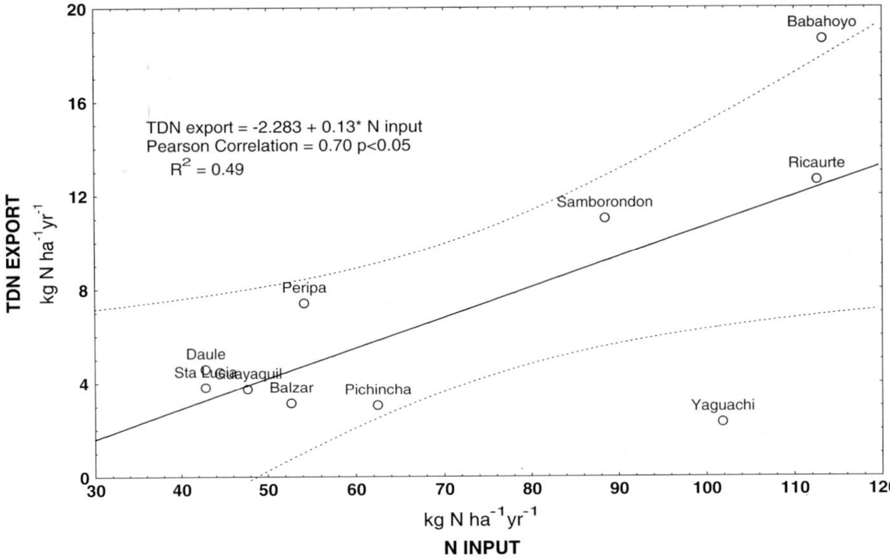

Figure 3. Export of total dissolved nitrogen (TDN) as a function of N input.

N concentrations can vary substantially with discharge in tropical rivers (McDowell and Asbury 1994), and our limited sampling of particulate N did not cover the full range of flows encountered at our site, we have not attempted

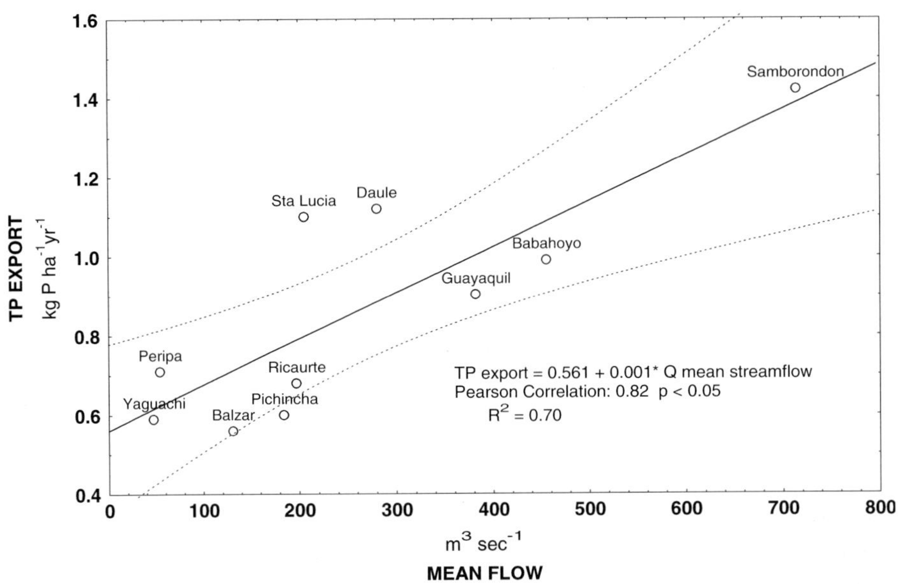

Figure 4. Export of total dissolved nitrogen (TDN) as a function of N input.

to estimate a specific value for particulate N export. However, our small data set suggests that the particulate N may be equal in magnitude to our measured TDN losses. Previous work has shown that larger watersheds tend to have higher losses of particulate N, and thus there may be proportionally more particulate N loss in the larger sub-watersheds of the Guayas Basin (Lewis et al. 1999).

We found that at the basin level the largest sink of N was via exports in food (43% of N input), followed by the N transported into the rivers (14% of N inputs), leaving 43% of N inputs unaccounted for in the landscape. We attempted to 'close' the N budget at the overall Guayas Basin scale by estimating the fate of the remaining N inputs, by considering fractions stored in soils or vegetation of landscape, or lost to the atmosphere by volatilization and denitrification processes (Jaworski et al. 1992; McMahan and Woodside 1997; Van Breemen et al. 2002). Though highly uncertain, we use literature values to establish a first guess at the overall fate of N in the basin. Denitrification in agricultural and tropical soils can be very significant, particularly in wet areas with high inputs of N fertilizers such the Guayas Basin (Velthof et al. 1997; Downing et al. 1999). Rates of denitrification in watersheds vary broadly and are dependent on pH, soil moisture, organic carbon availability and temperature (Van Breemen et al. 2002). Previous data on denitrification in tropical agricultural lands and tropical aquatic freshwater are very scarce. Losses of around 25% may be experienced following application of urea fertilizers (Vallis and Keating 1980), and up to 40% for urea application on sugar cane (McKee and Eyre 1999). We estimated denitrification in the terrestrial landscape assuming losses of 25% of applied nitrogen fertilizer in cropland, 3 kg ha^{-1} yr^{-1} from pastures, and 0.6 kg ha yr^{-1} from forest area based on literature values (McKee and Eyre 2000; Valiela and Bowen 2002). These rates of denitrification were applied to the land use map and then aggregated to obtain an average for the Guayas Basin. The exports of TDN above (7.4 kg N ha^{-1} yr^{-1}) and below (3 kg N ha^{-1} yr^{-1}) the Daule Peripa Dam suggest extensive denitrification is occurring in the reservoir. We have not considered denitrification in the sewage plants of the region. Given that only primary sewage treatment is used in the basin, and that sewage treatment is only available to 20% of the rural population and 60% of the urban population (INEC 2001), we suspect that denitrification during sewage treatment is unlikely to be significant N sink.

Ammonia volatilization is a major pathway for N loss from urea-based fertilizers (Harper and Stewart 1987). Volatilization is likely to be occurring in this tropical agricultural watershed where fertilization is based on urea application. Ammonia volatized from excreta has been estimated between 20 and 30% of the nitrogen deposited in the manure in intensive ranching areas (Vallis and Keating 1997; McKee and Eyre 2000). We estimated volatilization losses using the lowest reported rate (20%) because of the low intensity of ranching in the Basin.

Though admittedly uncertain, we estimate that 13.6 kg N ha^{-1} yr^{-1} (25% of N input) may be retained as storage in the soil, vegetation and forest Our estimate of denitrification in the overall landscape was 9.4 kg N ha^{-1} yr^{-1} (13% of N inputs) and volatilization was 6.7 kg N ha^{-1} yr^{-1} (10% of N input). Comparing our measurements of N exported to the rivers of 10 kg N ha^{-1} yr^{-1} (14% of N input), with the nutrient losses by erosion estimated by Koning (1999) of 20 N kg ha^{-1} yr^{-1} from agricultural land in the coastal region, suggests that our estimation of denitrification and volatilization in the landscape (about 15 N ha^{-1} yr^{-1}) may be an adequate first estimation.

We did not evaluate the amount of denitrification in the rivers, though it may be important in closing the N cycle. Rates of denitrification in aquatic systems are thought to be more rapid in the tropics than the temperate zone (Downing et al. 1998), hence riverine denitrification may be playing an important role in reducing N loading to the coastal zone in the Guayas estuary.

Nutrient balances and agricultural practices

Land use is the key factor contributing to the nutrient loadings to the landscape in this region. Agricultural land use in particular dominates nutrient cycling in the region, in terms of both inputs to the basin (via fertilizer applications and crop & animal production) and exports from the basin (in vegetables, meat, milk, and eggs). This highlights the importance of international trade in determining the nutrient budgets for the region. Ecuador is a net importer of fertilizers and agrochemicals and a net exporter of tropical food products. For example, of the bananas, coffee and shrimp produced each year, 60, 80, and 98% respectively are exported (FAO 2003). Fertilizers and agrochemicals represent between 30 and 40% of agricultural production costs.

Regional differences in nitrogen export in the two major rivers of the basin, the Babahoyo and Daule, are probably the result of several factors. Riverine export as TDN was higher in the Babahoyo River than in the Daule River (Table 6). Export of TDN in the Daule may be reduced by denitrification in the reservoir above the Daule-Peripa Dam, as denitrification is an important process in tropical reservoirs (Downing et al. 1999). The dam also regulates N flux by decreasing streamflow during the rainy season, thereby decreasing nitrogen loading to the rivers. Differences in land use patterns between the two basins may also explain differences in nutrient export. In the Daule, with its lower N export, cropland is 51% of land area, whereas in the Babahoyo River watershed it is 65%. Crops such as banana and sugar cane, which are more abundant in the Babahoyo watershed, also require more fertilizer additions than annual crops such as rice, maize and soybeans. The rate of fertilizer application to the permanent crops is high (>200 kg N ha^{-1} yr^{-1}), contributing to greater leaching of N to the rivers. Large leaching losses of nutrients also commonly occurs after sugar cane harvest and burning vegetation (Vallis and Keating 1997), and this may also contribute to the higher N loading in the Babahoyo River.

With respect to soil fertility, our nutrient budgets are a useful indicator of sustainability. In cases of negative balances where net nutrient outputs exceed the inputs, stocks of soil nutrients are probably declining, endangering agricultural production which may trigger land degradation (Stoorvogel 1993; De Koning 1999; Van den Bosch et al. 1988; Priess et al. 2001). On the other hand, if the balance is strongly positive, nutrient enrichment may lead to eutrophication in rivers and downstream in coastal areas. At the regional level there is a net accumulation of nutrients (both N and P) in the Guayas Basin. At the sub-watershed level there is a positive balance in most of the watersheds suggesting that N and P inputs introduced into the systems by farming practices introduce external inputs to offset nutrient removal in exported crops. If no losses had occurred, these basins may qualify as sustainable with respect to soil fertility, at least by these criteria (Stoorvogel 1993; De Koning 1999; Van den Bosch et al. 1988; Priess et al. 2001). However, direct assessments of soil fertility should be conducted to verify that nutrients are retained in the soil.

Two of our study basins with intensive agriculture (Balzar and Peripa) have net N balances near or below zero, indicating soil degradation and potential loss of productive capacity. Even the intensive addition of nutrients in fertilizers did not compensate for the nutrients leaving in the crops exported from these basins, which are devoted mainly to bananas and rice production. A previous assessment of the banana industry in Ecuador (Borbor 1999) showed that even though fertilizer application has increased from 150 kg ha^{-1} in 1985 to 550 kg ha^{-1} in 1998, banana yields have decreased from 30 to 20 Mt ha^{-1} (Mt – metric ton). In these watersheds that are losing N and P there is a considerable risk of a decline in soil fertility and long-term economic losses for farmers.

Despite the possibility of N depletion in some sub-watersheds, our assessment of the entire Guayas Basin shows that it is retaining a significant fraction of total N inputs (25%), and an even greater fraction of net N inputs. There is considerable uncertainty about the proportion of N and P retained in the soil due, however, due to the lack of empirical data on soil nutrient retention. Random soil samples analyzed by Instituto Investigaciones Agropecuarias (INIAP) suggest a great variability in nutrient content depending on the region and type of land use (unpublished data). In general, nutrient content is higher in the lower basin in the upper basin (INIAP, unpublished reports). Although we were unable to tightly constrain estimates of denitrification, we estimate that there were significant N losses via denitrification in this wet region, in particular due to those associated with the use of urea-based fertilizers. Rice paddies actually fix and retain N within the soil, but during wet and warm conditions they may also favor denitrification.

Land use decisions are linked to biophysical (local level) and socio-economic factors (regional and international level). In Ecuador, the fertile soils of the Guayas Basin, the national population growth, and the international demand for tropical crops have combined to create strong pressures on the land of this region. This has led to shorter fallow periods and the conversion of non-suitable agricultural land to agricultural crops and grasslands, often in marginal areas (Southgate and Whitaker 1994), resulting in greater potential for nutrient depletion especially in the highlands (De Koning 1999). We have identified areas within the Guayas basin where nutrient depletion is already evident, which suggests that agricultural productivity in the region may be affected in time as food production intensifies.

The sustainability of a land use system is comprised of ecological, agro-technical, and socio-economic dimensions (Van Ittersum et al. 1998). In the case of the Guayas Basin the main concerns are economic efficiency and the agro-technical performance. Thus, the export farming of this region has relied on large applications of fertilizers and agrochemicals to obtain high yields. However, it is important to differentiate between sustainable agricultural management and simply more energy intensive production processes. These may increase yield, but if the fertilizer is removed yields probably would fall to levels below their original pre-fertilization value because site quality has declined (Hall and Hall 1993; Hall et al. 1998). Since N fertilizers are very energy intensive this may become a much larger issue in the future if energy prices increase greatly, especially after Ecuador ceases to be a net producer of petroleum (Hall et al. 2003; Hallock et al. 2004).

Conclusions

Nutrient budgets in the Guayas watershed are driven largely by agricultural production. Nutrient inputs to the terrestrial landscape are dominated by fertilizer inputs, accounting for about 50% of N and P inputs. These nutrient inputs provide economic return in the form of agricultural commodities, most of which are exported out of the country. Although at the overall basin level there is a net accumulation or stable amounts of nutrients in the system, there are signs of nutrient loss and soil degradation in portions of the landscape that are subject to intensive fertilization, irrigation, and ranching activities. Further, there is a direct relationship between nutrient inputs to the landscape and nutrient loadings in surface waters of the region. However, only a small fraction of N (14%) and a larger fraction of P (38%) inputs was leached to rivers. Our analysis suggests that N river export is more related to land use and agricultural practices, while P is driven by runoff and erosion process. This highlights the need for research to understand the nutrient dynamics in the landscapes of this region. Understanding nutrient budgets is a useful analysis tool in order to incorporate better management practices that avoid soil degradation in the long run and mitigate N and P pollution problems in the coastal zone. Further research in this tropical agricultural watershed should be able to extend the comparisons to smaller watersheds where data acquisition can be more extensive.

Acknowledgements

We are grateful for funding to support this research from the Inter-American Institute for Global Change. We thank Dennis Swaney and Karin Limburg for their insights at various stages of this work.

References

U.S. Department of Agriculture, Agricultural Research Service 2003. USDA National Nutrient Database for Standard Reference, Release 5. Nutrient Data Laboratory Home Page, http://www.nal.usda.gov/fnic/foodcomp.

Almanaque Electrónico Ecuatoriano (AEE) 1998. Sistemas de Información Geográfica para Aplicaciones Agropecuarias en el Ordenamiento de Territorio y Manejo Integral de Cuencas. Alianza Jatun Sacha/CDC-Ecuador, Escuela Superior Politécnica del Ejercito (ESPE), y Centro Internacional de mejoramiento del maiz y trigo (CIMMYT-Ecuador).

Borbor M. 1999. A systems analysis of banana and shrimp production in Ecuador emphasizing the environmental impact on coastal ecosystems. M.S. thesis, State University of New York, College of Environmental Science and Forestry, Syracuse, NY, p. 73.

Borbor V.J. 1965. Principales pastos cultivados en las Haciendas ribereñas del rio Guayas. B.S. Thesis. Department of Agricultural Engineering. Universidad Guayaquil, Ecuador.

Boyer E.W. and Howarth R.W. (eds) 2002. The Nitrogen Cycle at Regional to Global Scales. Kluwer Academic Publishers, 518 pp.

Boyer E.W., Goodale C.L., Jaworski N.A. and Howarth R.W. 2002. Anthropogenic nitrogen sources and relationships to riverine nitrogen export in the northeastern USA. Biochemistry 57/58: 137–169.

Bruijnzeel L.A. 1996. Predicting the impacts of land cover transformation in the humid tropics: the need for integrated research. In: Gash J.H.C., Nobre C.A., Roberts J.M. and Victoria R.L. (eds.), Amazonian Deforestation and Climate, John Wiley & Sons, New York, NY, USA, pp. 15–55.

Caraco N.F. 1993. Disturbance of the phosphorus cycle: a case of indirect effects of human activity. Trends Ecol. Evolut. 7: 51–54.

Cole J.J., Peierls B.L., Caraco N.F. and Pace M.L. (1993). Nitrogen loading of rivers as a human-driven process. In: McDonnell M.J. and Pickett S.T.A. (eds), Humans as Components of Ecosystems: The Ecology of Subtle Human Effects and Populated Areas. Springer-Verlag, pp. 141–157.

Comision Asesora Ambiental de la Presidencia de la Republica del Ecuador (CAAM)1996. Desarrollo y problematica ambiental del area del Golfo de Guayaquil, 105–110.

Comision de Estudios para el desarrollo de la Cuenca del Rio Guayas (CEDEGE) 2000. Presented by Jose M. Martin at the Conference of 25th anniversary of CEDEGE. CEDEGE y la planificacion de recursos hidraulicos en Ecuador.

Comision de Estudios para el desarrollo de la Cuenca del Rio Guayas (CEDEGE) 2001. Plan integral de gestion socio-ambiental de la Cuenca del Guayas y Peninsula de Santa Elena. Resumen Ejecutivo, 149 pp.

Cleveland C.C., Townsend A.R., Schimel D.S., Fisher H., Howarth R.W., Hedin L.O., Perakis S.S., Latty E.F., Von Fischer J.C., Elseroad A. and Wasson M.F. 1999. Global patterns of terrestrial biological nitrogen (N_2) fixation in natural ecosystems. Global Biogeochem. Cycles 13: 623–645.

Cronan C.S., Piampiano J.T. and Patterson H.H. 1999. Influence of land use and hydrology on exports of carbon and nitrogen in a Maine river basin. J. Environ. Quality 28: 953–961.

Downing J.A., McClain R.T., Melack J.M., Elser J., Rabalis N.N., Lewis W.M. Jr., Turner R.E., Corredor S.D., Yanez-Arancibia A., Kopaska J.A. and Howarth R.W. 1999. The impact of accelerating land use change on the N-Cycle of tropical aquatic ecosystems: current conditions and projected changes. Biogeochemistry 46: 109–148.

ECLAC, Economic Comission for Latin American and the Caribbean 1999. Statistical Yearbook for Latin America and the Caribbean 1998, Chile. ISSN 10114–0697.

Falconi-Benitez Fander 2000. An integrated assessment of changes of land-use in Ecuador. Paper presented in the Advances in Energy Studies, Porto Venere 2000.

Food Agriculture Organization 2002. FAOSTAT, statistical database. http://www.apps.fao.org/.

Filoso S., Martinelli L.A., Williams M.R., Lara L.B., Krusche A., Ballester M.V., Victoria R. and de Camargo P.B. 2003. Land use and nitrogen export in the Piracicaba River basin, Southeast Brazil. Biogeochemistry 65: 275–294.

Galloway J.N., Schlesinger W.H., Levy H. II, Michaels A. and Schnoor J.L. 1995. Nitrogen fixation: anthropogenic enhancement-environmental response. Global Biogeochem. Cycles 9: 235–252.

Hall Charles A., Ko J.-Y., Lee C.-L. and Wang H.Q. 1998. Ricardo Lives: The Inverse Relation of Resource Exploitation Intensity and Efficiency in Costa Rican Agriculture and its Relation to Sustainable Development. Advances in Energy Studies Conference, Musis, Roma.

Hall C.A.S. and Hall M.H.P. 1993. The efficiency of land and energy use in tropical economies and agriculture. Agricult., Ecosyst. Environ. 46: 1–30.

Hallock J.L. Jr., Tharakan P.J., Hall C.A.S., Jefferson Michael and Wu W. 2004. Forecasting the limits to the availability and diversity of global conventional oil supply. Energy 30(10): 1673–1696.

Harper D.M. and Stewart W.D.P. 1987. The effects of land use upon water chemistry, particularly nutrient enrichment, in shallow lowland lakes: comparative studies of three lochs in Scotland. Hydrobiologia 148: 211–229.

Howarth R.W., Billen G., Swanney D., Townsed A., Jaworski N., Lajtha K., Downing J.A., Elmgreen R., Caraco N., Jordan T., Berendese F., Freney J., Kudeyarov V., Murdoch P. and Zhao-Liang Z. 1996. Regional nitrogen budgets and riverine N & P fluxes for the drainage to the North Atlantic Ocean: natural and human influences. Biogeochemistry 35: 75–139.

Houghton R.A. 1994. The worldwide extent of land-use change. BioScience 44: 305–313.

Tiessen H., Chacon P. and Cuevas E. 1994. Phosphorus and nitrogen status in soils and vegetation along a toposequence of Dystrophic Rainforests on the Upper Rio Negro. Oecologia 99: 145–150.

Instituto Nacional de Hidrologia y Meteorologia (INAMHI), Anuarios meteorológicos.

Instituto Nacional de Estadísticas y Censos (INEC) 2001. VI Censo Nacional de Población y Vivienda del 2000.

Instituto Nacional de Estadísticas y Censos (INEC) 2002. III Censo Nacional Agropecuario 2001.

Instituto Nacional de Investigaciones Agropecuarias (INIAP) 2000. Costos de las tecnologias de los principales cultivos del Ecuador, 124 pp.

Jaworski N.A., Groffman P.M., Keller A.A. and Prager J.C. 1992. A watershed nitrogen and phosphorus balance: the Upper Potomac River Basin. Estuaries 15: 83–95.

de Koning G.H.F. 1999. Spatially explicit analysis of land use change: a case study for Ecuador. Thesis. Wageningen University Agricultural University, 172 pp.

Lacerda L.D., Kremer H.H., Kjerfve B., Salomons W., Marshall J.I. and Crossland C.J. 2002. South American Basins: LOICZ Global Change Assessment and Synthesis of River Watershed – Coastal Sea Interaction and Human Dimensions. LOICZ Reports & Studies No. 21, Volume 1: ii + 95 pp, Volume 2: ii + 127 pp, LOICZ, Texel, The Netherlands.

Lewis W.M. Jr., Melack J.M., McDowell W.H., McClain M. and Richey J.E. (1999). Nitrogen yields from undisturbed watersheds in the Americas. In: New Perspectives on Nitrogen Cycling in the Temperate and Tropical Americas, pp. 149–162.

McDowell W.H. and Asbury C.E. 1994. Export of carbon, nitrogen, and major ions from three tropical montane watersheds. Limnol. Oceanogr. 39: 111–125.

Merriam J., McDowell W.H. and Currie W.S. 1996. A high-temperature catalytic oxidation technique for determining total dissolved nitrogen. Soil Sci. Soc. Am. J. 60: 1050–1055.

McKee L. and Eyre B.D. 2000. Nitrogen and phosphorus budget for the subtropical Richmond river catchment, Australia. Biogeochemistry 50: 207–239.

Ojima D.S., Galvin K.A. and Turner B.L. II 1994. The global impact of land-use change. BioScience 44: 300–304.

Organizacion Panamerica para la Salud (OPS) – Division de Salud y Ambiente 1994. Manual de evaluacion y manejo de sustancias toxicas en aguas superficiales. Seccion 3: Evaluacion preliminaria rapida, Lima, Peru.

Pierzynski G.M., Hettiarachi G.M. and Koelliker J.K., 1997. Methods for assessing the Impacts of Soil Degradation on Water Quality. Methods for assessment of soil degradation, pp. 513–545.

Skole D.L., Chomentowski W.H., Salas W.A. and Nobre A.D. 1994. Physical and human dimensions of deforestation in Amazonia. BioScience 44: 314–324.

Smaling E.M.A. and Oenema 1997. Estimating nutrient balances in agro-ecosystems at different spatial scales. Methods for assessment of soil degradation. pp. 229–250.

Southgate D. and Whitaker M. 1994. Economic Progress and the Environment: One Developing Country's Policy Crisis. Oxford University Press.

Sibbesen E. 1989. Phosphorus cycling in intensive agriculture with special reference to countries in the temperate zone of Western Europe. In: Tiessen H. (ed.), Phosphorus cycles in terrestrial and aquatic ecosystems. Regional workshop 1: Europe, SCOPE/UNEP Proceeding, University of Saskatchewan, Saskatoon, Canada, pp. 112–122.

Sibbesen E. and Rutger Metzger A. 1995. Phosphorus balance in the European Agriculture. In: Tiessen, H. (ed.), Phosphorus in the Global Environment: Transfers, cycles and management. SCOPE Vol 54, pp. 43–59.

Stoorvogel J.J. 1993. Optimizing land use distribution to minimize nutrient depletion: a case study for the Atlantic Zone of Costa Rica. Geoderma 60: 277–292.

Twilley R., Gottfried R., Rivera-Monroy V., Zhang W., Montano-Armijos M., Bodero A., 1998. An approach and preliminary model of integrating ecological and economic constraints of environmental quality in the Guayas River estuary, Ecuador. Environmental Science and Policy, 271–288.

USGS 2003. United States Geological Survey, Elevation derivative database. Online URL: http://www.edcdaac.usgs.gov/gtopo30/hydro/.

Valiela I. and Bowen J.L. 2002. Nitrogen sources to watersheds and estuaries: role of land cover mosaics and losses within watersheds. Environ. Pollut. 118: 239–248.

Vallis I. and Keating B.A. 1997. Uptake and loss of fertilizer and soil nitrogen in sugarcane crops. Proc. Aust. Soc. Sugar Cane Tech.

Van Breemen N., Boyer E.W., Goodale C.L., Jaworski N.A., Paustian K., Seitzinger S.P., Lajtha K., Mayer B., Van Dam D., Howarth R.W., Nadelhoffer K.J., Eve M. and Billen G. 2002. Where did all the nitrogen go? Fate of nitrogen inputs to large watersheds in the northeastern USA. Biogeochemistry 57/58: 267–293.

Van den Bosch H., de Jager A. and Vlaming J. 1988. Monitoring nutrient flows and economic performance in African farming systems (NUTMON) II. Tool development. Agric., Ecosyst. Environ. 71: 49–62.

Van Ittersum M.K., Rabbinge R. and Van Latesteijn H.C. 1998. Exploratory land use studies and their role in strategic policy making. Agricult. Syst. 58(3): 309–330.

Velthof G.L., Oenema O., Postma R. and van Beusichem M.L. 1997. Effects of type and amount of applied fertilizer on nitrous oxide fluxes from intensively manged grassland. Nutrient Cycling Agroecosyst. 46: 257–267.

Vitousek P.M., Aber J.D., Howarth R.W., Likens G.E., Matson P.A., Schindler D.W. and Tilman D.G. 1997. Human alteration of the global nitrogen cycle: sources and consequences. Ecol. Appl. 7(3): 737–750.

Webb B.W., Phillips J.M. and Walling D.E. 2000. A new approach to deriving 'best estimate' chemical fluxes for rivers draining the LOIS study area. Sci. Total Environ. 251/252: 45–54.

Yanez-ArancibiaA & Lara-Dominguez A.L. 1998. Mangrove Ecosystems in Tropical America: Structure, Function, and Management. EPOMEX Scientific Series 3, University of Campeche, Mexico.

Biogeochemistry (2006) 79: 163–186
DOI 10.1007/s10533-006-9010-1

The influence of climate on average nitrogen export from large watersheds in the Northeastern United States

R.W. HOWARTH[1,*], D.P. SWANEY[1], E.W. BOYER[2], R. MARINO[1], N. JAWORSKI[3] and C. GOODALE[1]
[1]*Department of Ecology & Evolutionary Biology, Cornell University, Ithaca, NY 14853, USA;* [2]*Department of Environmental Science, Policy & Management, University of California, Berkeley, CA, USA;* [3]*U.S. Environmental Protection Agency (retired), 27 Tarzwell Drive, Narragansett, RI 02882, USA; *Author for correspondence (e-mail: rwh2@cornell.edu)*

Key words: Nitrogen, Nutrients, Nitrogen flux, Nitrogen pollution, Denitrification, Climate change, Net anthropogenic nitrogen inputs, NANI, Watersheds, River, River basin

Abstract. The flux of nitrogen in large rivers in North America and Europe is well explained as a function of the net anthropogenic inputs of nitrogen to the landscape, with on average 20 to 25% of these inputs exported in rivers and 75 to 80% of the nitrogen retained or denitrified in the landscape. Here, we use data for average riverine nitrogen fluxes and anthropogenic inputs of nitrogen over a 6-year period (1988–1993) for 16 major watersheds in the northeastern United States to examine if there is also a climatic influence on nitrogen fluxes in rivers. Previous studies have shown that for any given river, nitrogen fluxes are greater in years with higher discharge, but this can be interpreted as storage of nitrogen in the landscape during dry years and flushing of this stored nitrogen during wet years. Our analyses demonstrate that there is also a longer-term steady-state influence of climate on riverine nitrogen fluxes. Those watersheds that have higher precipitation and higher discharge export a greater fraction of the net anthropogenic inputs of nitrogen. This fractional export ranges from 10 to 15% of the nitrogen inputs in drier watersheds in the northeastern United States to over 35% in the wetter watersheds. We believe this is driven by lower rates of denitrification in the wetter watersheds, perhaps because shorter water residence times do not allow for as much denitrification in riparian wetlands and low-order streams. Using mean projections for the consequences of future climate change on precipitation and discharge, we estimate that nitrogen fluxes in the Susquehanna River to Chesapeake Bay may increase by 3 to 17% by 2030 and by 16 to 65% by 2095 due to greater fractional delivery of net anthropogenic nitrogen inputs as precipitation and discharge increase. Although these projections are highly uncertain, they suggest a need to better consider the influence of climate on riverine nitrogen fluxes as part of management efforts to control coastal nitrogen pollution.

Introduction

Over the past several decades, eutrophication in coastal marine ecosystems has grown and is now considered the biggest pollution problem in the coastal waters of the U.S. (Howarth et al. 2000; NRC 2000). Eutrophication lowers biotic diversity, leads to hypoxic and anoxic conditions, increases the incidence and duration of some types of harmful algal blooms, degrades the habitat quality of seagrass beds or even destroys them, and can lead to changes in ecological food webs that reduce fish and shellfish production (NRC 2000). The Environmental Protection Agency's National Coastal Condition Report (EPA 2001) lists eutrophic condition as one of the three greatest threats to the health of the nation's estuaries, along with poor benthic condition (a result, in part, of eutrophication) and wetland loss. Some 40% of the estuarine area in the conterminous U.S. is severely degraded from eutrophication, and 67% is degraded to some extent (Bricker et al. 1999; EPA 2001). In the northeastern United States (defined as Chesapeake Bay north through Maine), some 60% of the estuarine area shows a high expression of eutrophic condition (EPA 2001). Eutrophication of coastal marine ecosystems is driven primarily by nitrogen inputs (Howarth 1988; Nixon 1995; NRC 2000; Howarth and Marino, 2006). From 1960 to 1980, average nitrogen fluxes in rivers to the coastal waters of the United States are estimated to have increased by 67% (Howarth et al. 2002a). During the 1980's, nitrogen fluxes increased little if at all. However, riverine nitrogen fluxes in the United States are estimated to have again increased steadily over the past 15 years, although less rapidly than during the 1960s and 1970s (Howarth et al. 2002a).

Climate variability and climate change are likely to have a profound effect on the delivery of nutrients to coastal marine ecosystems, but there is great uncertainty as to the detailed responses expected (Scavia 2002). This uncertainty results in part from divergent predictions for future climate change, for example with some global models predicting a drier climate and some a wetter climate in the northeastern United States as atmospheric carbon dioxide levels continue to rise over the next century (Wolock and McCabe 1999). Further uncertainty results from the non-linearity in response of riverine freshwater discharge to changes in climate, with some models suggesting discharge will increase disproportionately to increases in precipitation, and others suggesting increases in discharge will be less than increases in precipitation (Najjar 1999; Wolock and McCabe 1999; Najjar et al. 2000). Beyond these uncertainties in the physical climate system and the hydrologic responses of watersheds, the biogeochemical responses to changes in climate and hydrology are difficult to predict, particularly for nitrogen. However, sustained changes in nitrogen processing within the landscape are likely to have very significant effects on the health of coastal marine ecosystems.

Watersheds with greater precipitation and discharge will tend to have higher erosion rates, and this leads to higher fluxes of phosphorus from the landscape since most of the phosphorus in large rivers is particle bound (Howarth et al. 1995, 2002b; Moore et al. 1997). Nitrogen moves through the landscape primarily in dissolved forms, and nitrogen fluxes seem to be primarily controlled by the sources and sinks of nitrogen in the landscape. For disturbed landscapes in the temperate zone, an average of 20 to 25% of the nitrogen inputs resulting from human activity is exported in rivers (Howarth et al. 1996, 2002b; Boyer et al. 2002). Is there a climatic influence on this relationship? For examining global patterns of nitrate flux in large rivers, some models have assumed that the non-point-source contribution is controlled in part by area-specific discharge (Seitzinger and Kroeze 1998; Caraco and Cole 1999), but in a direct comparison among these and other models, those without discharge or other climatic parameters proved to be at least as accurate and precise in predicting multi-year average fluxes (Alexander et al. 2002). For the Mississippi River basin, McIsaac et al. (2001) demonstrated that during dry years, nitrogen accumulates in the soil or groundwater, and during wet years, this stored nitrogen is flushed out. The time scale of response in their study was only a few years. What would be the consequences of a sustained change in climate over a longer period of time? Over longer time scales, the primary issues are not short-term storage and flushing, but rather whether there are changes in nitrogen sinks in the landscape (storage in soils and biomass, or in rates of denitrification). In this paper, we further address the influence of climate on average riverine nitrogen flux by examining the relationship of net anthropogenic nitrogen inputs (NANI) on 6-year mean nitrogen fluxes in 16 major rivers across a climate gradient in the northeastern United States. By studying this climatic gradient, we can ascertain the longer-term steady-state effects of climate on riverine nitrogen fluxes.

Methods

We build upon the analysis of anthropogenic nitrogen sources and riverine nitrogen fluxes for 16 major watersheds in the northeastern U.S. done by Boyer et al. (2002) for the time period of 1988 through 1993. These watersheds are (moving north to south, from Maine to Virginia) the Penobscot, Kennebec, Androscoggin, Saco, Merrimack, Charles, Blackstone, Connecticut, Mohawk, upper Hudson, Delaware, Schuylkill, Susquehanna, Potomac, Rappahannock, and James. This is the same set of large watersheds used by Alexander et al. (2002) in their comparison of models for predicting nitrate and total nitrogen fluxes. It is important to note that both our work and that of Boyer et al. (2002) are based on the watershed areas upriver of defined USGS monitoring stations, and so do not generally include the heavily urbanized areas immediately along the coast. So defined, the watershed areas vary from 475 km^2 for the Charles River basin to over 70,000 km^2 for the Susquehanna (Table 1). The single largest land-use type in all 16 watersheds is forest, ranging from 48% of the area of the Schuylkill to 87% of the area of the Saco. Agricultural land use varies from 1.5% of the land area in the Penobscot River basin to 38% in the Schuylkill, and urban land use varies from 0.4% of the area in the Penobscot to 22% in the Charles River basin. Population densities vary from 8 individuals per km^2 in the Penobscot basin to 556 individuals per km^2 in the Charles River basin (Table 1). Further information on the watersheds is given in Boyer et al. (2002).

We estimated annual average river discharges using daily discharge data from river gauging stations located at the outlet of each watershed (USGS 2005). Annual discharge for all 16 rivers for the period 1950 through 2003 is

Table 1. Characteristics of the 16 major watersheds of the northeastern U.S. during the period 1988 to 1993. Watersheds are defined as the area upstream of USGS gauging stations, as in Boyer et al. (2002).

	Area (km^2)	Population density (# km^{-2})	Mean discharge (mm $year^{-1}$)	Mean precipitation (mm $year^{-1}$)	Mean temp. (°C)	Mean N export (kg N km^{-2} $year^{-1}$)
Penobscot	20,109	8	588	1075	4.3	320
Kennebec	13,994	9	566	1085	4.3	330
Androscoggin	8,451	17	640	1151	4.6	400
Saco	3,349	16	672	1218	5.8	390
Merrimack	12,005	143	589	1148	7.4	500
Charles	475	556	583	1207	9.7	1760*
Blackstone	1,115	276	651	1260	9.0	1140
Connecticut	25,019	65	642	1160	6.3	540
Hudson	11,942	32	622	1126	6.6	500
Mohawk	8,935	54	548	1142	6.8	800
Delaware	17,560	85	547	1131	8.7	960
Schuylkill	4,903	293	488	1134	10.6	1760
Susquehanna	70,189	54	487	1022	8.9	980
Potomac	29,940	63	328	985	11.3	900
Rappahannock	4,134	24	360	1045	12.6	470
James	16,206	24	407	934	10.1	310

*Includes nitrogen in wastewater flows diverted out of the watershed.

shown in Figure 1. The period of analysis for our study (1988–1993) is indicated by grey shading in the figure. Note that discharges during our period of analysis and the preceding several years are broadly representative of the longer time frame, without unusually high or low discharge years. Annual average riverine nitrogen exports were estimated for the 1988–1993 period from USGS data on total nitrogen concentrations (collected at the gauging stations at approximately monthly intervals) using the estimator approach described in Cohn et al. (1992). This regression-based method is a flow-weighted interpolation of the concentration measurements (Boyer et al. 2002). Mean estimates for precipitation and temperature for the 6-year period were obtained from the VEMAP-II historical climate reconstruction (Kittel et al. 1997; Boyer et al. 2002).

We determined the net anthropogenic nitrogen inputs (NANI) to each watershed using the approach of Howarth et al. (1996). In this method, NANI is the sum of fertilizer use, nitrogen fixation in agro-ecosystems, the net import of nitrogen in human food and animal feeds, and the atmospheric deposition of oxidized nitrogen (NO_y). Note that wastewater discharges are not considered explicitly in this analysis, since the nitrogen in wastewater originates from food (either imported or grown within the region, with the source nitrogen from fertilizer use or agricultural nitrogen fixation). Similarly, the deposition of ammonia and ammonium is not considered an input in this approach, as the large majority of the ammonia and ammonium deposited in a watershed is assumed to have originated from emissions within the same watershed

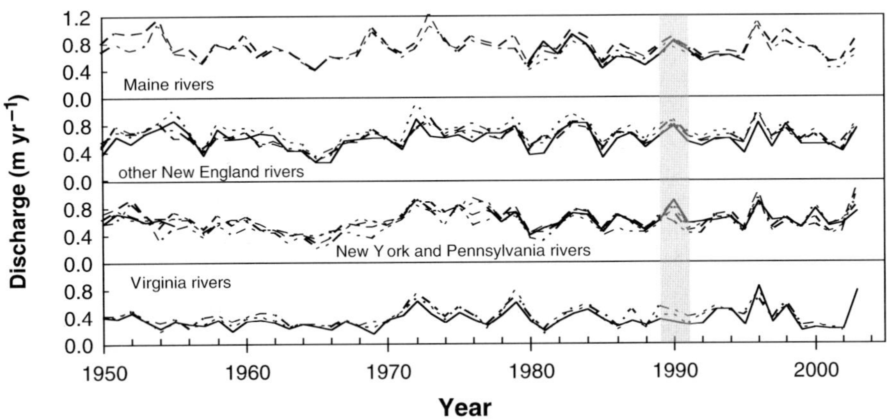

Figure 1. Annual freshwater discharge for each of the 16 northeastern US watersheds from 1950 to 2003. Top panel shows the 4 rivers that are mostly in Maine (Penobscot, Kennebec, Androscoggin, and Saco), the second panel from top shows the 4 other rivers in New England that are further to the west or south (Merrimack, Connecticut, Charles, and Blackstone), the third panel from the top shows the 5 rivers that are largely in New York State and Pennsylvania (Hudson, Mohawk, Delaware, Susquehanna, and Schuylkill), and the bottom panel shows the 3 rivers that are largely in Virginia and Maryland (Potomac, Rappahannock, and James). Our period of analysis for this study (1988–1993) is indicated by grey shade.

(Howarth et al. 1996). This is generally true in large watersheds and regions, but in smaller watersheds, there may be significant fluxes of ammonia and ammonium through atmospheric transport across different watersheds. Boyer et al. (2002) attempted to estimate these cross-boundary fluxes of ammonia and ammonium for the 16 major northeastern watersheds. However, this requires many highly uncertain assumptions, and in any case, the net ammonia/ammonium deposition due to cross-watershed transport in the atmosphere is small relative to NO_y deposition (Boyer et al. 2002). Therefore, we simply consider the NO_y term here. Note that in the Howarth et al. (1996) study, we used only the part of NO_y deposition estimated to originate from human activity in calculating NANI, rather than the total NO_y deposition we use here. On average for watersheds in the northeastern U.S., the total NO_y deposition is 2.3% greater than the anthropogenically derived NO_y deposition (Howarth et al. 1996).

We estimate the atmospheric deposition of NO_y (both wet and dry deposition) using the approach of Ollinger et al. (1993), based on a spatial model that extrapolates data from depositional monitoring networks (such as NADP) with a consideration of topographic effects. For most watersheds, we used the Ollinger et al. (1993) model, updated with more recent depositional velocities for dry deposition (Lovett and Rueth 1999). A few of the watersheds (Potomac, Rappahannock, and James) are outside of the geographic range for the regression equations used in Ollinger et al. (1993); for those three watersheds, we used the regression relationships put forth by Lovett and Lindberg (1993) that relate dry deposition to wet deposition (Boyer et al. 2002).

We estimate fertilizer use in each watershed using county-based sales data (Battaglin and Goolsby 1994) from 1991, scaled to the watersheds by weighting by the percentage of county area in each watershed (Boyer et al. 2002). Nitrogen fixation associated with agricultural crops is estimated from the area of particular types of crops (soybeans, alfalfa, snap beans, and hay and pasture) multiplied by literature-derived estimates of fixation rates associated with those individual crop types (Boyer et al. 2002). The net import of nitrogen in human food and animal feeds is estimated from a mass balance of needs versus production; that is, the difference between per capita estimates of the nitrogen in food and feed needs for humans and domestic animals and the nitrogen in foods and feeds produced within a watershed (Boyer et al. 2002). These estimates are somewhat sensitive to the assumed efficiency of nitrogen use in animal production; we use the values of van Horn (1998), which are based on U.S. agricultural practices.

Results

The average riverine nitrogen export from the 16 watersheds over the 6 year period from 1988 through 1993 ranged from a low of 310 to 330 kg N km^{-2} $year^{-1}$ for the James, Penobscot, and Kennebec River basins to a high of ~1760 kg N km^{-2} year^{-1} for the Charles and Schuylkill River basins (Table 1). In comparison, without human disturbance average watersheds in the north temperate zone are estimated to export approximately 100 kg N km^{-2} year^{-1} (Howarth et al. 1996, 2002b; NRC 2000). The fluxes from the Charles and Schuylkill basins are quite high, and in fact exceed the average flux from the watersheds of the highly populated, heavily industrialized and agriculturally intensive watersheds that drain to the North Sea in Europe (1450 kg N km^{-2} year^{-1}; Howarth et al. 1996). Seven out of the 16 watersheds in the northeastern U.S. have nitrogen fluxes that exceed the average flow down the Mississippi River basin (570 kg N km^{-2} year^{-1}; Howarth et al. 1996).

The 16 watersheds vary in the relative importance of the various nitrogen inputs to the overall NANI estimate (Table 2). The majority of NANI comes from NO_y deposition in the 4 watersheds in Maine (the Penobscot, Kennebec, Androscoggin, and Saco River basins). In the watersheds further south, the NO_y deposition rates are higher than in Maine, but other sources increase even more (Table 2). The net importation of nitrogen in food and feed is quite important in watersheds with higher population densities, and this makes up more than half of NANI in the Charles and Blackstone River basins. In many watersheds, agricultural inputs from fertilizer use and nitrogen fixation are dominant, and these make up 50% or more of NANI in the Mohawk, Delaware, Potomac, Rappahannock, and James River basins (Table 2). Overall, for

Table 2. Average annual nitrogen inputs from anthropogenic sources to the 16 major watersheds of the northeastern U.S. for the period 1988 to 1993 (kg N km^{-2} year^{-1}).

	NO_y deposition	N fertilizer use	Agricultural N fixation	Net N import in foods and feeds	Total Net anthropogenic N inputs (NANI)
Penobscot	360	90	70	40	560
Kennebec	430	50	160	150	790
Androscoggin	500	80	150	240	970
Saco	570	40	100	100	810
Merrimack	610	150	210	710	1680
Charles	670	200	190	2090	3150
Blackstone	710	310	310	1500	2830
Connecticut	630	270	360	570	1830
Hudson	660	200	370	270	1500
Mohawk	710	410	1240	620	2980
Delaware	810	530	680	350	2370
Schuylkill	890	1210	1230	1950	5280
Susquehanna	820	620	1150	1100	3690
Potomac	710	1020	1170	1450	4350
Rappahannock	620	1030	1440	610	3700
James	650	360	700	400	2110
Area-weighted mean	680	560	740	740	2720
northeastern US mean (Howarth et al. 1996)	1200	600	750	1000	3550

these 16 watersheds, the area-weighted mean nitrogen inputs to the watersheds are reasonably evenly distributed between NO_y deposition (680 kg N km^{-2} year^{-1}), fertilizer use (560 kg N km^{-2} year^{-1}), nitrogen fixation in agro-ecosystems (740 kg N km^{-2} year^{-1}), and the net importation of nitrogen in foods and feeds (740 kg N km^{-2} year^{-1}; Table 2).

The analysis presented here is similar to that presented in Howarth et al. (1996) for the northeastern U.S. as a whole in terms of average riverine nitrogen fluxes, and the mean value for the entire northeastern U.S. presented in Howarth et al. (1996) sits in the center of, and is bracketed nicely by, the riverine nitrogen flux values for the 16 major watersheds. The analysis here also is similar in terms of the agricultural sources to the regions (fertilizer use and agricultural nitrogen fixation), but the estimates given in Howarth et al. (1996) are substantially greater for NO_y deposition and for the net importation of nitrogen in food and feeds (Table 2). For the importation of nitrogen in food and feeds, we attribute this difference to the inclusion of the heavily populated coastal margin cities (New York City, Boston, Washington, Providence, Philadelphia, etc.) within the area included in the Howarth et al. (1996) analysis but excluded from Boyer et al. (2002) and this study. For the most part, these urban centers in the northeastern U.S. are down-river from the USGS gauging stations which define the watershed areas used by Boyer et al. (2002) and this study.

For NO_y deposition, the mean value for the 16 watersheds reported in Boyer et al. (2002) and used here is 680 kg N km^{-2} year^{-1}, while for the entire northeastern U.S. Howarth et al. (1996) used an estimate of 1200 kg N km^{-2} year^{-1} (Table 2). As was the case with the net importation of nitrogen in food and feeds, the difference in these estimates may reflect the different geographic boundaries, with higher deposition in the more urbanized areas. Deposition in the more rural areas represented by the 16 watersheds (as defined up-river of the USGS gauging stations) may not reflect the potentially high levels of deposition that occur near emission sources in urban areas (Holland et al. 1999; Lovett et al. 2000; Howarth et al. 2002a). However, the different estimates also may be due in part to different methodologies. As stated above, Boyer et al. (2002) used depositional monitoring data for their estimate. The core data are from the National Acid Deposition Program (NADP), whose stations are purposefully located in rural areas where urban and agricultural influences are minimal (NADP 2005). Spatial coverage is sparse, and scaling the point observations over space and time is difficult (Meyers et al. 2001). Further, only wet deposition is measured at the NADP stations, and challenges remain in how to estimate contributions from dry deposition, given the complexity of factors controlling deposition velocities (Ollinger et al. 1993, Meyers et al. 2001). The estimate used in Howarth et al. (1996) comes from the GCTM model, which predicts depositional patterns globally at a relatively course spatial scale using emission sources as inputs and modeling atmospheric transformations and transport (Prospero et al. 1996). A similar, more recent model (TM3) used by Galloway et al. (2004) for their global and regional nitrogen budgets yields a comparable estimate for the northeastern U.S. as did the GCTM model. These emission-based models are attractive, in that at least at very course spatial scales, they are as accurate as the emission data. However, they cannot easily be applied at a spatial scale fine enough to give estimates for the individual 16 northeastern watersheds. For the analysis in this paper, we therefore relied on the estimates from Boyer et al. (2002), which may well be robust for the rural areas represented by these watersheds. However, it is important to note that the actual total NO_y deposition to the northeastern U.S. may be substantially higher. If so, much of this additional NO_y deposition likely falls on the more urbanized landscape near the coast, where retention is low, and so it is likely to have a high percentage export to coastal waters (Howarth et al. 2002b).

The average annual riverine nitrogen fluxes from the 16 watersheds are highly correlated with NANI to each watershed (Figure 2). The relationship is very similar to that observed when comparing the large regional areas that drain into the North Atlantic Ocean, both from North America and from Europe (Howarth et al. 1996). Note in both cases the y-intercept of the linear regression is approximately 100 kg N km^{-2} year^{-1}, which has been used to provide an estimate of what the nitrogen flux off the landscape for temperate watersheds might be, absent human inputs of nitrogen (that is, NANI = 0; NRC 2000; Howarth et al. 2002b). Here, the slope of the regression is 0.26, indicating that on average only 26% of the human inputs of nitrogen to the landscape (NANI) are exported in downstream river export, and that 74% must be retained in the landscape or lost through denitrification. This is a similar slope to that observed in the coarser spatial-scale analysis of the North

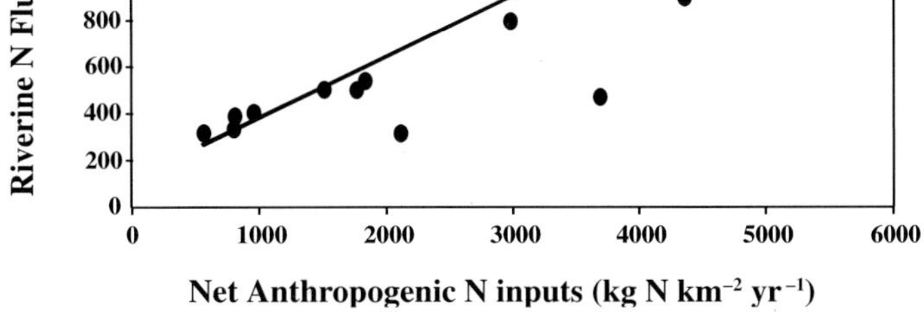

Figure 2. Average annual riverine nitrogen fluxes are strongly related to the net anthropogenic nitrogen inputs (NANI) to the watershed ($Y = 0.26X + 107$; $R^2 = 0.62$; $p = 0.0003$). Note that this relationship is similar to that reported in Boyer et al. (2002), although the latter was based on total nitrogen inputs, and not just the anthropogenic sources.

Atlantic drainage basin (Howarth et al. 1996). Note also that Boyer et al. (2002) presented a somewhat different analysis (in their figure 6). There, the x-axis is for total nitrogen inputs, and so includes an estimate for the natural rate of nitrogen fixation in forests. The Boyer et al. (2002) figure also includes some estimated net input of nitrogen from deposition of ammonia and ammonium which is not included in the NANI estimate here (see methods, above). The relationship shown in Boyer et al. (2002) looks very similar to that here, except that the intercept (7 kg N km^{-2} $year^{-1}$) was much closer to 0. This is consistent with the idea that a watershed that consistently receives no nitrogen inputs (from natural or anthropogenic sources) would export little or no nitrogen, and thus gives us greater confidence in using the intercept from Figure 2 (107 kg N km^{-2} $year^{-1}$) as an estimate of the riverine nitrogen flux for temperate watersheds which have only natural inputs of nitrogen.

The average riverine nitrogen flux from the 16 watersheds is fairly well explained just from NANI ($R^2=0.62$, $p=0.0003$; Figure 2), and one could easily believe that much of the scatter results from quality of data or from differences among the watersheds in characteristics such as soil type and topography. However, we note that the points lying above the regression line tend to be watersheds with higher discharge and precipitation, while those below it are from "less wet" watersheds (Table 1). To evaluate whether some aspect of climate has an influence on the long-term average flux of nitrogen from these watersheds (in addition to the influence of NANI), we examined the fractional delivery of NANI and examined its relationship to climatically related parameters (precipitation, temperature, and discharge). We define the fractional delivery as the riverine nitrogen flux that is above the natural background flux expected absent any anthropogenic nitrogen inputs, divided by NANI for that watershed. That is,

$$F = (R - 107)/\mathrm{NANI} \qquad (1)$$

where F is the fractional delivery of NANI, R is the long-term average riverine flux of nitrogen (kg N km^{-2} $year^{-1}$), 107 represents the natural background riverine nitrogen flux in the absence of human activity (kg N km^{-2} $year^{-1}$), and NANI is the net anthropogenic nitrogen input (kg N km^{-2} $year^{-1}$).

The fractional delivery of NANI is well correlated with both precipitation ($R^2=0.53$; $p=0.0015$; Figure 3a) and discharge ($R^2=0.48$; $p=0.003$; Figure 3b). Note that precipitation (P) and discharge (Q) are themselves correlated ($R^2=0.66$; plot not shown). Clearly, watersheds with greater precipitation and higher discharge have higher fractional deliveries, ranging from a high of 0.2 to 0.43 for watersheds with precipitation greater than 1,100 mm $year^{-1}$ and discharges greater than 500 mm $year^{-1}$ to 0.1 to 0.18 for watersheds with less precipitation and lower discharge. Temperature is not as good a predictor of fractional delivery of NANI, and the relationship, while suggestive, is at best marginally significant ($p=0.11$) and has a lower R^2 value (0.17; Figure 3c). Note however the suggestion of an inverse relationship, with

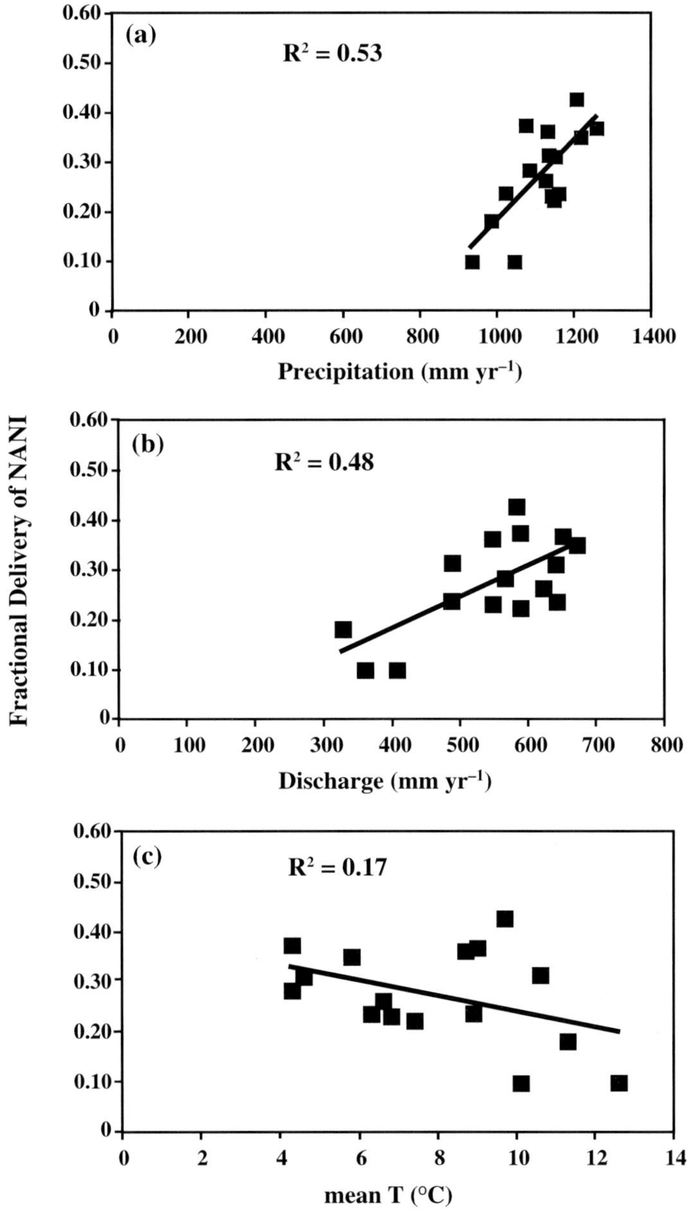

Figure 3. The fractional delivery of net anthropogenic nitrogen inputs (NANI) for the 16 watersheds plotted as a function of mean precipitation (a, top panel; $R^2 = 0.53$, $p = 0.0015$), mean discharge (b, middle panel; $R^2 = 0.48$, $p = 0.003$), and mean temperature (c, bottom panel; $R^2 = 0.17$, $p = 0.11$).

greater fractional delivery of NANI in the colder watersheds. While temperature is correlated with discharge ($R^2 = 0.56$; plot not shown), temperature is not well correlated with precipitation in these watersheds ($R^2 = 0.11$; $p = 0.20$; plot not shown).

To develop a predictive equation for the riverine nitrogen flux (R), we can re-arrange Equation (1) and substitute single variable relationships for F discussed above in an equation of the form:

$$R = F * \text{NANI} + 107 \quad (2)$$

For example, we could use precipitation as a predictor of F (Figure 3a) since that relationship had the best explanatory power. Substituting the regression parameters for Figure 3a into Equation (2) yields:

$$R = (0.0008 * P - 0.62) * \text{NANI} + 107 \quad (3)$$

where P is precipitation (mm year^{-1}). This equation can be written in the following form:

$$R = (0.0008 * P * \text{NANI}) - (0.62 * \text{NANI}) + 107 \quad (4)$$

Alternatively, to get the best parameter fit for an equation of the form of Equation (4), we can obtain coefficients that relate R to (P * NANI) and NANI by using a 2-variable linear regression with interacting terms. This yields the equation:

$$R = (0.00095 * P * \text{NANI}) - (0.762 * \text{NANI}) + 55 \quad (5)$$

The relationship is highly significant ($p < 0.000001$) and has an R^2 value of 0.875, or an adjusted R^2 of 0.855 (Table 3). Both the interaction term (P * NANI) and the NANI term contribute significantly to this relationship ($p = 0.0002$ and $p = 0.0024$, respectively; Table 3). Re-arranging Equation (5) into the form of Equation (3) yields:

$$R = (0.00095 * P - 0.762) * \text{NANI} + 55 \quad (6)$$

The intercept of 55 kg N km^{-2} year^{-1} is lower than the 107 kg N km^{-2} year^{-1} determined from the NANI vs. riverine nitrogen flux regression (Figure 2), but not significantly so. The 95% confidence interval for the intercept determined in Equation (6) extends from −155 to +255 kg N km^{-2} year^{-1}. Note that the term (0.00095 * P–0.762) expresses the fractional delivery of NANI, or F.

We also tested a more complex model, including not only (P * NANI) and NANI as input terms but also P (that is, the complete interaction model for NANI and P). This 3-variable, interacting-term regression model is also significant (Table 4), but less so than the simpler model using just (P*NANI) and NANI. The addition of P alone does not contribute significantly ($p = 0.613$; Table 3), and its inclusion in the regression lessens the significance of the other two terms in comparison to the simpler 2-term model ($p = 0.036$ for (P*NANI) compared to $p = 0.0002$, and $p = 0.091$ for NANI, compared to $p = 0.0024$;

Table 3. Summary statistics for several linear regression models that predict riverine nitrogen flux (R) based on precipitation (P) and net anthropogenic nitrogen inputs (NANI).

	Interacting-term model with NANI and P *NANI	Interacting-term model with P, NANI, and P *NANI	non-interacting term model with P and NANI
Regression statistics			
R^2	0.875	0.877	0.821
Adjusted R^2	0.855	0.847	0.793
Standard error	179.5	184.8	749.3
Observations	16	16	16
ANOVA			
df for regression	2	2	2
df for residual	13	13	13
F	45.36	28.63	29.75
P	$<$0.00001	0.00001	0.00001
Intercept			
Coefficient	55	920	−2710
Standard error	93	1670	749
t-statistic	0.59	0.55	−3.61
p-value	0.56	0.59	0.003
NANI			
Coefficient	−0.762	−1.03	0.287
Standard error	92.7	0.561	0.040
t-statistic	−3.76	−1.84	7.24
p-value	0.0024	0.091	0.00001
P			
Coefficient	–	−0.77	2.47
Standard error	–	1.49	0.65
t-statistic	–	−0.520	3.80
p-value	–	0.61	0.002
P *NANI			
Coefficient	0.00095	0.00095	–
Standard error	0.00018	0.00018	–
t-statistic	5.12	5.12	–
p-value	0.0002	0.036	

Table 3). This adds to our confidence in the approach we used to derive a predictive equation for riverine nitrogen flux by combining the relationship between NANI and riverine N flux (Equation (2); Figure 2) with the relationship which best predicts the fractional delivery of NANI as a function of precipitation (Figure 3a).

A regression that relates riverine discharge to P and NANI without an interaction of P and NANI yields the following equation:

$$R = (2.47 * P) + (0.29 * \text{NANI}) - 2710 \tag{7}$$

This relationship, too, is highly significant ($p = 0.00001$) with both the P and NANI terms contributing significantly to the regression ($p = 0.002$ and $p = 0.0001$, respectively; Table 3). However, the adjusted R^2 value (0.79) and F-

Table 4. Summary statistics for two interacting-term linear regression models that predict riverine nitrogen flux (R), based on discharge (Q) and net anthropogenic nitrogen inputs (NANI) or based on population density (D) and precipitation (P).

	Interacting-term model with NANI and Q*NANI	Interacting-term model with D and D*P
Regression statistics		
R^2	0.874	0.828
Adjusted R^2	0.855	0.801
Standard error	179.7	210.4
Observations	16	16
ANOVA		
df for regression	2	2
df for residual	13	13
F	45.28	31.25
P	<0.000001	0.00001
Intercept		
Coefficient	−101	374
Standard error	101	73.5
t-statistic	−1.00	5.09
p-value	0.33	0.0002
NANI		
Coefficient	−0.096	–
Standard error	0.077	–
t-statistic	−1.24	–
p-value	0.24	–
D		
Co efficient	–	25.0
Standard error	–	8.6
t-statistic	–	2.90
p-value	–	0.012
Interacting term		
Coefficient	0.00087	−0.18
Standard error	0.00017	0.0071
t-statistic	5.11	−1.58
p-value	0.0002	0.023

ratio (28.9) are slightly lower than for equations 5 and 6, where *P* and NANI interact (compare Tables 3 and 5). As discussed below, predictors that do not include interaction terms, such as that in Equation (7), lead to dramatically different mechanistic interpretations than do the predictive equations which include such interactions, such as equations 5 and 6.

The slope of the regression line when riverine nitrogen fluxes predicted using NANI and precipitation (Equation 6) are plotted against the actual observed nitrogen fluxes is very close to 1:1, with a very good linear fit (Figure 4a). However, an exponential fit of the regression looks reasonable as well (Figure 4b), and statistically, the two fits are indistinguishable. The exponential fit indicates the possibility of a bias in Equation (6), with it under-predicting riverine nitrogen fluxes at both the low and high end of the relationship. At the low end, this would be consistent with a "pristine" riverine nitrogen flux

(NANI = 0) that is greater than the 55 kg N km^{-2} $year^{-1}$ predicted from Equation (6). The exponential fit instead suggests a "pristine" riverine nitrogen flux of 255 kg N km^{-2} $year^{-1}$ (Figure 4b). At the high end of the relationship, the exponential fit would be consistent with the concept of nitrogen saturation (Aber et al. 1998, 2003). That is, the percentage of nitrogen exported from the landscape may increase disproportionately with nitrogen loading to the landscape above a certain point. Aber et al. (2003) have shown that nitrogen losses from forests in the northeastern US increase dramatically and non-linearly as atmospheric deposition exceeds ~700 kg N km^{-2} $year^{-1}$, as occurs in several of the watersheds included in our data analysis.

Discharge is not quite as good a predictor of the fractional delivery of NANI as is precipitation, but it is still significant (Figure 3a and b). We can use

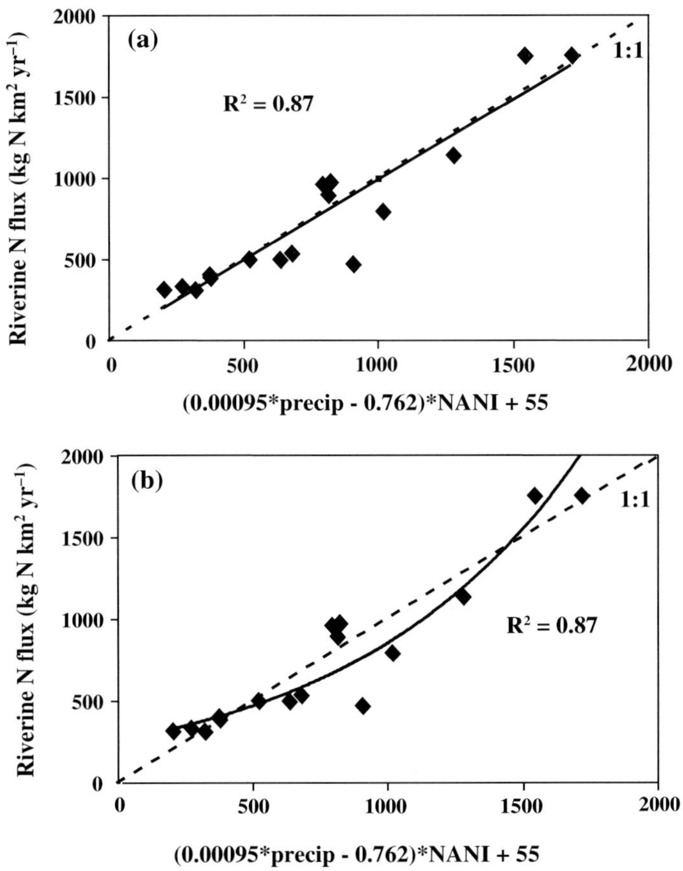

Figure 4. Equation (6) in the text ($R = (0.00095 * P - 0.762) * \text{NANI} + 55$) is an excellent predictor of riverine nitrogen fluxes. Note the similarity of the regression line for the prediction and the 1:1 line (a, top). There is some suggestion of an exponential fit (b, bottom; $Y = 255\ e^{0.0012X}$), which may indicate nitrogen saturation in the landscape at higher inputs of anthropogenic nitrogen.

discharge instead of precipitation to develop a predictive equation for riverine nitrogen flux following the steps outlined above, using a 2-variable, interacting-term regression with Q and (Q*NANI). This also gives a highly significant relationship (Table 4) and yields the equation:

$$R = (0.00087 * Q - 0.096) * \text{NANI} - 101 \quad (8)$$

The regression statistics are very similar to those for the 2-variable model using precipitation (compare Table 4 with Table 3), with one exception: the NANI term alone does not contribute significantly to this regression ($p = 0.24$; Table 4), and the regression is driven largely by the interactive (Q*NANI) term. Note also that the intercept is negative (−101 kg N km^{-2} $year^{-1}$). This intercept corresponds to the predicted riverine nitrogen flux in the situation where there was no human disturbance (NANI = 0), and a negative nitrogen flux from rivers is of course nonsensical. The 95% confidence limits for the intercept, however, extend from −319 to +116 kg N km^{-2} $year^{-1}$. Plots of the riverine nitrogen fluxes predicted from Equation (8) (using the NANI plus the interaction term Q*NANI) are shown in Figure 5a and b. These are very similar to those plots showing predictions based on (P*NANI) and NANI (Figure 4a and b).

It is also of interest to know whether simpler "proxy variables" for NANI have as much explanatory power. Population density is an example of such a variable, and has been used in many studies as an explanatory variable for nutrient discharge (Peierls et al. 1991; Smith et al. 2003). We tested a relationship of the same form Equation as 4, but substituting population density (D) for NANI, and obtained:

$$R = (-0.018 * P * D) + (25.0 * D) + 374 \quad (9)$$

With an adjusted R^2 of 0.80, it is a good relationship that is highly significant (Table 4), but not as predictive as using NANI (Table 3). Both the D term and the interaction term are significant ($p = 0.012$ and $p = 0.023$, respectively), but it is interesting to note that the signs are opposite those of the corresponding terms in Equation (5), suggesting that population density is not behaving as a simple proxy for NANI. As discussed above, equations 5 and 6 suggest that the slope of the relationship between R and NANI is itself a positive linear function of P; Equation (9) indicates that the corresponding slope of the relationship between R and D is a negative linear function of P.

Discussion

Given that many of the statistical models we explore do a good to excellent job of predicting riverine nitrogen flux, one must interpret them with care. Nonetheless, our analysis indicates a greater fractional export of NANI from the watersheds with greater precipitation (Figure 3a) and discharge (Figure 3b). Note that this greater fractional export of NANI is not due to

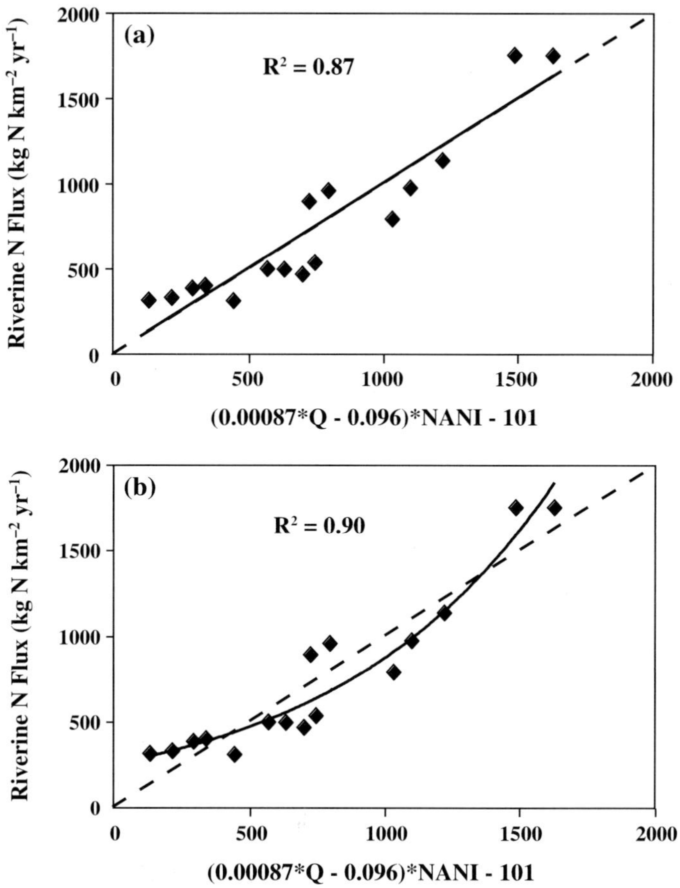

Figure 5. Equation (8) in the text ($R=(0.00087 * Q-0.0.096) * \text{NANI}-101$) is also an excellent predictor of riverine nitrogen fluxes. Note the similarity to Figure 4, including the close fit to the 1:1 line (a, top) and the suggestion of an exponential fit (b, bottom; $Y=251\ e^{0.0021X}$).

flushing during wet years of nitrogen stored in the landscape during preceding dry years, as observed by McIsaac et al. (2001). Our analysis is based on mean fluxes of nitrogen over a 6-year period in watersheds across a climatic gradient, and the discharge from these watersheds during the period of analysis (1988–1993) is typical of the longer time scale observed over the last half century (Figure 1). We therefore believe that our results reflect the long-term influence of climate on the fate of NANI. The most plausible interpretation mechanistically is that the sinks for nitrogen in the landscape are smaller in watersheds with greater precipitation and discharge.

In general, the sinks for reactive nitrogen in the environment are poorly known (Galloway et al. 2004), but they are as well estimated for these 16 northeastern U.S. watersheds as for any other region on Earth. The best estimates are that for the NANI not exported in rivers from these 16 watersheds, roughly one third accumulates in soils or biomass or is exported from the watersheds in wood, while approximately two thirds is denitrified (van Breemen et al. 2002). It may seem paradoxical that a wetter climate would lead to either less storage of nitrogen in soils and biomass or less denitrification, as one might actually predict greater accumulation of organic matter in the soils of wetter environments, and greater rates of denitrification in wetter environments where soils are perhaps more likely to be waterlogged. We suggest that the major influence of climate on the nitrogen sinks is for less denitrification in the watersheds with greater precipitation and discharge, due to faster flushing of water through riparian wetlands and low-order streams. These riparian wetlands and low-order streams are likely to be sites of significant denitrification, and the amount of nitrogen that can be removed from these systems is directly related to the water residence time (Howarth et al. 1996; van Breemen et al. 2002; Seitzinger et al. 2002).

Lewis et al. (1999) have demonstrated that the nitrogen fluxes from undisturbed catchments in the tropics are greater where discharge is higher, a result which could be explained by higher rates of biological nitrogen fixation in the wetter environments, lower sinks for nitrogen in the wetter environments, or both. Lewis (2002) found a very similar relationship for small catchments in the United States where rates of deposition were relatively low (mean of 280 kg N km^{-2} $year^{-1}$, but note that deposition was greater than 400 to 500 kg N km^{-2} $year^{-1}$ in many of the catchments). For these catchments, it is unlikely that higher rates of nitrogen fixation can explain the pattern: assuming that only 20 to 25% of the nitrogen inputs (atmospheric deposition plus natural biological nitrogen fixation) are exported in stream flow (as for average NANI and total nitrogen inputs in larger temperate-zone watersheds; Howarth et al. 1996; Boyer et al. 2002), rates of biological nitrogen fixation would have to exceed 2,000 kg N km^{-2} $year^{-1}$ in the wetter catchments to support the observed nitrogen exports in streams. Such rates have not been observed in temperate-zone terrestrial ecosystems (Cleveland et al. 1999) and are an order of magnitude higher than estimates for the forests of the 16 major northeastern watersheds (Boyer et al. 2002). We suggest that the most likely explanation for the observation of Lewis (2002) is that the nitrogen sinks are smaller in the wetter environments, and that a higher fraction of NANI (and perhaps natural nitrogen fixation) is exported, as suggested in our analysis.

The statistical models we present in this paper fall into two general types: those that have an interactive term between NANI and precipitation or discharge (equations 5, 6, and 8) and those where there is no multiplicative interaction between NANI and the climate variable (Equation 7). These have very different physical interpretations. For the models shown in equations 6 and 8, the influence of climate is on the fractional export of NANI, and a greater fraction of NANI is exported in watersheds with more precipitation (Equation 6) and higher discharge (Equation 8). This sort of model indicates that the background flux associated with natural sources of nitrogen in the landscape is small, and that the climate might have a major influence on riverine nitrogen flux by altering the fractional delivery of NANI. For the non-interactive type of model (Equation 7), there is still a strong influence of climate on riverine nitrogen flux, with higher fluxes in wetter environments. However, there is no influence of climate (precipitation) on the amount of NANI that is exported, which remains constant at 29% (see Equation 7), but rather only on the background or "natural" flux of nitrogen. This flux must originate with the natural rate of biological nitrogen fixation. According to this model, then, the higher riverine nitrogen fluxes in the watersheds with more precipitation are due to higher rates of biological nitrogen fixation in forests. These rates of fixation are not well known, but in our earlier estimates, the rates of fixation are in fact lower in the wetter watersheds (Boyer et al. 2002). The physical interpretation of Equation 7 suggests that the rates of nitrogen fixation in the wetter watersheds must be of the magnitude of 2,800 kg N km^{-2} $year^{-1}$ higher than in the driest of the 16 northeastern watersheds. As noted above, these rates would be at least an order of magnitude higher than likely for temperate-zone forests (Cleveland et al. 1999; Boyer et al. 2002). Thus, while both classes of models provide very good to excellent statistical fits to the data, only the interacting-term models (such as equations 6 and 8) lead to realistic interpretations. We conclude that the effect of climate on riverine nitrogen export is very likely to be on the fractional delivery of NANI (as illustrated in Figure 3a and b) rather than the background natural flux of nitrogen.

We can use these interacting-term models (equations 6 and 8) to begin to estimate how future climate change might affect riverine nitrogen fluxes. Najjar et al. (2002) provide estimates for future changes in precipitation and discharge for the mid-Atlantic coastal region, which includes roughly half of the 16 watersheds in our study. Their estimates, based on both the Hadley Centre and Canadian Climate Centre models for global climate change, suggest a mean likely increase in precipitation of 4% and of discharge of 2% by 2030, with increases in precipitation and discharge by 2095 of 15 and 11% respectively. Such estimates are quite uncertain (Table 5), but they provide a context for

Table 5. Predicted consequences of climate change on riverine nitrogen flux in the Susquehanna River to Chesapeake Bay. Estimates rely on the range and mean projections of change in precipitation and discharge by 2030 and by 2095 from Najjar et al. (2000) and on our equations 6 and 8, which relate riverine nitrogen flux to NANI and either precipitation or discharge. NANI is assumed not to change. Mean projected values are shown, with the range given in parentheses.

	2030	2095
Change in precipitation	+4% (−1% to +8%)	+15% (+6% to +24%)
Change in discharge	+2% (−2% to +6%)	+11% (−4% to +27%)
Change in riverine nitrogen flux, based on precipitation	+17% (−4% to +35%)	+65% (+26% to +200%)
Change in riverine nitrogen flux, based on discharge	+3% (−3% to +8%)	+16% (−6% to +38%)

examining the consequences on nitrogen fluxes. In Table 5, we illustrate the potential magnitude of changes in riverine nitrogen fluxes for the Susquehanna River basin due to future climate change, using both precipitation and discharge as predictors, and using both the mean projections and the range of projections given by Najjar et al. (2002). We chose the Susquehanna River for this analysis both because it is the largest of the watersheds in the northeastern United States and because it is the major input of nitrogen to main stem of Chesapeake Bay (Hagy et al. 2004), one of the most nutrient-sensitive estuaries in the country (NRC 2000). Further, the range of predicted future values for discharge and precipitation in the Susquehanna River basin is within the range of values currently observed across the climate gradient for the 16 northeastern U.S. rivers, so we need not extrapolate our models beyond the observational data upon which they are based. This is critical, particularly for precipitation, where the relationship between precipitation and fractional delivery of nitrogen inputs is quite steep (Figure 3a), and undoubtedly is not linear when the precipitation is less than 800 or greater than 1,300 mm year^{-1}.

For these predictions of the consequences of climate change on nitrogen fluxes in the Susquehanna, we assume that NANI remains constant into the future. Note that the estimate based on NANI and discharge (Equation 8) over-predicts the flux from the 1988 to 1993 period by 13%, while the estimate based on NANI and precipitation (Equation 6) underestimates this flux by 16%. Note also that all of the nitrogen projections given in Table 5 respond in a non-linear way to climate forcing, with the nitrogen increases or decreases larger than the respective changes in either precipitation or discharge. This nonlinearity is particularly pronounced for the estimates based on precipitation. While our projections obviously carry a great deal of uncertainty, they suggest that compared to the 1988–1993 period, nitrogen fluxes down the Susquehanna in 2030 may be 3 to 17% greater and in 2095 may be 16 to 65% greater in response to climate change (Table 5, using mean estimates based on the discharge and precipitation models). Such changes would obviously make it much more difficult to achieve nitrogen reduction for Chesapeake Bay.

While riverine nitrogen fluxes for the northeastern watersheds are very well explained on the basis of nitrogen inputs to the landscape and climate (Figures 5 and 6), it must be noted that a variety of management options are available for greatly reducing nitrogen fluxes in rivers without necessarily decreasing the net anthropogenic nitrogen inputs (NANI; see Howarth in press and Howarth et al. 2006 for recent reviews of some of these options). For example, planting winter cover crops on agricultural fields or switching from annual to perennial crops can greatly reduce nitrogen losses from the fields even when there is no reduction in fertilizer application (Randall et al. 1997; Staver and Brinsfield 1998; Randall and Mulla 2001). That NANI so well explains the riverine nitrogen fluxes, therefore, suggests that farming and other nitrogen management practices (such as wastewater disposal) during the 1988–1993 period of our study were relatively uniform across the watersheds. Great opportunity exists to improve these management practices, and thereby help reduce nitrogen pollution in coastal waters. Nonetheless, our results indicate that climate plays a significant role in determining the magnitude of the flux in rivers of nitrogen from human-dominated landscapes.

Given the uncertainties in our analysis, our conclusions must be tempered, and our projections must be qualified. Nonetheless, the analysis of the 16 northeastern watersheds illustrates that climate probably has a pronounced, sustained influence on the flux of nitrogen in large rivers. The percentage of nitrogen inputs to the landscape that is exported to coastal ecosystems by rivers is greater in the watersheds with wetter climates, probably because the nitrogen sinks in the landscape (primarily denitrification) are less. We believe that the relationships developed here should be tested in a wider set of large watersheds across as broad a climate gradient as possible. Should the relationships prove robust, then the influence of future climate change must be an important consideration in any management plans to control nitrogen inputs to coastal marine ecosystems.

Acknowledgements

We thank Steven Ellner and Åsa Danielsson for helpful discussions of the statistical aspects of our analyses and Alan Townsend, Greg McIsaac, Sybil Seitzinger, and Ken Staver for comments on earlier drafts of the manuscript. This research was funded by grant R830882 from the EPA STAR program, by a USDA Hatch grant, and by an endowment provided to Cornell University by David R. Atkinson.

References

Aber J., McDowell W., Nadelhoffer K., Magill A., Berntson G., Kamakea M., McNulty S., Currie W., Rustad L. and Fernandez I. 1998. Nitrogen saturation is temperate forest ecosystems. BioScience 48: 921–934.

Aber J., Goodale C., Ollinger S.V., Smith M., Magill A.H., Martin M.E., Hallett R. and Stoddard J.L. 2003. Is nitrogen deposition altering the nitrogen status of northeastern forests? BioScience 53: 375–389.

Alexander R.B., Johnes P.J., Boyer E.W. and Smith R.A. 2002. A comparison of models for estimating the riverine export of nitrogen from large watersheds. Biogeochemistry 57/58: 295–339.

Battaglin, W.A. and Goolsby, D.A. 1994. Spatial Data in Geographic Information System Format on Agricultural Chemical Use, Land Use, and Cropping Practices in the United States. USGS Water-Resources Investigations Report 94-4176 Report available online at http://water.usgs.gov/pubs/wri944176/. Data available on-line at http://water.usgs.gov/GIS/metadata/usgswrd/nit91.html.

Boyer E.W., Goodale C.L., Jaworski N.A. and Howarth R.W. 2002. Effects of anthropogenic nitrogen loading on riverine nitrogen export in the northeastern US. Biogeochemistry 57&58: 137–169.

Bricker S.B., Clement C.G., Pirhalla D.E., Orland S.P. and Farrow D.G.G. 1999. National Estuarine Eutrophication Assessment: A Summary of Conditions, Historical Trends, and Future

Outlook. National Ocean Service, National Oceanic and Atmospheric Administration, Silver Springs, MD.

Caraco N.F. and Cole J.J. 1999. Human impact on nitrate export: An analysis using major world rivers. Ambio 28: 167–170.

Cleveland C.C., Townsend A.R., Schimel D.S., Fisher H., Howarth R.W., Hedin L.O., Perakis S.S., Latty E.F., von Fischer J.C., Elseroad A. and Wasson M.F. 1999. Global patterns of terrestrial biological nitrogen (N_2) fixation in natural systems. Global Biogeochem. Cycles 13: 623–645.

Cohn T.A. et al. 1992. The validity of a simple statistical model for estimating fluvial constituent loads: an empirical study involving loads entering the Chesapeake Bay. Water Resour. Res 28: 2353–2363.

Environmental Protection Agency 2001. National Coastal Condition Report. EPA-620/R-01/005, Office of Research and Development and Office of Water, U.S. Environmental Protection Agency, Washington, DC.

Galloway J.N., Dentener F.J., Capone D.G., Boyer E.W., Howarth R.W., Seitzinger S.P., Asner G.P., Cleveland C., Green P.A., Holland E., Karl D.M., Michaels A., Porter J.H., Townsend A. and Vorosmarty C. 2004. Nitrogen cycles: past, present, and future. Biogeochemistry 70: 153–226.

Hagy J.D., Boynton W.R., Keefe C.W. and Wood K.V. 2004. Hypoxia in Chesapeake Bay, 1950–2001: Long-term change in relation to nutrient loading and river flow. Estuaries 27: 634–658.

Holland E., Dentener F., Braswell B. and Sulzman J. 1999. Contemporary and pre-industrial global reactive nitrogen budgets. Biogeochemistry 4: 7–43.

Howarth R.W. 1988. Nutrient limitation of net primary production in marine ecosystems. Annual Rev. Ecol. Systemat. 19: 89–110.

Howarth, R.W. In press. The development of policy approaches for reducing nitrogen pollution to coastal waters of the USA. China Science.

Howarth R.W., Jensen H., Marino R. and Postma H. 1995. Transport to and processing of phosphorus in near-shore and oceanic waters. In: Tiessen H. (ed.), Phosphorus in the Global Environment, SCOPE #54, Wiley & Sons, Chichester, pp. 323–345.

Howarth R.W., Billen G., Swaney D., Townsend A., Jaworski N., Lajtha K., Downing J.A., Elmgren R., Caraco N., Jordan T., Berendse F., Freney J., Kudeyarov V., Murdoch P. and Zhu Zhao-liang. 1996. Riverine inputs of nitrogen to the North Atlantic Ocean: fluxes and human influences. Biogeochemistry 35: 75–139.

Howarth R.W., Anderson D., Cloern J., Elfring C., Hopkinson C., Lapointe B., Malone T., Marcus N., McGlathery K., Sharpley A. and Walker D. 2000. Nutrient pollution of coastal rivers, bays, and seas. Issues Ecol. 7: 1–15.

Howarth R.W., Boyer E.W., Pabich W.J. and Galloway J.N. 2002a. Nitrogen use in the United States from 1961–200 and potential future trends. Ambio 31: 88–96.

Howarth R., Walker D. and Sharpley A. 2002b. Sources of nitrogen pollution to coastal waters of the United States. Estuaries 25: 656–676.

Howarth R.W. and Marino R. 2006. Nitrogen as the limiting nutrient for eutrophication in coastal marine ecosystems: Evolving views over 3 decades. Limnol. Oceanogr. 51: 288–295.

Howarth R.W., Ramakrishna K., Choi E., Elmgren R., Martinelli L., Mendoza A., Moomaw W., Palm C., Boy R., Scholes M., and Zhu Zhao-Liang. 2006. Chapter 9: Nutrient Management, Responses Assessment. Ecosystems and Human Well Being. Vol. 3, Policy Responses. The Millenium Ecosystem Assessment. Island Press, Washington, DC, pp. 295–311.

Kittel T.G.F., Royle J.A., Daly C., Rosenbloom N.A., Gibson W.P., Fisher H.H., Schimel D.S., Berliner L.M., and VEMAP2 Participants. 1997. A gridded historical (1895–1993) bioclimate dataset for the conterminous United States. In: Reno N.V. (ed.), Proceedings of the 10th Conference on Applied Climatology.

Lewis W.M. 2002. Yield of nitrogen from minimally disturbed watersheds of the United States. Biogeochemistry 57/58: 375–385.

Lewis W.M., Melack J.M., McDowell W.H., McClain M., and Richey J.E. 1999. Nitrogen yields from undisturbed watersheds in the Americas. Biogeochemistry 46: 149–162.

Lovett G. and Lindberg S.E. 1993. Atmospheric deposition and canopy interactions of nitrogen in forests. Can. J. For. Res 23: 1603–1616.

Lovett G.M. and Rueth H. 1999. Potential N mineralization and nitrification in American beech and sugar maple stands along a N deposition gradient in the northeastern US. Ecol. Appl. 9: 1330–1344.

Lovett G.M., Traynor M.M., Pouyal R.V., Carreiro M.M., Zhu W.X. and Baxter J.W. 2000. Atmospheric deposition to oak forests along an urban-rural gradient. Env. Sci. Tech 34: 4294–4300.

McIsaac G.F., David M.B., Gertner G.Z. and Goolsby D.A. 2001. Net anthropogenic N input to the Mississippi River basin and nitrate flux to the Gulf of Mexico. Nature 414: 166–167.

Meyers T., Sickles J., Dennis R., Russell K., Galloway J. and Church T. 2001. Atmospheric nitrogen deposition to coastal estuaries and their watersheds. In: Valigura R.A., Alexander R.B., Castro M.S., Meyers T.P., Paerl H.W., Stacey P.E. and Turner R.E. (eds), Nitrogen Loading in Coastal Water Bodies: An Atmospheric Perspective, American Geophysical Union, Washington, DC, pp. 53–76.

Moore M.H., Pace M., Mather J., Murdoch P.S., Howarth R.W., Folt C.L., Chen C.Y., Hemond H.F., Flebbe P.A. and Driscoll C.T. 1997. Potential effects of climate change on the freshwater ecosystems of the New England/mid-Atlantic region. Water Resources 11: 925–947.

NADP. 2005. National Atmospheric Deposition Program/National Trends Network. NADP Program Office, Illinois State Water Survey, 2204 Griffith Dr., Champaign, IL 61820. [online] URL: http://nadp.sws.uiuc.edu/nadpdata.

Najjar R.G. 1999. The water balance of the Susquehanna River basin and its response to climate change. J. Hydrol. 219: 7–19.

Najjar R.G., Walker H.A., Anderson P.J., Barron E.J., Bord R.J., Gibson J.R., Kennedy V.S., Knight C.G., Megonigal J.P., O'Connor R.E., Polsky C.D., Psuty N.P., Richards B.A., Soreson L.G., Steele E.M. and Swanson R.S. 2000. The potential impacts of climate change on the mid-Atlantic coastal region. Climate Res. 14: 219–233.

NRC 2000. Clean Coastal Waters: Understanding and Reducing the Effects of Nutrient Pollution. National Academies Press, Washington, DC.

Nixon S.W. 1995. Coastal marine eutrophication: a definition, social causes, and future concerns. Ophelia 41: 199–219.

Ollinger S.V., Aber J.D., Lovett G.M., Millham S.E., Lathrop R.G. and Ellis J.M. 1993. A spatial model of atmospheric deposition for the northeastern U.S.. Ecol. Appl. 3: 459–472.

Prospero J.M., Barrett K., Church T., Dentener F., Duce R.A., Galloway J.N., Levy H., Moody J. and Quinn P. 1996. Atmospheric deposition of nutrients to the North Atlantic Basin. Biogeochemistry 35: 27–73.

Peierls B., Caraco N., Pace M. and Cole J.J. 1991. Human influence on river nitrogen. Nature 350: 386–387.

Randall G.W. and Mulla D.J. 2001. Nitrate nitrogen in surface waters as influence by climatic conditions and agricultural practices. J. Environ. Qual. 30: 337–344.

Randall G.W., Huggins D.R., Russelle M.P., Fuchs D.J., Nelson W.W. and Anderson J.L. 1997. Nitrate losses through subsurface tile drainage in CRP, alfalfa, and row crop systems. J. Environ. Qual 26: 1240–1247.

Scavia D., Field J.C., Boesch , Buddemeier R., Burkett V., Canyan D., Fogarty M., Harwell M.A., Howarth R.W., Mason C., Reed D.J., Royer T.C., Sallenger A.H. and Titus J.G. 2002. Climate change impacts on US coastal and marine ecosystems. Estuaries 25: 149–164.

Seitzinger S.P. and Kroeze C. 1998. Global distribution of nitrous oxide production and N inputs in freshwater and coastal marine ecosystems. Global Biogeochem. Cycles 12: 93–113.

Seitzinger S.P., Styles R.V., Boyer E.W., Alexander R., Billen G., Howarth R., Mayer B. and van Breemen N. 2002. Nitrogen retention in rivers: model development and application to watersheds in the northeastern US. Biogeochemistry 57&58: 199–237.

Smith S.V., Swaney D., Talaue-McManus L., Bartley J.D., Sandhei P.T., McLaughlin C.J., Dupra V.C., Crossland C.J., Buddemeier R.W., Maxwell B.A. and Wulff F. 2003. Humans, hydrology, and thedistribution of inorganic nutrient loading to the ocean. BioScience 53: 235–245.

Staver K.W. and Brinsfield R.B. 1998. Use of cereal grain winter cover crops to reduce groundwater nitrate contamination in the Mid-Atlantic coastal plain. J. Soil Water Conserv. 53: 230–240.

USGS. 2005. National Water Information System Data Retrieval [online] URL: http://waterdata.usgs.gov/nwis-w/US/.

Van Breemen N., Boyer E.W., Goodale C.L., Jaworski N.A., Paustian K., Seitzinger S., Lajtha K., Mayer B., van Dam D., Howarth R.W., Nadelhoffer K.J., Eve M. and Billen G. 2002. Where did all the nitrogen go? Fate of nitrogen inputs to large watersheds in the northeastern USA. Biogeochemistry 57&58: 267–293.

Van Horn H.H. 1998. Factors affecting manure quantity, quality, and use. Proceedings of the Mid-South Ruminant Nutrition Conference, Dallas-Ft. Worth, May 7–8, 1998. Texas Animal Nutrition Council, pp 9–20.

Wolock D.M. and McCabe G.M. 1999. Simulated effects of climate change on mean annual runoff in the conterminous Unites States. J. Am. Wat. Res. Assoc. 35: 1341–1350.

Biogeochemistry (2006) 79: 187–208
DOI 10.1007/s10533-006-9011-0

Coastal eutrophication assessment in the United States

DONALD SCAVIA[1,]* and SUZANNE B. BRICKER[2]
[1]*University of Michigan, Ann Arbor, MI, USA;* [2]*National Oceanic and Atmospheric Administration, Silver Spring, MD, USA; *Author for correspondence (e-mail: scavia@umich.edu; phone: +1-734-615-4860; fax: +1-734-763-8965)*

Key words: Assessment, Coastal, Eutrophication, Gulf of Mexico, Hypoxia

Abstract. Recent national assessments document that nitrogen-driven coastal eutrophication is widespread and increasing in the United States. This significant coastal pollution problem includes impacts including increased areas and severity of hypoxic and anoxic waters; alteration of food webs; degradation and loss of sea grass beds, kelp beds and coral reefs; loss of biodiversity; and increased incidences and duration of harmful algal blooms. In this paper, we review two complementary approaches to assessing the causes and consequences of these trends, as well as potential remedies for them. The first is a national-scale assessment, drawn primarily from expert knowledge of those most familiar with the individual estuaries and integrated into a common analysis framework. The second approach, focused on the Mississippi/Atchafalaya basin – the largest US drainage basin – draws upon both quantitative and qualitative analyses within a comprehensive framework, Integrated Assessment.

Recent national assessments (e.g., Bricker et al. 1999; NRC 2000; CENR 2003) document that coastal eutrophication is widespread and increasing in the United States and that, among the diverse issues confronting US coastal systems, the impact of excess nutrients is the most important pollution problem (Howarth et al. 2000; NRC 2000; Ocean Commission 2004). These impacts include increased areas and severity of hypoxic and anoxic waters; alteration of food webs; degradation and loss of sea grass beds, kelp beds and coral reefs; loss of biodiversity; and increased incidences and duration of harmful algal blooms. While both nitrogen and phosphorus contribute to nutrient pollution, nitrogen is the more significant driver of eutrophication in most US coastal areas and humans have increased the average flux of nitrogen to the coastal waters of the United States by four to fivefold; in some regions the increase has been as large as 10-fold (NRC 2000; Howarth et al. 2002). Population growth, expanded land development, and intensified agriculture are likely to increase nitrogen loads substantially in the future.

Efforts to address these issues most often require actions at local levels; however, there is value in providing both national and watershed perspectives on the issue. In this paper, we review two complementary approaches to assessment. The first is a national assessment, drawn primarily from expert knowledge of those most familiar with the individual estuaries and integrated into a common analysis framework. The second approach – an Integrated Assessment of the Causes and Consequences of Hypoxia in the Northern Gulf of Mexico – addresses the largest US drainage basin and draws upon both quantitative and qualitative analyses within a comprehensive framework that considers causes and consequences of nutrient inputs, and examines alternatives to reduce, mitigate and control nutrient related impacts. The sections below outline both approaches and illustrate how they produce information at complementary scales and for different audiences.

National estuarine eutrophication assessment

In the early 1990s in response to the knowledge that some estuaries were showing signs of nutrient related degradation as evidenced by hypoxia in Long Island Sound, Chesapeake and Mobile bays (Welsh 1991) and the concern that this might be a wide spread problem, NOAA conducted a nationwide assessment to discern the magnitude, severity, and location of eutrophic conditions. The intent was to learn whether these problems were local, regional, or national in scale, to determine probable causes, and to provide this information to managers such that observed problems could be addressed at the appropriate level.

The National Estuarine Eutrophication Assessment (NEEA) was carried out in three phases. First, questionnaires were used nationally to collect data and information for 16 nutrient related water quality variables (NOAA 1996, 1997a–c, 1998). Second, results for six of the 16 variables were selected to provide an assessment of overall conditions within the water bodies (Bricker et al. 1999). Finally, additional databases were used to evaluate the probable causes of observed conditions and to make projections about future outlook.

The original method is described here in brief, as are recent improvements and modifications made by the Assessment of Estuarine Trophic Status group (ASSETS, Bricker et al. 2003). A full description of the original method can be found in Bricker et al. (1999) and details for modifications can be found in Bricker et al. (2003). Additionally, a brief description is provided of the continuation of this work through the NEEA Update Program (Bricker et al. 2004). This program is intended to provide results to inform managers, researchers, and politicians about the success of legislation and management measures designed to address eutrophication issues. The update, anticipated for release in late 2006, (http://ian.umces.edu/neea), will provide case studies that examine the success of management since the early 1990s in order to inform recommendations for analysis and application of appropriate management measures in systems nationwide. It is meant to be a companion program to a National Research Program for Nutrient Pollution in Coastal Waters (Howarth et al. 2003) and interactive with European Commission efforts such as the Water Framework Directive 2000/60/EC (WFD e.g. OSPAR 2002, 2003; Coast 2003).

Table 1. Indicator parameters and rationale, thresholds and justification for primary and secondary symptoms of estuarine eutrophication (from Bricker et al. 2003).

Indicator and rationale	Thresholds and ranges	Threshold justification
Algal Blooms: Chl *a* is used as an indicator of phytoplankton primary productivity. Highest concentrations in an estuary during the annual bloom period were recorded. High levels cause dieoff of SAV and low bottom water dissolved oxygen.	Hypereutrophic: $> 60 \mu g$ Chl*a* l^{-1} High: > 20 but $\leq 60 \mu g$ Chl*a* l^{-1} Medium: > 5 but $\leq 20 \mu g$ Chl*a* l^{-1} Low: > 0 but $\leq 5 \mu g$ Chl*a* l^{-1}	• Estuaries with highest annual Chl *a* < than $5 \mu g\, l^{-1}$ appear unimpacted (Nixon 1983), however, this level is detrimental to survival of corals (Lapointe and Matzie 1996). • At $20 \mu g\, l^{-1}$ SAV shows declines (Stevenson et al. 1993) and community shifts from diverse mixture to monoculture (Twilley et al. 1985). • At $60 \mu g\, l^{-1}$ high turbidity and low bottom water dissolved oxygen are observed (Jaworski 1981).
Macroalgae and Epiphytes: Excessive macroalgal and epiphyte growth is known to suffocate bivalves and cause dieoff of SAV.	Problem: detrimental impact to biological resources (e.g. dieoffs of SAV) No Problem: no apparent impacts on biological resources	There is no standard measure or threshold above which macroalgae and/or epiphytes are considered to be a problem to the biological resources, and it is rare to find quantitative information. However some studies show that: • Macroalgae (*Ulva* or *Enteromorpha)* above 100 g dry wt m^{-2}bottom causes SAV dieoff (Dennison et al. 1992). • Epiphyte colonizing SAV at a dry weight equal to the dry wt cm^{-2} of the host plant will cause dieoff of the host plant (Dennison et al. 1992). • In the absence of a standard concentration determinations were heuristic.

Table 1. Continued.

Indicator and rationale	Thresholds and ranges	Threshold justification
Nuisance and Toxic Blooms: Problem conditions for toxic blooms result from the production of toxin by the organism. For nuisance blooms, excessive abundance of small organisms that clog filter feeders siphons.	Problem = detrimental impact to biological resources (e.g. dieoffs of filter feeding bivalves and fish, respiratory irritation) No Problem – no apparent impacts on biological resources	Nutrient input increases cause changes in nutrient ratios that promote growth of nuisance and toxic algae (Rabalais et al. 1996). • Threshold determination is difficult because toxicity of chemicals produced by the different species vary, e.g., some dinoflagellates become toxic at cell counts in excess of 10^6 cells l^{-1}, others are a problem at 10^5 cells l^{-1}; *Pfiesteria piscicida* is toxic at levels below 10^2 cells l^{-1} (Burkholder et al. 1992a). • In the absence of a standard concentration determinations were heuristic.
Dissolved Oxygen Concentrations: Bottom water dissolved oxygen concentration has become a standard measurement to assess the general condition of a water body due to its importance to the survival of benthic organisms.	Anoxia = 0 mg l^{-1} Hypoxia = >0 but ≤ 2 mg l^{-1} Biologically Stressful = >2 but ≤ 5 mg l^{-1}	• Bottom water concentrations of 2 mg l^{-1} or less, have significantly reduced benthic macroinfauna and infauna, and success of trawling for demersal species (Rabalais and Harper 1992). • The range of 2–5 mg l^{-1} is included in this survey since field and laboratory observations have also shown oxygen stress responses in invertebrate and fish fauna at these concentrations (Rabalais and Harper, 1992.)

Submerged Aquatic Vegetation (SAV): The measure of SAV is spatial coverage since this is the most common data available, though diversity and density of plants is available for some estuaries.	High = $\geq 50\%$ and $\leq 100\%$ estuarine surface water area Medium = $\geq 25\%$ but $< 50\%$ of estuarine surface water area Low = $\geq 1\%$ but $< 25\%$ estuarine surface water area.	Submerged vascular plants such as *zostera marina* and *Potamogeon perfoliatus*, are thought to play a vital role in the ecology of nearshore environments to depths of 1–2 m. These plants attenuate variable inputs of nutrients and sediment, and are thought to be invaluable nursery areas. In relatively pristine waterbodies, SAV thrive while die-offs and absence of SAV is generally believed to be an indication of a eutrophic condition, associated with high turbidity caused by increased nutrient and Chl *a* concentrations (Orth and Moore 1984; Stevenson et al. 1993; Boynton et al. 1996). Additionally, high nutrient concentrations may cause an imbalance in nutrient supply ratios leading to dieoff of SAV (Burkholder et al. 1992b).

Indicator selection and characterization

Sixteen nutrient-related water quality indicators were selected to characterize eutrophic conditions across very different systems (Hinga et al. 1991). This suite of variables was broad enough to assess all estuarine types and to provide clear distinction of conditions among estuaries. The data and information collected included magnitude, timing, and frequency of occurrence of extreme conditions during the annual cycle. For example, Chlorophyll *a* data considers the concentrations during the largest bloom (e.g. winter-spring bloom), the spatial area over which the highest concentrations occur, and the frequency with which blooms recur (e.g., annual, persistent, episodic).

Response ranges were developed from U.S. estuarine data and were selected to be simple to use and to distinguish the magnitude of eutrophic symptoms among estuaries. For example, Chlorophyll a and dissolved oxygen concentrations were assigned to High, Medium and Low categories based on observations as well as discussions with local investigators. Chlorophyll a bloom concentrations were assigned as Low = 0–5 $\mu g\ l^{-1}$, Medium = > 5–20 $\mu g\ l^{-1}$, High = > 20–60 μg^{-1} and Hypereutrophic = > 60 $\mu g\ l^{-1}$. Dissolved oxygen conditions were assigned as Anoxic = 0 mg l^{-1}, Hypoxic = > 0–2 mg l^{-1}, Biologically Stressful = > 2–5 mg l^{-1} (Table 1; for full detail of criteria see Bricker et al. 1999 or Bricker et al. 2003). Although some response criteria may not distinguish among estuaries within a region, they are intended to distinguish among estuaries on a broad geographic basis. However, these ranges did not work universally for the 138 systems and the NEEA Update Program is working to re-evaluate the ranges and develop criteria that will more accurately characterize conditions by type of estuary (where type classification is determined primarily by similar physical and hydrologic characteristics).

For epiphyte and macroalgal abundances, and nuisance and toxic algal blooms, a different approach was used because there is no standard measure for these variables. For these indicator variables, NEEA participants were asked simply if they were or were not a problem in their system (e.g., nuisance and toxic blooms causing fish kills, macroalgae causing losses of SAV or smothering of bivalves).

Data for conditions and trends in 138 U.S. estuaries and the Mississippi/Atchafalaya River Plume were collected in a series of questionnaires, site visits, and regional workshops (NOAA 1996, 1997a–c, 1998). Data were collected by salinity zone (Tidal Fresh = 0–0.5 ppt, Mixing Zone = 0.5–25 ppt, Seawater Zone $\geq$ 25 ppt) for each system, providing a basis for comparison among the highly varied systems. About 400 participants from academia, and state, federal, and local agencies provided information and data. A reliability assessment (self-assigned) ranging from 'highly confident' to 'speculative' was offered for each response since the information varies from statistically tested scientific data to general observations. The final assessment also has an associated reliability assessment based on the reliability of data and information used in the analysis.

The NEEA/ASSETS assessment methodology

The NEEA model (Bricker et al. 1999), and recent modifications described in the Assessment of Estuarine Trophic Status (ASSETS; Bricker et al. 2003), uses a Pressure-State-Response framework to assess eutrophication in three component parts:

- Overall human influence (OHI) on development of conditions (Pressure),
- Overall eutrophic conditions (OEC) within a water body (State), and
- Determination of future outlook (DFO) for conditions within the system (Response).

Aspects of each component use a decision logic approach to combine data and information into single multi-dimensional descriptors and matrices are used to combine two components into a single descriptor for each of the three components.

Pressure – overall human influence (OHI)

The 'Pressure' component of the assessment is designed to determine the influence of human related inputs relative to the natural tendency of a system to either retain or flush nutrients (i.e. susceptibility). This component is determined by combining in a matrix an estimation of susceptibility of a system, and the level of nutrient inputs from the watershed. Participants in the NEEA used watershed nutrient model estimates (SPARROW; Smith et al. 1997), watershed population density and other demographic data in the Coastal Assessment and Data Synthesis (CADS 1999) to estimate inputs. These were divided into High, Medium, and Low categories. Hydrologic and physical data from CADS (1999) was used to determine susceptibility. In a logic decision approach, the dilution potential takes into account vertical stratification and dilution volume which are dependent upon the stratification status. Similarly, the flushing potential is estimated from tide range and the ratio of freshwater inflow to the volume of the estuary. The final susceptibility estimate of High, Medium, or Low is determined in a matrix combining the dilution and flushing values (Bricker et al. 1999).

In ASSETS, improvements were made to the original methodology by applying a simple model to better estimate the level of human related nutrient inputs to the system. The model compares anthropogenic nutrient loading and natural background concentrations and also factors in potential nutrient inputs from oceanic sources thus addressing the question of whether management measures would be successful. The results of the model calculation, essentially a ratio of land or human related inputs to oceanic inputs, are assigned one of five categories: High, Moderately High, Moderate, Moderately Low, and Low and are used in the matrix with the susceptibility measure in place of the nutrient load estimates that were used in the NEEA. For a full description of model development and use of the matrix to estimate the level of human influence see Bricker et al. (2003).

State – overall eutrophic condition (OEC)

To determine overall eutrophic condition, six variables were selected from the original 16 that were characterized in the NEEA (Table 1; Bricker et al. 1999). These were divided into two groups: primary or early stage symptoms (chlorophyll a, epiphytes, macroalgae), and secondary or well developed eutrophication symptoms (dissolved oxygen, Submerged Aquatic Vegetation (SAV) loss, harmful algal bloom occurrence). In the original NEEA, a logic decision approach was used to determine the level for each variable within each estuarine salinity zone. The level for an indicator is a combined value of the extreme concentration or condition of the variable (e.g. bloom concentration of Chl *a*, or lowest concentration of dissolved oxygen), the spatial area over which the extreme conditions occur, and the frequency with which it is observed (e.g. annually, periodically, episodically; Table 2). The separate salinity zone results are then combined to give a weighted average value for the

Table 2. Logic decision approach for Chlorophyll *a* level of expression (Bricker et al. 1999).

IF	AND	AND	THEN	
Concentration	*Spatial coverage*	*Frequency*	*Expression*	*Value*
Hypereutrophic or High	High	Periodic	High	1
	Moderate	Periodic	High	1
	Low	Periodic	Moderate	0.5
	Very Low	Periodic	Moderate	0.5
	High	Episodic	High	1
	Moderate	Episodic	Moderate	0.5
	Low/Very Low	Episodic	Low	0.25
	Any spatial coverage	Unknown	Flag A	0.5
	Unknown	Any frequency	Flag A	0.5
Medium	High	Periodic	High	1
	Moderate	Periodic	Moderate	0.5
	Low/Very Low	Periodic	Low	0.25
	High	Episodic	Moderate	0.5
	Mod/Low/Very Low	Episodic	Low	0.25
	Any spatial coverage	Unknown	Flag A	0.5
	Unknown	Any frequency	Flag A	0.5
Low	Any spatial coverage	Any frequency	Low	0.25
Unknown	Unknown	Unknown	Not included in calculation at zone level	

Chlorophyll *a* Level of Expression Determination.
Spatial coverage and frequency of occurrence are used to determine the level of expression for each salinity zone and are then aggregated up to the estuary level.

estuary which is given a numerical value that is then converted to a categorical rating (i.e. High, Moderate, Low).

The overall primary symptom level is determined by averaging the values for Chl*a*, epiphytes, and macroalgae; whereas the highest of the three secondary symptoms (dissolved oxygen, loss of SAV, nuisance and/or toxic bloom occurrences) is selected based on the assumption that these symptoms indicate a well developed problem. These values are combined to determine an overall rating of eutrophic conditions for the estuary (Figure 1). Assessment results show that nutrient related water quality problems occur on a national basis (Figure 2).

Recent modifications have been made that allow the use of data rather than 'expert knowledge' for Chl *a* and dissolved oxygen. These variables are measured in a standard manner and statistical criteria were developed to quantify them in a more robust manner while staying true to the intent of

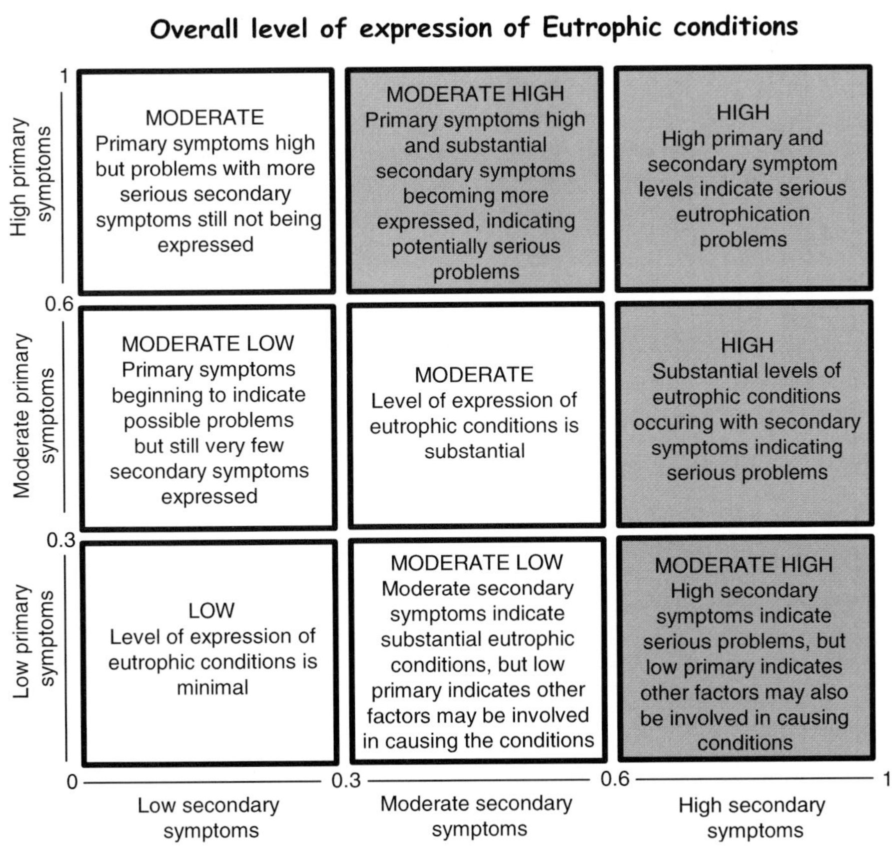

Figure 1. Matrix for determination of overall eutrophic condition from primary and secondary symptom levels (from Bricker et al. 2003).

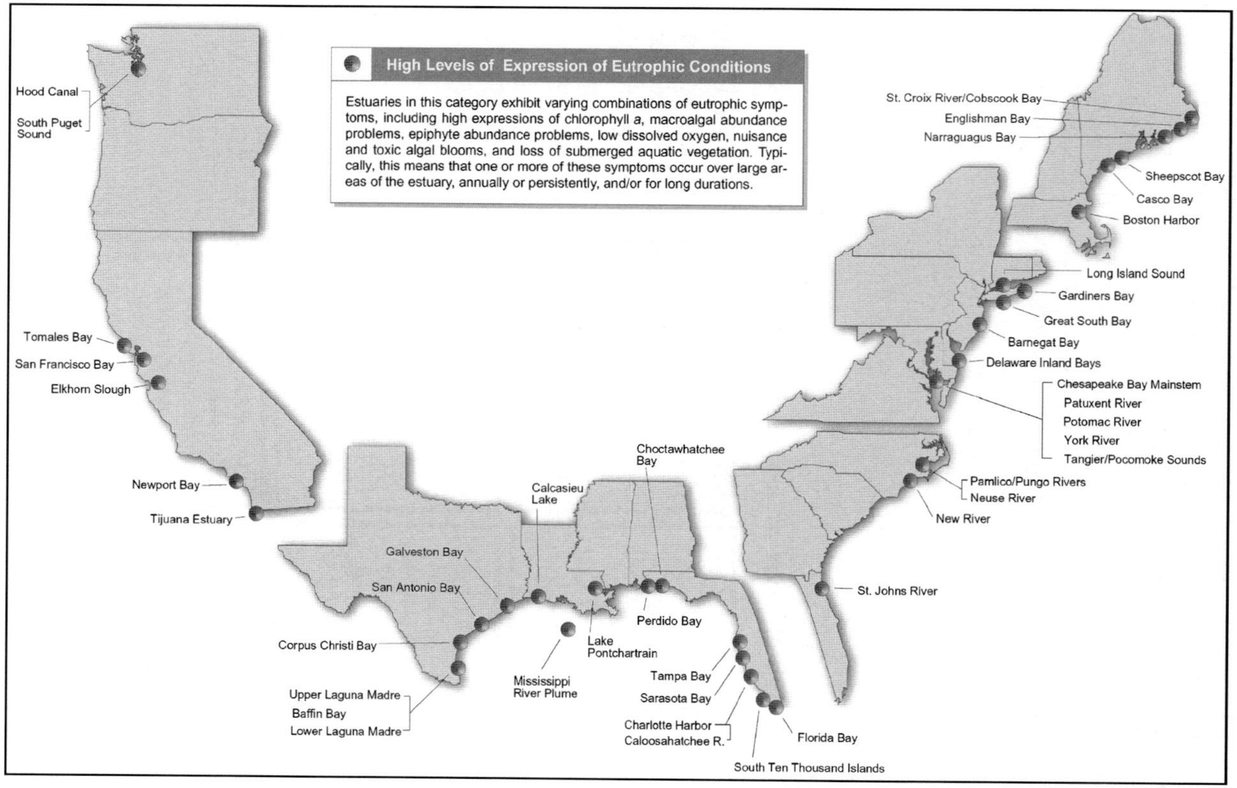

Figure 2. High level eutrophic conditions in estuaries of the United States (from Bricker et al. 1999).

reporting extreme concentrations. The approach used is to determine 90th percentile concentration during the annual cycle for Chlorophyll *a* and the 10th percentile for dissolved oxygen (Bricker et al. 2003). These are then converted to the categorical ratings using the thresholds of the NEEA (i.e. for Chl *a*; Low = 0–5 μg l^{-1}, Medium $\geq$ 5–20 μg l^{-1}, High = $\geq$ 20–60 μg $^{-1}$ and Hypereutrophic $\geq$ 60 μg l^{-1}) and used in the same manner as described above. Additional improvements to the original 'expert knowledge' methodology have also been proposed for macroalgae, and submerged aquatic vegetation based on comparison of potential area of colonization and effective colonized area. Presently these are still determined heuristically. Due to the lack of data on a national basis, epiphytes are no longer used as an indicator.

Response – determination of future outlook (DFO)

The Response component or future outlook is designed to estimate changes that might occur given predicted changes in nutrient input to a system. Like the other components this is determined by a matrix that combines susceptibility of the system with expected future changes in nutrient loads. Predictions of nutrient loading (increase, decrease, unchanged) are based on predicted population increase, planned management actions and expected changes in watershed uses. Results show that 86 systems were expected to become worse and only eight to improve from the early 1990s to the year 2020 (Bricker et al. 1999). The NEEA Update anticipated for late in 2006 will provide an interim report of changes over a decade, from the early 1990s to the early 2000s.

Synthesis – grouping pressure, state and response indicators

An additional modification to the original method (ASSETS; Bricker et al. 2003) combines the OEC, OHI, and DFO into a single overall score falling into one of five categories: high, good, moderate, poor, or bad. These categories conform to the EU Water Framework Directive (EUWFD; 2000/60/EC) and the framework provides a scale for setting eutrophication related reference conditions for different types of systems (e.g., Bettencourt et al. 2004).

Additional modifications: NEEA update program

Further modifications that are presently being pursued in the NEEA update program include the development of a type classification based on physical and hydrologic characteristics using the Deluxe Integrated System for Clustering Operations (DISCO) tool (Smith and Maxwell 2002). Preliminary results are promising (Smith et al. 2004) and will be used to determine type specific reference conditions and thresholds for desirable/undesirable conditions for indicator variables. Additionally, indicator variables are being evaluated by type to ensure that all types of estuaries are assessed with indicators that are relevant. For instance, in types where there is no SAV under natural conditions, an alternative indicator will be used.

A socio-economic/human use indicator is being developed where changes in fish catch rate are related to changes in water quality in the manner of Lipton and Hicks (1999, 2003) and Mistiaen et al. (2003). Preliminary analysis of Long Island Sound data shows that as nitrogen inputs decrease, dissolved oxygen and recreational catch of Striped Bass increase. The increase in catch is shown to be related to changes in oxygen when other influences (e.g. fishermen avidity and experience, temperature, changes in fish stock) are accounted for (Mason et al. 2004).

Finally, an online tool is being developed that can be used by scientists and managers to assess eutrophication and compare with other systems of similar types from the US and internationally (http://www.eutro.org). Results will be stored so that trends can be developed and tracked online with successive assessments. Additionally, the site provides publications that describe: the legislative context nationally and internationally that drive assessment and management, the general concept of the eutrophication issue, the development of the NEEA/ASSETS methodology, and programs supporting monitoring, assessment, management and research of nutrient related eutrophication in US and EU estuaries and coastal waters.

An integrated assessment of the causes and consequences of hypoxia in the northern Gulf of Mexico

The integrated assessment (IA) of the causes and consequences of hypoxia in the northern Gulf of Mexico (CENR 2000) is an example of a watershed approach to evaluating options for management action. While this approach is proving to be effective on this massive scale, it can be as effective, in fact more tractable, on smaller scales more typical of estuarine watersheds. The Mississippi River system ranks among the world's top 10 rivers in length, freshwater discharge, and sediment delivery and drains 41% of the contiguous United States (Figure 3). This massive river system discharges to the Louisiana/Texas continental shelf producing the largest zone of oxygen-depleted coastal waters (hypoxia) in the western Atlantic Ocean. Recurring summer hypoxia in the northern Gulf of Mexico has received considerable scientific and policy attention because of potential ecological and economic impacts from this very large zone of low oxygen, and because of the implications for management within its massive watershed (CENR 2000; Mitsch et al. 2001; Task Force 2001; Rabalais et al. 2002). In 1998, the Congress passed and the President signed into law the Harmful Algal Bloom and Hypoxia Research and Control Act (HABHRCA), which among other things called for an Integrated

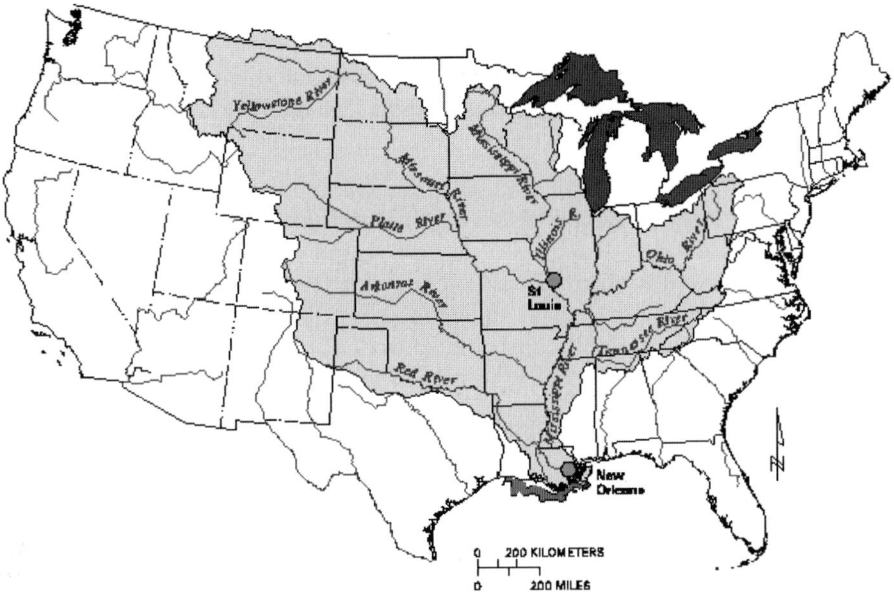

Figure 3. Mississippi/Atchafalaya River Basin and Gulf of Mexico Hypoxia (Rabalais et al. 2002).

Assessment of its causes and consequences in support of an Action Plan to reduce the size of the hypoxic zone. The analysis followed the Integrated Assessment approach:

1. Define the policy relevant question around which the assessment is to be performed. This was established early in the HABHRCA language – 'assess the causes and consequences of hypoxia in the northern Gulf of Mexico.'
2. Document the status and trends of appropriate environmental, social, and economic conditions related to the issue. This is a value-independent description of current conditions and, to the extent possible, the historical trends in those properties.
3. Describe the environmental, social, and economic causes and consequences of those trends. This often includes simulation, statistical, and other explanatory models and analyses. Again, these descriptions are fact-based although subject to analysis and interpretation.
4. Provide forecasts of likely future conditions under a range of policy and/or management actions. This can be quantitative forecasts from models or other trend analysis tools. These are subject to considerable scientific evaluation and interpretation.
5. Provide technical guidance for the most cost effective means of implementing each of those options. These efforts are designed to provide those who are responsible for implementation the menu of approaches available

to them, along with some evaluation of their potential for success and cost-effectiveness.
6. Provide an assessment of the uncertainties associated with the information generated for the above steps and outline key monitoring, research, and modeling needs to improve future assessments in this area.

To form a solid basis for the IA, six technical reports were commissioned (Brezonik et al. 1999; Diaz and Solow 1999; Doering et al. 1999; Goolsby et al 1999; Mitch et al. 1999; Rabalais et al. 1999) and subjected to independent peer review, followed by formal public comment (Rabalais et al 2002). These reports documented the ecological and economic extent, characteristics, causes, and effects of Gulf hypoxia; the flux and sources of nutrients in the Mississippi River system; the effects of reducing nutrient loads on waters within the basin and in the Gulf; methods to reduce nutrient loads; and the social and economic costs and benefits of methods to reduce nutrient loads. Description of the production, peer review, and public comment processes, along with the six reports, the Integrated Assessment, public comments, and responses to those comments are available at: http://www.nos.noaa.gov:80/Products/pubs_hypox.html.

Regions of oxygen concentrations below 2 mg l^{-1} (hypoxia) that form off the Louisiana coast each spring and summer increased from an average of 8300 km^2 in 1985–1992 to over 16,000 km^2 in 1993–2001 (Rabalais et al. 2002), and reached a record 22,000 km^2 in 2002 (Figure 4). The IA concluded that the almost threefold increase in nitrogen load to the Gulf (Goolsby et al. 1999) has been the primary external driver of increased hypoxia (Brezonik et al. 1999;

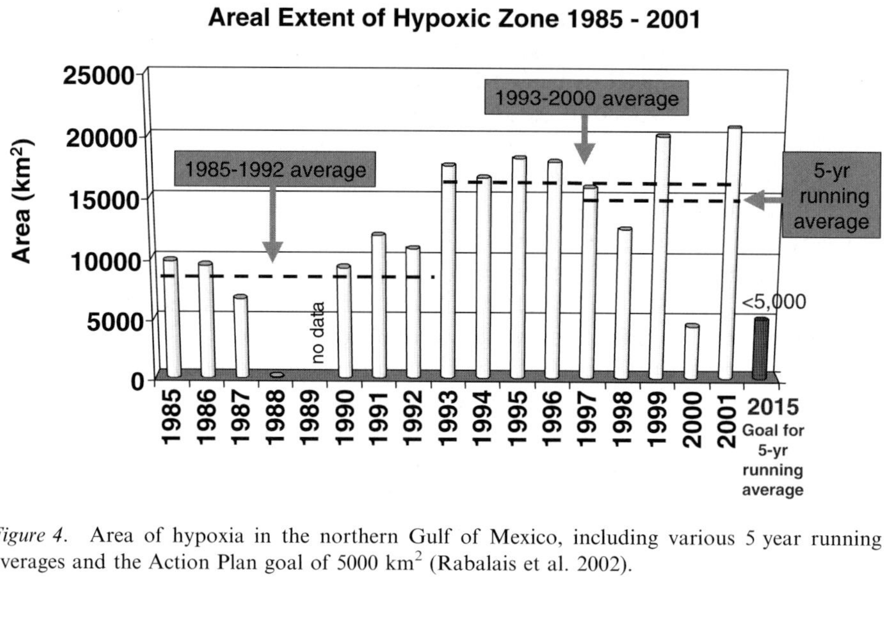

Figure 4. Area of hypoxia in the northern Gulf of Mexico, including various 5 year running averages and the Action Plan goal of 5000 km^2 (Rabalais et al. 2002).

Rabalais et al. 1999). This riverine nitrogen input stimulates coastal algal production and the subsequent settling of organic matter below the pycnocline. Because the pycnocline inhibits vertical oxygen flux, decomposition of organic matter below the pycnocline consumes oxygen faster than it is replenished, resulting in declining oxygen concentrations during the period of stratification. Two key questions were asked during development of policies to reduce, mitigate, and control Gulf hypoxia (Task Force 2001; Rabalais et al. 2002). The first was: 'When did large-scale hypoxia start in the Gulf of Mexico?' Knowing the answer to this question is important both for understanding its underlying causes and for identifying reasonable and practical goals for reducing its size. During development of the initial IA, the answer to this question was based on sporadic historical data and selected sediment core records, and it could only suggest that significant changes in bottom water oxygen conditions likely took place after the 1950s. The second question debated during development of the Action Plan was: 'What nitrogen load reduction would be needed to reach the societal goal set for hypoxia?' Model analysis during the development of the initial IA indicated that reducing N loads by 30–50% should increase oxygen concentrations in the bottom waters by 35–50% (Bierman et al. 2001; Brezonik et al. 1999).

The IA also identified the most significant nitrogen sources within the basin as being agricultural non-point sources originating in midwestern states. Ninety percent of the nitrate inputs to the Gulf come from non-point sources; 74% from agricultural non-point sources; and 56% of the nitrate enters the system north of Ohio River. Thus, the IA and resulting Action Plan focused on addressing these non-point sources.

Mitsch et al. (1999, 2001) reviewed the range of methods and technologies available to reduce N loads, including those from point and non-point sources, and concluded that the most effective methods are those that keep nitrogen on agricultural lands and those that encourage denitrification within the watershed (Figure 5). Of the 2.5 million metric tons of N that could potentially be prevented from reaching the Gulf, 24% could potentially result from denitrification losses in wetlands and riparian buffers; whereas 73% could potentially result from improved nitrogen management on farm lands. Only 2–3% would be reduced from upgrading all sewage treatment plants.

Doering et al. (1999) evaluated costs associated with the various nitrogen control and mitigation options. For example, they concluded that agricultural practices (other than reduced fertilizer use) to reduce the loss of nitrogen at the edge of the field would cost $0.88 per kg of Nitrogen for a 20% loss reduction and $3.37 per kg of nitrogen for a 40% loss reduction. The higher costs per unit N loss reflect the fact that achieving 40% loss reductions would require retiring more productive lands than a 20% reduction. The costs associated with achieving a comparable 20% N load reduction from reduced fertilizer applications is $0.69/unit N load, from construction and operation of 5 million acres of wetlands is $8.90/unit N load, and from construction of 19 million acres of riparian buffers is $26.03/unit N load.

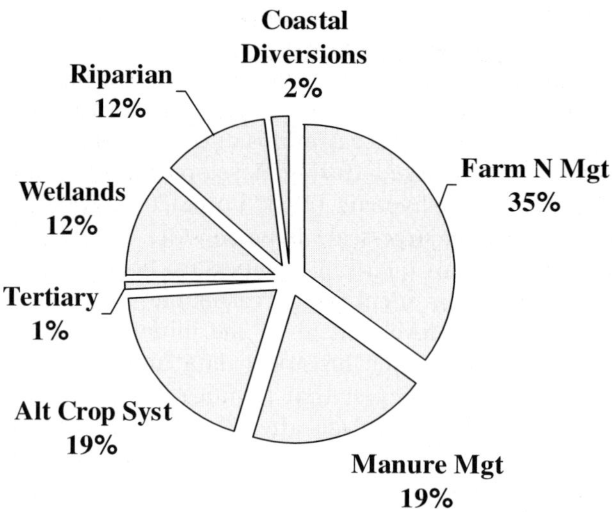

Figure 5. Potential sources of N load reduction. Total potential reduction from all sources is 2.5 million metric tons per year. Data Source: Mitsch et al. (1999, 2001), CENR (2000).

It is interesting to note that, while gross costs per unit N load reduction vary considerably from \$0.69 to \$0.88 to \$8.90 for reduced fertilizer use, other improved agricultural practices, and wetland construction respectively, the net costs (\$0.67, \$0.80, \$1.00, respectively) are closer to each other. These net costs include economic benefits such as reduced fertilizer purchases and wetland benefits such as flood control, wildlife protection, and recreation. These relatively comparable unit costs across the three major control approaches gave the Task Force and related agencies flexibility in implementation. Fortunately, recent reanalysis (McIsaac et al. 2001, 2002) of the effects of potential management actions on nitrogen loads from the Mississippi River Basin suggest that those reductions might be even easier to achieve than estimated in the original studies supporting the Action Plan (Doering et al. 1999).

Analysis and summary

Clark and Majone (1985) outlined four criteria for evaluating integrated assessments – Technical adequacy, Value, Legitimacy, and Effectiveness. The Gulf hypoxia IA and the National assessment (NEEA) score well on all four criteria. 'Technical adequacy' was achieved in the IA through the independent peer review of the six background technical reports and subsequent publications in the primary literature. NEAA was built from the peer reviewed literature, a formal process for integrating expert opinion, and peer review of the overall product. 'Value' in the IA was established by responding directly to directives of the HABHRCA statute and the guidance and expectations of the Task Force responsible for creating an action plan. The NEEA, published in 1999 effectively raised the profile of this national issue in Congress, with federal and state agencies, and in stimulation of a comprehensive evaluation by the National Research Council (NRC 2000). 'Legitimacy' in the IA was established by engaging a wide range of stakeholders in the development and review of the IA and background reports, as well as through formal public comment, facilitated and open science meetings, and participation in seven public meetings of the Task Force. The participation of most U.S. scientists and many policy makers with expertise and experience in coastal eutrophication ensured legitimacy for the NEEA.

According to Clark and Majone (1985), 'Effectiveness' is best measured by two factors – did the assessment make a difference in a policy outcome and/or did it influence how policy makers understand the problem. The first and most obvious measure of effectiveness of the IA is that it led to an Action Plan (Task Force 2001) that was endorsed by eight Federal Agencies, nine Basin States, and two Tribes and delivered to the President and the Congress in 2001. The second measure is that the IA has changed the way scientists and policy makers think about and discuss options for dealing with excess nutrient loads to the Gulf of Mexico, and to some extent, the rest of agriculturally dominated coastal watersheds. NEEA similarly changed the focus of how U.S. agencies and scientists viewed the problem of coastal eutrophication, as well as stimulating continued dialog in both Congress and the administration (e.g., HAB-HRCA, Ocean Commission 2004).

Morgan and Dowlatabadi (1996) suggest that assessments should be iterative and include new information as it becomes available. Just as the National Estuarine Eutrophication Assessment is being modified, updated, and improved, new information is becoming available for the Gulf hypoxia assessment, in time for the 5 year reassessment called for in the Action Plan (Task Force 2001). While modeling and other analyses were sufficient to allow the Task Force to establish a goal of reducing the average size of the hypoxic region to below 5000 km^2 by reducing N loads by 30%, subsequent modeling efforts helped refine those estimates. These more recent analyses (Scavia et al. 2003, 2004) compared three very different models to reach the consensus that large-scale hypoxia likely did not start in the Gulf of Mexico until the mid-1970s and that the 30% nitrogen load reduction called for in the Action Plan may not be sufficient to reach the plan's goal. Load reductions of 35–45% are likely to be needed to reach the hypoxia goals in most years.

These two examples provide templates for different approaches to assessing coastal eutrophication. The national assessment was able to establish the overall state of the nation with respect to coastal eutrophication even though data were not uniformly available or comprehensively obtained. The use of expert knowledge within a consistent and comprehensive framework produced information upon which Federal policy makers could form a basis for action. It also provided a sense of status and trends, with tentative notions of futures, for specific estuaries and regions, and it formed a basis for a more comprehensive assessment that addressed causes and potential remedies at similar national scales (CENR 2003). Continued improvements to the approach including development of type specific indicators and thresholds and identification and apportionment of watershed nutrient sources will result in more accurate assessments of conditions and more targeted recommendations for management.

The second approach focused on a particular watershed, *albeit* the largest in the United States. This focus on the specific causes and consequences of nutrient pollution for this watershed and receiving water enabled targeted options for restoration, as opposed to the more general recommendations of the national assessment. Similar approaches at smaller scales, such as individual estuarine watersheds, are likely to provide a more effective basis for managing coastal eutrophication.

References

Bettencourt A., Bricker S.B., Ferreira J.G., Franco A., Marques J.C., Melo J.J., Nobre A., Ramos L., Reis C.S., Salas F., Silva M.C., Simas T. and Wolff W. 2004. Typology and Reference Conditions for Portuguese Transitional and Coastal Waters Development of guidelines for the application of the European Union Water Framework Directive. Instituto da Agua (INAG)-Institute of Marine Science (IMAR), Lisbon. http://www.ecowin.org/ticor/documents/TICOR%20book.pdf.

Bierman Jr. V.J., Hinz S.C., Wiseman Jr. W.J., Rabalais N.N. and Turner R.E. 2001 Forecasting hypoxia in the Gulf of Mexico: Responses to nutrient loadings from the Mississippi River Basin. Presentation at Coastal Zone 01, July 18, 2001, Cleveland.

Boynton W.R., Murray L., Hagy J.D., Stokes C. and Kemp W.M. 1996. A comparative analysis of eutrophication patterns in a temperate coastal lagoon. Estuaries 19(2B): 408–421.

Brezonik P.L., Bierman Jr. V.J., Alexander R., Anderson J., Barko J., Dortch M., Hatch L., Hitchcock G.L., Keeney D., Mulla D., Smith V., Walker C., Whitledge T. and Wiseman Jr. W.J. 1999. Effects of Reducing Nutrient Loads to Surface Waters within the Mississippi River Basin and the Gulf of Mexico. Topic 4 Report for the Integrated Assessment of Hypoxia in the Gulf of Mexico. NOAA Coastal Ocean Program Decision Analysis Series No. 18, National Oceanic and Atmospheric Administration, Coastal Ocean Program, Silver Spring.

Bricker S., Matlock G., Snider J., Mason A., Alber M., Boynton W., Brock D., Brush G., Chestnut D., Claussen U., Dennison W., Dettmann E., Dunn D., Ferreira J., Flemer D., Fong P., Fourqurean J., Hameedi J., Hernandez D., Hoover D., Johnston D., Jones S., Kamer K., Kelty R., Keeley D., Langan R., Latimer J., Lipton D., Magnien R., Malone T., Morrison G., Newton J., Pennock J., Rabalais N., Scheurer D., Sharp J., Smith D., Smith S., Tester P., Thom R., Trueblood D. and Van Dolah R. 2004. National Estuarine Eutrophication Assessment Update: Workshop summary and recommendations for development of a long-term monitoring and assessment program. Proceedings of a workshop September 4–5 2002, Patuxent Wildlife Research Refuge, Laurel, Maryland. National Oceanic and Atmospheric Administration, National Ocean Service, National Centers for Coastal Ocean Science. Silver Spring. http://www.eutro.org/documents/dldo.pdf.

Bricker S.B., Clement C.G., Pirhalla D.E., Orlando S.P. and Farrow D.R.G. 1999. National Estuarine Eutrophication Assessment. Effects of Nutrient Enrichment in the Nation's Estuaries. NOAA, National Ocean Service, Special Projects Office and National Centers for Coastal Ocean Science, Silver Spring. http://www.spo.nos.noaa.gov/projects/cads/nees/Eutro_Report.pdf.

Bricker S.B., Ferreira J.G. and Simas T. 2003. An integrated methodology for assessment of estuarine trophic status. Ecol. Model. 169: 39–60. http://www.coastalscience.noaa.gov/documents/assets.pdf.

Burkholder J.M., Noga E.J., Hobbs C.H. and Glasgow H.B. Jr. 1992a. New 'phantom' dinoflagellate is the causative agent of major estuarine fish kills. Nature 358: 407–410.

Burkholder J.M., Mason K.M. and Glasgow H.B. Jr. 1992b. Water-column nitrate enrichment promotes decline of eelgrass *Zostera marina*: evidence from seasonal mesocosm experiments. Mar. Ecol. Prog. Ser. 81: 163–178.

Coastal Assessment and Data Synthesis System (CADS). 1999. NOAA, National Ocean Service, Special Projects Office, Silver Spring. http://www.cads.nos.noaa.gov.

CENR 2003. An Assessment of Coastal Hypoxia and Eutrophication in U.S. Coastal Waters. National Science and Technology Council Committee on Environment and Natural Resources, Washington. http://www.nccos.noaa.gov/publications/hypoxia.pdf.

CENR 2000. Integrated Assessment of Hypoxia in the Northern Gulf of Mexico. National Science and Technology Council Committee on Environment and Natural Resources, Washington.

Clark W.C. and Majone G. 1985. The critical appraisal of scientific enquiries with policy implications. Science, Technology and Human Values 10(3): 6–19.

COAST. 2003. European Union Common Implementation Strategy Working Group 24 Final Draft Guidance on Coastal and Transitional Waters. European Union Water Directors, Copenhagen. http://www.eutro.org/documents/wfd%20cis2.4%20(coast)%20guidance%20on%20tcw.pdf.

Dennison W.C., Moore K.A. and Stevenson J.C. 1992. Chapter III: SAV habitat requirements development. In: Batuik R.A., Orth R.J., Moore K.A., Dennison W.C., Stevenson J.C., Staver L.W., Carter V., Rybicki N.B., Hickman R.E., Kollar S., Bieber S. and Heasly P. (eds), Chesapeake Bay Submerged Aquatic Vegetation Habitat Requirements and Restoration Targets: A Technical Synthesis. US Environmental Protection Agency, Chesapeake Bay Program, Annapolis.

Diaz R.J. and Solow A. 1999. Ecological and economic consequences of hypoxia. Topic 2. Gulf of Mexico Hypoxia Assessment. NOAA Coastal Ocean Program Decision Analysis Series. National Oceanic and Atmospheric Administration, Coastal Ocean Program, Silver Spring.

Doering, O.C. et al. 1999. Evaluation of the economic costs and benefits of methods for reducing nitrogen loads to the Gulf of Mexico. National Oceanic and Atmospheric Administration, Coastal Ocean Program Decision Analysis Series No. 20, Silver Spring.

European Commission Water Framework Directive 2000/60/EC (WFD) http://www.europa.eu.int/comm/environment/water/water-framework/index_en.html.

Goolsby D.A., Battaglin W.A., Lawrence G.B., Artz R.S., Aulenbach B.T., Hooper R.P., Keeney D.R. and Stensland G.J. 1999. Flux and Sources of Nutrients in the Mississippi–Atchafalaya River Basin, Topic 3 Report for the Integrated Assessment of Hypoxia in the Gulf of Mexico. NOAA Coastal Ocean Program Decision Analysis Series No. 17. National Oceanic and Atmospheric Administration, Coastal Ocean Program, Silver Spring.

Hinga K.R., Stanley D.W., Klein C.J., Lucid D.T. and Katz M.J. (eds) 1991. The National Estuarine Eutrophication Project: Workshop Preceedings. National Oceanic and Atmospheric Administration and the University of Rhode Island Graduate School of Oceanography, Rockville http://www.eutro.org/documents/hinga%201991.pdf.

Howarth R.W., Anderson D., Cloern J., Elfring C., Hopkinson C., Lapointe B., Malone T., Marcus N., McGlathery K., Sharpley K. and Walker D. 2000. Nutrient pollution of coastal rivers, bays, and seas. Issues Ecol. 7: 1–15.

Howarth R., Marino R. and Scavia D. 2003. Nutrient Pollution in Coastal Waters – Priority Topics for an Integrated National Research Program for the United States. United States Department of Commerce, National Oceanic and Atmospheric Administration, National Ocean Service, National Centers for Coastal Ocean Science. Silver Spring.

Jaworski N.A. 1981. Sources of nutrients and the scale of eutrophication problems in estuaries. In: Neilson B.J. and Cronin L.E. (eds.), Estuaries and Nutrients, Humana Press, Clifton, pp. 83–110.

Lapointe B.E. and Matzie W.R. 1996. Effects of stormwater nutrient discharges on eutrophication processes in nearshore waters of the Florida Keys. Estuaries 19(2B): 422–435.

Lipton D.W. and Hicks R. 2003. The cost of stress: low dissolved oxygen and recreational striped bass (*Morone saxatilis*) fishing in the Patuxent River. Estuaries 26: 310–315.

Lipton D.W. and Hicks R. 1999. Linking water quality improvements to recreational fishing values: the case of Chesapeake Bay striped bass. In: Evaluating the Benefits of Recreational Fisheries. Fisheries Centre Research Reports 7(2). University of British Columbia.

Mason A., Lipton D.W. and Bricker S.B. 2004. Improving indicators of water quality degradation impacts for management of estuarie and coastal waters, p. 102. *In* American Society of Limnology and Oceanography and The Oceanography Society 2004 Ocean Research Conference Abstract Book. February 15–20, 2004. Hawaii Convention Center, Honolulu. http://www.aslo.org/honolulu2004/files/aslo-tos-2004-abstracts.pdf.

McIssac G.F., David M.B., Gertner G.Z. and Goolsby D.A. 2001. Nitrate flux in the Mississippi River. Nature 414: 166–167.

McIssac G.F., David M.B., Gertner G.Z. and Goolsby D.A. 2002. Relating net nitrogen input in the Mississippi River basin to nitrate flux in the Lower Mississippi River: a comparison of approaches. J. Environ. Quality 31: 1610–1622.

Morgan M.G. and Dowlatabadi H. 1996. Learning from integrated assessment of climate change. Climatic Change 34: 337–368.

Mitsch W.J., Day Jr. J.W., Gilliam J.W., Groffman P.M., Hey D.L., Randall G.W. and Wang N. 1999. Reducing Nutrient Loads, Especially Nitrate-Nitrogen, to Surface Water, Ground Water, and the Gulf of Mexico. Topic 5 Report for the Integrated Assessment of Hypoxia in the Gulf of Mexico. NOAA Coastal Ocean Program Decision Analysis Series No. 19. National Oceanic and Atmospheric Administration, Coastal Ocean Program, Silver Spring.

Mitsch W.J., Day J.W. Jr., Gilliam J.W., Groffman P.M., Hey D.L., Randall G.W. and Wang N. 2001. Reducing nitrogen loading to the Gulf of Mexico from the Mississippi River basin: strategies to counter a persistent ecological problem. BioScience 15: 373–388.

Mistiaen J.A., Strand I.E. and Lipton D.W. 2003. Effects of environmental stress on blue crab (*Callinectes sapidus*) harvests in Chesapeake Bay tributaries. Estuaries 26: 316–322.

National Oceanic and Atmospheric Administration (NOAA). 1996. NOAA's Estuarine Eutrophication Survey. Volume 1: South Atlantic Region. National Oceanic and Atmospheric Administration, National Ocean Service, Office of Ocean Resources Conservation and Assessment. Silver Spring http://www.eutro.org/documents/south%20atlantic%20regional%20report.pdf.

National Oceanic and Atmospheric Administration (NOAA). 1997a. NOAA's Estuarine Eutrophication Survey. Volume 2: Mid-Atlantic Region. National Oceanic and Atmospheric Administration, National Ocean Service, Office of Ocean Resources Conservation and Assessment. Silver Spring http://www.eutro.org/documents/mid-atlantic%20regional%20report.pdf.

National Oceanic and Atmospheric Administration (NOAA). 1997b. NOAA's Estuarine Eutrophication Survey. Volume 3: North Atlantic Region. National Oceanic and Atmospheric Administration, National Ocean Service, Office of Ocean Resources Conservation and Assessment. Silver Spring http://www.eutro.org/documents/north%20atlantic%20regional%20 report.pdf.

National Oceanic and Atmospheric Administration (NOAA). 1997c. NOAA's Estuarine Eutrophication Survey. Volume 4: Gulf of Mexico Region. National Oceanic and Atmospheric Administration, National Ocean Service, Office of Ocean Resources Conservation and Assessment. Silver Spring http://www.eutro.org/documents/gulf%20of%20mexico%20regional%20 report.pdf.

National Oceanic and Atmospheric Administration (NOAA). 1998. NOAA's Estuarine Eutrophication Survey. Volume 5: Pacific Coast Region. National Oceanic and Atmospheric Administration, National Ocean Service, Office of Ocean Resources Conservation and Assessment. Silver Spring http://www.eutro.org/documents/pacific%20regional%20report.pdf.

National Research Council (NRC) 2000. Clean Coastal Waters: Understanding and Reducing the Effects of Nutrient Pollution. National Academy Press, Washington.

Nixon S.W. 1983. Estuarine Ecology: A Comparative and Experimental Analysis using 14 Estuaries and the MERL Ecosystems. Final Report to the U.S. Environmental Protection Agency. Chesapeake Bay Program, Washington.

Ocean Commission. 2004. An Ocean Blueprint for the 21st Century. Final Report. U.S. Commission on Ocean Policy. Washington, DC.

Oslo Paris Convention for the Protection of the North Sea (OSPAR). 2003. OSPAR Integrated Report 2003 on the Eutrophication Status of the OSPAR Maritime Area Based Upon the First Application of the Comprehensive Procedure. OSPAR Commission for the protection of the marine environment of the North-East Atlantic. http://www.eutro.org/documents/p00189_Eutrophication%20Status%20Report%202003.pdf.

Oslo Paris Convention for the Protection of the North Sea (OSPAR). 2002. Common Assessment Criteria, their Assessment Levels and Area Classification within the Comprehensive Procedure of the Common Procedure. OSPAR Commission for the protection of the marine environment of the North-East Atlantic. http://www.ospar.org/eng/html/welcome.html.

Orth R.J. and Moore K.A. 1984. Distribution and abundance of submerged aquatic vegetation in Chesapeake Bay: An historical perspective. Estuaries 7: 531–540.

Rabalais N.N., Turner R.E., Justic D., Dortch Q., Wiseman W.J. Jr. and Sen Gupta B.K. 1996. Nutrient changes in the Mississippi Rover and system responses on the adjacent continental shelf. Estuaries 19(2B): 386–407.

Rabalais N.N. and Harper Jr. D.E. 1992. Studies of benthic biota in area affected by moderate and severe hypoxia. In: National Oceanic and Atmospheric Administration, Coastal Ocean Program Office, Nutrient Enhanced Coastal Ocean Productivity: Proceedings of a Workshop. Louisiana Universities Marine Consortium, October 1991 (pp. 150–153). Sea Grant Program, Texas A & M University, Galveston.

Rabalais N.N., Turner R.E., Justić D., Dortch Q. and Wiseman Jr. W.J. 1999. Characterization of Hypoxia: Topic 1 Report for the Integrated Assessment of Hypoxia in the Gulf of Mexico. NOAA Coastal Ocean Program Decision Analysis Series No. 15. National Oceanic and Atmospheric Administration, Coastal Ocean Program, Silver Spring.

Rabalais N.N., Turner R.E. and Scavia D. 2002. Beyond science into policy: Gulf of Mexico Hypoxia and the Mississippi River. BioScience 52: 129–142.

Scavia D., Rabalais N.N., Turner R.E., Justić D. and Wiseman W.J. Jr. 2003. Predicting the response of Gulf of Mexico hypoxia to variations in Mississippi River nitrogen load. Limnol. Oceanogr. 48: 951–956.

Scavia D., Justić D. and Bierman V.J. Jr. 2004. Reducing hypoxia in the Gulf of Mexico: advice from three models. Estuaries 27: 419–425.

Smith C.A. and Maxwell B.A. 2002. Deluxe Integrated System for Clustering Operations (DISCO). http://www.narya.engin.swarthmore.edu/disco/.

Smith S.V., Buddemeier R.W., Bricker S.B., Maxwell B.A., Pacheco P. and Mason A. 2004. Estuarine Typology: Perturbations and eutrophication responses. In: American Society of Limnology and Oceanography and The Oceanography Society 2004 Ocean Research Conference Abstract Book, p. 150. February 15–20, 2004. Hawaii Convention Center, Honolulu. http://www.aslo.org/honolulu2004/files/aslo-tos-2004-abstracts.pdf.

Smith R.A., Schwarz G.E. and Alexander R.B. 1997. Spatially referenced regressions on watershed attributes (SPARROW). Regional interpretation of water quality monitoring data. Water Resour. Res. 33: 2781–2798.

Stevenson J.C., Staver L.W. and Staver K.W. 1993. Water quality associated with survival of submersed aquatic vegetation along an estuarine gradient. Estuaries 16(2): 346–361.

Task Force. 2001. Mississippi River/Gulf of Mexico Watershed Nutrient Task Force. 2001. Action Plan for Reducing, Mitigating, and Controlling Hypoxia in the Northern Gulf of Mexico. Mississippi River/Gulf of Mexico Watershed Nutrient Task Force, Washington.

Twilley R.R., Kemp W.M., Staver K.W., Stevenson J.C. and Boynton W.R. 1985. Nutrient enrichment of estuarine submersed vascular plant communities. I. Algal growth and effects on production of plants and associated communities. Marine Ecol. Prog. Series 23: 179–191.

Welsh B. 1991. Anoxia and Hypoxia in Long Island Sound, Chesapeake Bay, and Mobile Bay: A comparative assessment. In: Hinga K.R., Stanley D.W., Klein C.J., Lucid D.T. and Katz M.J. (eds), The National Estuarine Eutrophication Project: Workshop Proceedings. National Oceanic and Atmospheric Administration and the University of Rhode Island Graduate School of Oceanography, Rockville, pp. 35–40 http://www.eutro.org/documents/hinga%201991.pdf.

Biogeochemistry (2006) 79: 209–237
DOI 10.1007/s10533-006-9006-x

Nitrogen cycling in tropical and temperate savannas

M.M.C. BUSTAMANTE[1,*], E. MEDINA[2], G.P. ASNER[3], G.B. NARDOTO[4] and D.C. GARCIA-MONTIEL[5]
[1]*Departamento de Ecologia, Universidade de Brasília, Brasília, DF, Brazil;* [2]*Instituto Venezolano de Investigaciones Cientificas, Caracas, Venezuela;* [3]*Department of Global Ecology, Carnegie Institution of Washington, Stanford, CA 94305, USA;* [4]*Centro de Energia Nuclear na Agricultura, Universidade de São Paulo, Avenida, Piracicaba, SP, Brazil;* [5]*Departamento de Ecologia, Universidade de Brasília, Brasília, DF, Brazil; *Author for correspondence (e-mail: mercedes@unb.br)*

Key words: Cerrado, Fire, Llanos, Mesquite, Nitrogen, Savannas, Trace gases

Abstract. Savannas are the most common vegetation type in the tropics and subtropics, ranging in physiognomy from grasslands with scattered woody plants to woodlands with heterogeneous grass cover. Productivity and organic matter turnover in savannas are controlled by interactions between water and nutrient availability, and this basic environmental structure is modified by fire frequency and land management practices. We compared temperate and tropical savannas in order to understand the strength of nitrogen (N) limitation of productivity. American tropical and temperate savannas are N limited systems, and the N cycle differs according to the woody plant density, fire frequency, land use change, N deposition and N fixation. Grazing and conversion to pasture have been the predominant land-use changes in most savannas. In the Cerrado and the Llanos tropical savannas, intensified use of fire for pasture management is leading to decreased woody plant density. Oppositely, in the Chaco and North American temperate savannas, fire suppression and grazing are leading to increases in woody density. In addition, the higher soil P availability in the Gran Chaco and the higher N deposition in North American savannas may be contributing to increases of N cycling and net productivity rates. Some aspects of the N budget for savannas of the American continent are still unclear and require further analysis to determine rates of N fixation, and to understand how spatial and temporal soil heterogeneity control N fluxes through soil solution and into streams.

Introduction

Savannas are the most common vegetation type in the tropics and subtropics (Solbrig 1991; Scholes and Archer 1997). They cover extensive areas of South America, Africa and Australia, and also occur in Central America, India, and North America (Figure 1). In a broad sense, savannas can be defined as ecosystems with a near continuous grass/herbaceous stratum, a discontinuous layer of trees and shrubs of variable density, and where growth patterns are closely associated with alternating wet and dry seasons (Bourlière and Hadley 1983 in Mistry 2000). The dry season can last from 2 to 9 months, and there is also substantial climate variability between years (Frost et al. 1986).

Savanna ecosystems are controlled by the interactions between water and nutrient availability, as savanna vegetation often occurs either on weathered

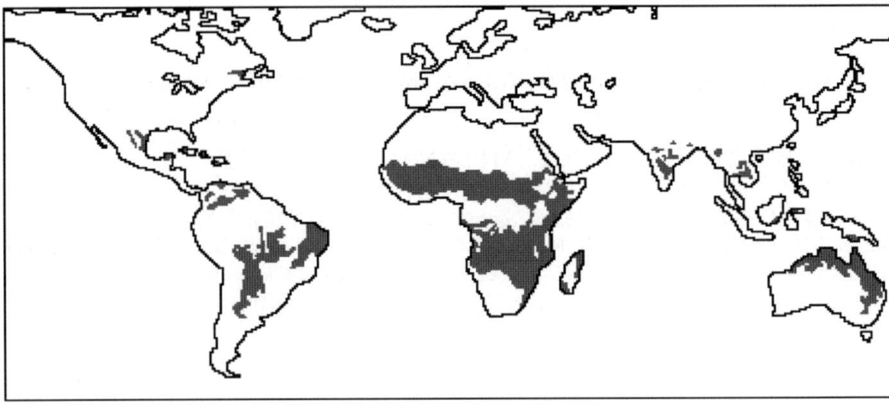

Figure 1. The approximate global distribution of tropical grasslands, savannas and woodlands (derived from Olson et al. (1983) by Scholes and Hall (1996).

soils with low nutrient availability, in regions of limited rainfall, or both (Medina 1987; Sarmiento 1996; Reatto et al. 1998). This basic environmental structure is modified by additional factors such as fire frequency and land management practices. Almost one fifth of the world's population lives in savanna regions (Frost et al. 1986), and these are probably the oldest ecosystems used by people, as they have supported hunting and pastoral practices for millennia. Presently savannas are experiencing many transformations due to agricultural intensification, over-grazing and changes in fire regimes. However, there is little understanding of the strength of N limitation of productivity, especially when comparing temperate and tropical savannas. This understanding requires data on inputs, outputs and internal cycling. Medina (1993) presented an extensive review of the mineral nutrition in tropical savannas containing considerable information concerning the N cycling. In this paper, we present a comparison between temperate and tropical American savannas based on the inputs, outputs and internal cycling of nitrogen. The interactions between N and other elements controlling the N cycle, as well as the effects of disturbance and land use changes, are presented.

In South America, savannas occur in two large patches north and south of the Equator (Solbrig 1996). The principal tropical savanna region south of the equator occurs entirely within Brazil, mostly in the central region of the country, is locally known as the Cerrado and covers approximately 2 millions km^2. This wet seasonal savanna is a mosaic of plant formations with a gradient of woody plant density ranging from open grasslands (campo limpo) through open scrubland (campo sujo and cerrado *sensu stricto*) to dense woodlands (cerradão) (Eiten 1972). The tropical savannas of northern South America (the Orinoco Llanos) extend from the Guaviare river in Colombia to the eastern coast of Venezuela covering approximately 500,000 km^2 (Sarmiento 1983). Vegetation physiognomy can also vary widely across the Llanos: from tree-less savanna grasslands to savanna woodlands with up to 80% tree cover (Sarmiento 1984).

In the subtropical region of South America, there are different ecosystems dominated by grasses, generally with sparse trees that can be considered as savannas due to the structure and function, but most of them differ remarkably from tropical savannas (Sarmiento 1996). Flooding and fire are key environmental factors that may differ, and drought is associated with low temperatures during the winter. While dystrophic acid soils dominate in the tropical savannas, soils are mesotrophic in the subtropical systems, and alkalinity and salinity can be important edaphic factors. These subtropical savannas reach a higher diversity in the Gran Chaco region in Bolivia, Paraguay, Argentina and small areas in Brazil. They are complex vegetation mosaics that vary throughout the Chaco region.

In North America, sub-tropical savannas dominated by the genera *Prosopis* (mesquite), *Acacia* and *Andropogon* are found throughout much of Texas and northern Mexico, or more than 500,000 km^2 (Bailey 1996). Areas of North American savannas are now dominated by introduced grasses of African origin (D'Antonio and Vitousek 1992; Asner et al. 2004). These savannas support a wide range of vegetation physiognomies, from open grasslands with sparse trees to dense mesquite woodlands having a nearly continuous understory of herbaceous cover. Rainfall in the sub-tropical savannas of North America, Australia and Africa ranges from less 400 mm to more than 1200 mm, and mean annual temperatures vary from 9 to 27 °C. Soils also vary substantially from oxic Ultisols on exposed Permian surfaces throughout Northern Texas to sandy alluvial Inceptisols in Northern Mexico and Arizona (SCS 1962).

In large regions of North and South American and Australian savannas, the vegetation has been heavily impacted by overgrazing following the introduction of large-scale cattle management practices (Asner et al. 2004). For example, in the Argentine Chaco region, grasslands disappeared and were replaced by a dense and thorny shrubland following years of heavy grazing (Morello and Saraiva-Toledo 1959; Bucher and Schofield 1981). In the Cerrado region, Indians and farmers used fire for the past several thousand years. The analysis of a sediment core of 6000 years revealed the presence of charcoal particles in all depths, indicating the occurrence of burning throughout this period (Vicentini 1999). However, in recent times the use of fire has become much more intensive with the conversion of extensive areas of Cerrado to pastures and grasslands. In other areas, fire suppression along with overgrazing and climate change are implicated in a widely observed shift from open savanna grasslands to ecosystems now densely populated by trees and shrubs (Archer 1995). In Texas grasslands and savannas, encroaching mesquite (*Prosopis glandulosa* var. glandulosa) has caused an increase in above-ground biomass, which, in turn, has increased soil carbon (C) and nitrogen (N) storage (Hibbard et al. 2001; Asner et al. 2003; Asner and Martin 2004).

Because global savannas cover an area of about 19.3 million km^2 and are typically burned at a frequency of 5 years or faster, they are important for regional and global atmospheric chemistry (Crutzen and Andreae 1999). The nitrogen cycle is particularly sensitive to changes induced by frequent fires due to substantial loss of this element through volatilization. In addition to the impacts of frequent fires on the N cycle, recent increases in anthropogenic deposition of N in natural ecosystems are more pronounced in grasslands and savannas (Asner et al. 2001). These authors pointed out that the primary controls over the fate of N in such regions are likely to be quite different than in forested systems, which have been the traditional focus of N pollution studies. The quantification of inputs and outputs of N in savanna ecosystems demands a more intensive research effort, as scarce information is available on N fixation rates and on both gaseous and solution N losses.

N inputs, outputs and internal cycling

Fire-nitrogen interactions

Nitrogen fluxes in savanna ecosystems are strongly influenced by the fire regime to which they are subjected. This is true of all savannas, independent of their geographical location (e.g., tropical or temperate). Fire affects the dynamics of the vegetation, particularly the grass/woody biomass ratio (Scholes and Archer 1997). Surface fires, which consume the fine fuel of the herbaceous layer, are the most common in savannas but fine fuel load varies with the degree of woodiness. The fine fuel of the herbaceous layer ranged from 85 to 97% of the fuel load from woodland savannas to open savannas in Central Brazil (Miranda et al. 2003). As fire frequency increases, the grass-tree ratio tends to increase (Sato et al. 1998), and the vegetation physiognomy shifts to a more open form, which subsequently favors the occurrence of intense fires (Miranda et al. 1996).

Greater losses of C, N and S of the aboveground biomass pools have been observed after fires from grassland formations than from cerrado savannas and cerradão woodlands (Kauffman et al. 1994). The differences in nutrient stocks and the higher fuel consumption during fires in more open areas of cerrado may explain the lower nutrient losses with greater density of woody plants. The woody components of the vegetation are the major pool of nutrients, and generally they do not burn during surface fires. Additionally, the nutrient stock in the leaves of the woody plants (Silva 1990) is smaller than the nutrient contained in the live biomass of the herbaceous layer (Batmanian 1983).

Nitrogen loss occurs through volatilization and via particulate matter formation during combustion and during subsequent wind-born transport (Kauffmann et al. 1994). Pivello and Coutinho (1992) estimated that, during a cycle of six prescribed fires in a *campo sujo* area (grass-dominated vegetation type), about 95% of the nitrogen in plant biomass was released to the atmosphere. In addition to the losses of nitrogen through biomass burning, several authors have shown that fire increased soil emissions of NO and N_2O

(Ward et al. 1992; Neff et al. 1995; Weitz et al. 1998; Wahlen et al. 2000). The emissions of N oxides in savannas are discussed further in Section 'Emissions of NO and N_2O' (below).

Atmospheric deposition

A study conducted in a cerrado area protected from fire in Central Brazil estimated total N input from atmospheric deposition at 4.2 kg ha^{-1} yr^{-1}, with a contribution of 2.2 kg ha^{-1} yr^{-1} as inorganic N (Resende 2001). This value of total atmospheric deposition corresponds to 3.4% of the litter stock but to less than 0.1% of the total N in the soil (0–100 cm). The same author determined that the N leached from the canopy and accumulated litter amounted to 5.1 kg N ha^{-1} yr^{-1}. Canopy and litter leachate was enriched in organic N.

A similar value of 2.2 kg ha^{-1} yr^{-1} was reported for the input of inorganic N from bulk precipitation and dry deposition in a *Trachypogon* savanna in Venezuelan Llanos (Montes and San Jose 1989). The input of NH_4-N was 2.0 kg ha^{-1} yr^{-1}, but organic nitrogen input was not measured. The reported inputs of inorganic N in the South American savannas are lower than the values found in an African savanna (~5 kg ha^{-1} yr^{-1}) (Villecourt and Roose 1978 cited by Abbadie et al. 1992). Comparing atmospheric deposition inputs and the amount of N in above- and below-ground plant biomass in a *Trachypogon* savanna, Montes and San Jose (1989) suggested that 19.5% of the N required for maximum biomass of 482 g m^{-2} during growth season was supplied by atmospheric deposition. They pointed out that the distribution of rains seemed to determine essential differences in the nutrient inputs, as the ratio of the nutrient input during the rainy and the dry season ranged from 2.3 to 7.8. Unusual differences in dry deposition chemistry between years seemed to be related to length of the dry season, occurrence of sporadic rains during the dry season, and the quality and quantity of fuel material. However, it cannot be assumed that all atmospheric deposition would be readily available for plant uptake. A recent report from Pacheco et al. (2004) for the Llanos indicated that soluble organic N inputs may be an important contribution, making up 80% of the total soluble N in rainfall. It remains to be established if this soluble organic N constitutes a net N input to the soil–vegetation system.

Nitrogen deposition has been monitored in temperate and sub-tropical savannas of Texas, USA as part of a national monitoring network (NADP 2003). NO_3 deposition varies from 6 to 10 kg N ha^{-1} yr^{-1}, depending upon geographic location, local and regional wind and rainfall patterns, and land use. Deposition of NH_4 ranges from 2 to 4 kg N ha^{-1} yr^{-1}, resulting in a total wet deposition range of about 8–14 kg N ha^{-1} yr^{-1}. These values are considered high, as they rival values observed in areas of the northeastern USA thought to be undergoing N pollution and saturation (Asner et al. 2001).

Nitrogen fixation

Global biological N fixation (BNF) in terrestrial ecosystems has been estimated at 128 Tg N yr^{-1}, supplying ~15% of the N requirement across all biome types (Galloway et al. 2004). According to this estimate, about 70% of BNF occurs in regions with warmer climates – Africa (25.9 Tg N yr^{-1}), Latin America (26.5 Tg N yr^{-1}) and Asia (21.4 Tg N yr^{-1}). Cleveland et al. (1999) compared global patterns of N fixation in natural ecosystems, and they included Cerrado vegetation in the xeromorphic woodlands category with BNF values of 9–34 kg N ha^{-1} yr^{-1}. The same study presented the N-fixation in tropical and wet savannas ranging from 16 to 44 kg N ha^{-1} yr^{-1}, concluding that this process is the major source of N in these systems.

An early attempt to estimate nitrogen balance in the savannas of northern South America was published by Medina (1982). This review concluded that N losses caused by burning grassland biomass amounted to less than 1% of the total soil N inventory down to 30 cm depth. However, the preliminary balance was strongly negative without including inputs through BNF. A more recent attempt to balance the N budget for the Orinoco savannas was published by Sanhueza and Crutzen (1998). In their budget (summarized in Table 1), the most uncertain numbers are those of N input via biological fixation, particularly in nodulated legumes. Without improved BNF values, as well as output data such as NH_3 volatilization, the calculated N balance ranges from −1.7 to −11.1 kg N ha^{-1} yr^{-1}. These values are small considering that Orinoco savannas contain between 4000 and 5000 kg N ha^{-1} (Medina 1982). However, until the uncertainties in the biological N fluxes are clarified, the ecological significance of these losses on ecosystem function in the long-term remains unclear.

Data on stocks and fluxes of N in a typical cerrado area (approx. 50% woody cover) protected from fire for 28 years compiled from several studies carried out on the same system are shown in Table 2. The total stock of N in the area is 1357 kg ha^{-1} considering the N stocks in soil (= 1116 kg N ha^{-1} in the 0–10 cm depth – this layer is supposed to be more influenced by the 28 years of fire suppression), in the biomass of green leaves of woody species (= 24 kg ha^{-1} – most of the N is allocated in leaves for photosynthesis), in the belowground biomass (= 95 kg ha^{-1} – roots 2 mm in 0–100 cm depth) and in fine litter (= 123 kg ha^{-1}). The inputs range from 20 to 48 kg ha^{-1} yr^{-1} considering 4 kg ha^{-1} yr^{-1} through atmospheric deposition and 16–44 kg ha^{-1} yr^{-1} through N fixation. The outputs in the absence of fire are very low as the N_2O emissions are negligible and NO emissions represent a loss of only 0.4 kg ha^{-1} yr^{-1}. Although data on N fluxes in streams are not yet available, they might also be negligible, as concentrations of NH_4+ and NO_3- in cerrado streams are 5–10 times lower than in streams of the eastern Amazon (Markewitz et al. 2001; Parron 2004). At an input rate of 20–48 kg ha^{-1} yr^{-1}, the time to accumulate a stock of 1357 kg ha^{-1} ranges from 28 to 68 years, which encompasses the time of fire suppression in the area (~28 years).

Table 1. General nitrogen budget for the Orinoco savannas (modified from Sanhueza and Crutzen 1998).

Process	Flux (kg N ha^{-1} yr^{-1})
N fixation	
Blue-green algae	1–2
Azospirillum associations	1.3–8
Nodulated legumes	(10–20) (estimated)
Atmospheric deposition	
Wet deposition	
NO_3	0.65–1.3
NH_4	0.56–3.3
Dry deposition	
$NO_2 + NO_3 + HNO_3$	0.42–0.87
$NH_4 + NH_3$	4.1–9.7
Total input	8–26 (9–46) considering estimated legume contribution
Denitrification	
N_2O	0.18–0.63
N_2	0.36–6.3
Biomass burning	
NO_x	1.2–3.6
NH_3	0.75–2.3
N_2O	0.05–0.15
RCN	0.35–1.1
N_2	6.5–20
NO emissions from soils	0.3–3
NH_3 volatilization	(5.5–8.5) (estimated)
Total output	9.7–37.1 (20–45.3) considering NH_3 volatilization

In temperate savannas of North America, estimates of biological N fixation range from 5 to 35 kg ha^{-1} yr^{-1} (Rundell et al. 1982). With higher N deposition (8–14 kg ha^{-1} yr^{-1}), the total inputs are estimated at 13–49 kg ha^{-1} yr^{-1} (Table 3). In comparison to outputs via N oxide trace gas emissions (0.3–3.4 kg ha^{-1} yr^{-1}; Martin et al. 2003) and low biomass burning fluxes due to widespread fire suppression (Archer 1995), the inputs are enormous and likely decreasing the strength of nitrogen limitation of primary production in many regions. For example, Asner and Martin (2004) used field and published data to show that total plant N requirement for a Texas savanna was about 21 kg N ha^{-1} yr^{-1} on oxic Ultisols with low woody and herbaceous plant cover, well within the range of estimates of N input via deposition and biological fixation. While woody encroachment in these regions increases the demand for nitrogen, soil organic N and N mineralization rates increased significantly, which suggests that N fixation increases with woody encroachment to accommodate N demand (Table 3). At this time, biomass burning

Table 2. General nitrogen budget for a cerrado *sensu stricto* protected from fire for 28 years.

Compartments	Stock (kg ha^{-1})	References
Total N – soil (0–100 cm)	4576	Resende (2001)
Total N – soil (0–10 cm)	1116	Resende (2001)
Aboveground biomass (woody species)	37,787	Silva (1990)
Belowground biomass (0–800 cm) (roots >2 mm)	27,649	Klink et al. (unpublished data)
Belowground biomass (0–100 cm) (roots >2 mm)	25,638	Klink et al. (unpublished data)
Fine litter production (kg ha^{-1} yr^{-1})	2300	Nardoto et al. (2006)
N in the biomass of leaves of woody species	24	Considering that leaves represent 5% of total aboveground biomass (Silva 1990) and that the mean N concentration in the leaves is 12.5 g kg^{-1} (Nardoto 2006)
N in the belowground biomass (0–800 cm) (roots >2 mm)	103	Resende (2001)
N in the belowground biomass (0–100 cm) (roots >2 mm)	95	Resende (2001)
N in the fine litter	123	
Process	Flux (kg ha^{-1} yr^{-1})	
Internal cycling		Nardoto and Bustamante (2003)
N mineralization (0–5 cm)	14	
N fixation	16–44	Cleveland et al. (1999)
Atmospheric deposition (wet + dry deposition)	4	Resende (2001)
Total input	20–48	
NO emissions from soils	0.4	Pinto (2003)

estimates of N losses or redistribution are scarce because fire suppression has been much more common throughout many temperate savannas (Scholes and Archer 1997).

N_2-fixation by Legume–Rhizobia associations

A characteristic feature of South American savannas is the high diversity of herbaceous and woody leguminous species. In fact, many of the forage legume species cultivated today in the tropical world originate from South America (Winter et al. 1989). Most legumes in tropical savannas live in association with native *Rhizobia* strains. In a large survey of legume species of the savannas in Venezuela, Barrios and Gonzalez (1971) found 109 nodulated species out of a total of 127 species investigated. Most of the nodulated species were within the subfamily Papilionoidae (73 species). The frequency of legume species in Venezuelan savannas has been associated with low levels of exchangeable Al and high levels of exchangeable Ca (these two parameters are inversely correlated; Medina and Bilbao 1991). In natural savannas, legumes have leaf

Table 3. General nitrogen budget for temperate mesquite savannas of North America.

Compartment	Stock (kg N ha^{-1}) without woody encroachment	With woody encroachment	References
Soil organic N (0–30 cm)	1900 on clays 2700 on loams	3200 on clays 3100 on loams	Asner and Martin (2004)
Process	Flux (kg N ha^{-1} yr^{-1})		
Internal N cycling N mineralization (0–30 cm)	16.2 (9.6) on clays 117.6 (19.7) on loams	29.6 (8.2) on clays 167.5 (26.9) on loams	Asner and Martin (2004)
	60 (1)	140 (30)–220 (20)	Hibbard et al. (2001)
N fixation			
Legumes	5–35		Rundell et al. (1982)
Atmospheric deposition			
Wet deposition			
NO_3	6–10		NADP (2002)
NH_4	2–4		NADP (2002)
Total input	13–49		
N_2O emissions from soils	<0.5	<0.5	Martin et al. (2003)
NO emissions from soils	0.3 (0.2) on clays 2.4 (0.4) on loams	0.6 (0.2) on clays 3.4 (0.5) on loams	Martin et al. (2003)

nitrogen concentrations above the critical level for normal growth; however, the levels of P and of K are clearly deficient. This nutrient status may lead to a decrease in plant growth due to the inhibition of effective nodulation (Medina and Bilbao 1991). Although the density and diversity of legume species in Cerrado areas are high (Felfili et al. 1992), there are very few reports about the activity of nodules in legume species occurring in this region (Leitão 1997). Low nodule activity was measured in Cerrado legume species of southeast Brazil (Faria et al. 1984), while another series of studies stressed the nodulation deficiency of the species with such capabilities (Campelo 1976; Döbereiner and Campelo 1977; Magalhães et al. 1982). Besides the need for N fixation by the plant, the magnitude of nodulation depends on the density of bacteria populations, and on the physical and chemical properties of soil. Factors such as P and Ca limitation, high soil acidity, Al saturation, and seasonal water stress may constrain BNF in tropical savannas of the Llanos and Cerrado. In the Chaco, Mazzarino et al. (1991a, b) worked with the dominant tree species *Prosopis flexuosa* DC (Fabaceae) and *Aspidosperma quebracho-blanco* Schlecht (Apocinaceae), and pointed out that N availability in soil under the legume species is higher than under the non-legume species as a consequence of the legume capacity to biologically fix nitrogen.

Very little is known about the spatial and temporal dynamics of BNF in North American savannas. The geographic extent of the major leguminous genera (e.g., *Prosopis*, *Acacia*) have been determined by Johnson and Mayeux (1990) who observed a large percentage of surveyed *Prosopis glandulosa* plants with nodules, with maximum occurrences in southern Texas and New Mexico.

Rates of BNF have been estimated in *Prosopis*, *Acacia* and other genera in Southern Texas (Zitzer et al. 1996; Archer et al. 2001). Based on Zitzer et al. (1996), we estimate BNF rates of up to 25 kg N ha^{-1} yr^{-1}. However, they emphasized that substantial variations are caused by water availability and the degree of woody cluster development, both of which affect soil N accumulation rates and total stocks. Rundell et al. (1982) estimated BNF among *Prosopis* plants at about 6–35 kg N ha^{-1} yr^{-1}. Like tropical systems, estimates of the temporal and spatial variability of BNF must be improved if the N balance of temperate savannas is to be understood.

Despite the few studies directly documenting nitrogen fixation by native legumes under natural conditions in South American savannas, there is some indirect evidence suggesting N fixation for a few species. These indirect evidences are based on the natural abundance of ^{15}N and the concentration of ureids and α-amino compounds in the xylem sap of young shoots (Medina and Bilbao 1991; Izaguirre-Mayoral et al. 1992; Sicardi de Mayorca and Izaguirre-Mayoral 1993).

The relative abundance of ureid (RAU%) was used by Izaguirre-Mayoral et al. (1992) as an index for nitrogen fixation capacity of herbaceous and suffrutescent legume species in central Venezuela. This index is based on the assumption that ureids are the form in which fixed nitrogen is exported from the nodules in the majority of tropical legumes. Soluble nitrogen compounds are extracted with alcohol from young stems and the total N content and its distribution in inorganic-N, α-amino-N and ureid-N is determined. RAU values above 60% are recorded in species with high nitrogen fixation activity, while values below 40% are indicative of low N_2 fixation activity. In Figure 2, the results of Sicardi de Mayorca and Izaguirre-Mayoral (1993) have been drawn to show the differences in N_2-fixation capacity in species separated in subfamilies, and measured during the peak of the rainy and dry seasons. The Faboideae contains most of the species studied. Besides, the best N_2-fixers are also within this subfamily. Drought strongly affects N_2-fixation capacity, as expressed by the reduction in RAU.

RAU values indicate short term N_2-fixation activity. To evaluate long-term fixation, the use of the natural abundance of ^{15}N, expressed as $\delta^{150}N$ in , appears to be the method of choice. The rationale is that within the same environmental conditions, N_2-fixing species have more negative values than non-fixing species. A few papers have reported large variations in $\delta^{15}N$ in species of South American savannas (Medina and Bilbao 1991; Sprent et al. 1996; Bustamante et al. 2004a; Medina and Izaguirre 2004). These papers show that, on average, the legumes have lower $\delta^{15}N$ (1–4) and higher N concentrations (~1%) than non-legumes. However, trees are consistently more negative than shrubs, irrespective of plant family (Bustamante et al. 2004a). Soils both in central Brazilian Cerrados and central Venezuelan savannas have strong positive $\delta^{15}N$ values (>3), and for the Venezuelan data set (Medina and Izaguirre 2004), the plant–soil difference of $\delta^{15}N$ values ranges from −4 to −6 for legumes and −0.4 to −2.1 for non-legumes. The Brazilian study

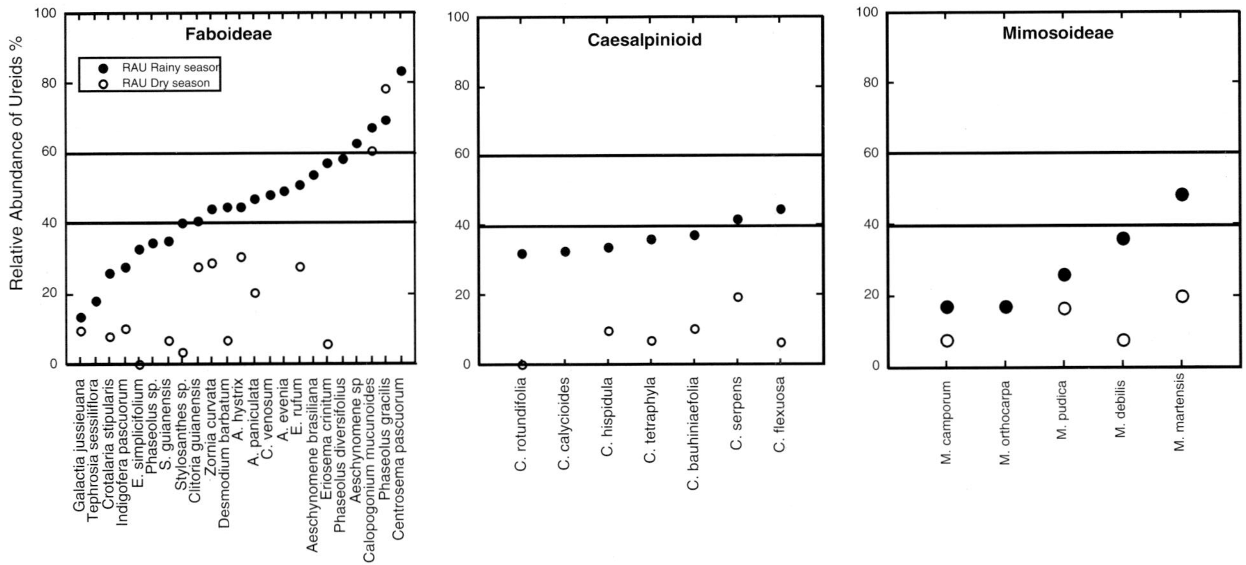

Arrangement of legumes species according to subfamily and Nitrogen Fixation Capacity (NFC) measured by the Relative SAbundance of Ureids (RAU%)
(Modified from Sicardi de Mallorca and Izaguirre-Mayoral 1993)

NFC	RAU %
HIGH	>60
INTERMEDIATE	40-60
LOW	<40

Figure 2. Arrangement of legumes species according to subfamily and NFC measured by the RAU% (modified from Sicardi de Mallorca and Izaguirre-Mayoral 1993).

(Bustamante et al. 2004a) reports strongly positive soil $\delta^{15}N$ values (>10) at depths of 1–2 m. It seems clear that most legumes, particularly those of the subfamily Faboideae, are active N_2-fixers under natural conditions but to have a reliable assessment of the contribution of legumes to the N budget of savannas it is necessary to understand controls over legume density and seasonal nitrogen fixation activity.

N_2-fixation by free-living microorganisms

Cyanobacteria can make important contributions to soil N in most humid savannas. In the *Trachypogon*-savannas of Central Venezuela, with 10–12% cyanobacterial coverage, N_2-fixation rates during the rainy season varied between 0.4 and 0.7 mg N m^{-2} day^{-1} (Santaella 1985). According to Chacón et al. (1991), these N-fixation rates would be enough to compensate potential N losses that can occur through fire.

Free-living bacteria in the rhizosphere of grass roots have been reported to fix significant amounts of atmospheric nitrogen (Boddey and Döbereiner 1995). The reports are somewhat inconsistent, and disagreement persists regarding the significance of this activity for the nitrogen budget of natural savannas. van Berkum and Day (1980) reported rates ranging from 14.7 to 51.4 g N ha^{-1} day^{-1} in field grown Brazilian grasses. It seems more likely that rhizospheric nitrogen fixation can occur as a slow but continuous process contributing small amounts of organic nitrogen to savanna ecosystems. These amounts, however, could become significant in areas where plant productivity is severely limited by nitrogen availability (Giller and Day 1985) or where legumes are not numerous (Abbadie et al. 1992). Nonsymbiotic nitrogen fixation has been reported to be a significant contribution to the nitrogen budget in savannas from West African. For these savannas, Robertson and Rosswall (1986) have reported N fixation from free-living bacteria of about 12 kg ha^{-1} yr^{-1}. Abril and Bucher (1999) also concluded that nitrogen fixation by free-living microorganisms is an important process in the Chaco, particularly in degraded areas. They measured changes in soil characteristics, nutrient availability and microbial activity in a gradient of grazing intensity (highly restored, moderately restored and highly degraded). Nitrogen fixation was more intense at the moderately restored site (no grazing for 8 years), followed by the highly degraded site (overgrazing), with the lowest values at the highly restored site (no grazing for 20 years). The authors suggested that high nutrient availability (particularly phosphorous) in the Chaco allows BNF (by both free-living and symbiotic organisms) and organic matter production to shape vegetation structure and function, including response to over-grazing. The Western Chaco soil nutrient contents are relatively high when compared to other savannas of the world (Table 4), most of which are located in highly weathered soils of much older origin (Huntley and Walker 1982). Effects of P limitation on BNF are discussed further in this paper. Additionally, the comparison of soil properties of the different savannas indicates that soil organic N is also relatively

Table 4. Comparison of organic carbon content, soil total nitrogen, C/N ratio and available in different savanna regions.

Ecosystem	Organic carbon %	Total nitrogen %	C/N	Extractable P (μg g^{-1})	References
Chaco (restored site)	4.7	0.28	16.8	52.5	Abril and Bucher (1999)
Llanos	1.2	0.07	17.1	2.3	Medina (1982)
Brazilian Cerrado	3.2	0.17	18.8	0.2	Resende (2001)
South African Savanna	4.1	0.18	22.8	32	Scholes and Walker (1993)
Australian Savannas	1.7	0.08	21.3	10	McKeon et al. (1991)
North America Mesquite Savannas	1.2	0.18	6.7		Martin et al. (2003)

high in mesquite savannas of Texas relative to the organic C these soils contain (Tables 4 and 5).

Mineralization/nitrification

In addition to N losses from frequent fires and environmental limitations on biological fixation, high C:N of litter (~60:1) may contribute to low rates of decomposition and mineralization in Cerrado areas, thus maintaining low N availability. The decomposition rate of the litter in a cerrado site was estimated at 2.2 years (Resende 2001). The mean biomass loss was 32% after about 1 year, and immobilization was observed for N, P and S, as their losses of these elements were lower than the biomass loss.

Nutrient use efficiency (NUE) can be estimated as the annual increment of biomass divided by the concentration of nutrients in the litter. A low NUE is usually associated with relatively high soil fertility, whereas a high efficiency may be linked to higher resorption in living biomass (Vitousek 1982). Compared to *Terra-firme* forests, cerrado sites showed higher efficiency for N, suggesting a greater limitation to productivity by nitrogen (Bustamante et al. 2004b).

The rate of net N mineralization in a cerrado site protected from fire was 14.7 N kg ha^{-1} yr^{-1} (Table 5), while in a burned cerrado site it was only 3.8 kg N ha^{-1} yr^{-1}, 1 year after the burning (Nardoto and Bustamante 2003). NH_4-N increased after fire, but no significant changes were observed for NO_3-N. NO_3-N accumulation occurred in short periods during the rainy season in both sites. Data from the same areas after a fire event 2 years later indicated a mineralization of 14.0 kg ha^{-1} yr^{-1} in the unburned cerrado and of 8.9 kg ha^{-1} yr^{-1} in the burned area (Viana 2002). The rates of N mineralization were lower in savannas sites with lower woody density (campo sujo) in the same location. Siqueira (2001) measured rates of 6.6 kg ha^{-1} yr^{-1} in an unburned campo sujo and of 3.8 kg ha^{-1} yr^{-1} in a burned campo sujo.

Table 5. Comparison of nitrogen stocks and fluxes in Cerrado, Llanos and Mesquite savannas of North and South America.

Compartments and fluxes	Cerrado[a]	Llanos[b]	Mesquite[c]
Major stocks ($kg\ ha^{-1}$)			
Total soil N (0–100 cm)	4576	5850 [i]	1900–3200
		4420[b] (0–40 cm)	
Aboveground biomass N	54.7[f]	23.7[b]–80.0 [i]	30–112
Belowground biomass N (0–100 cm)	94.5	14.6 [b] (0–30 cm) –40.0[i]	
Major fluxes ($kg\ ha^{-1}\ yr^{-1}$)			
N mineralization	14.0		60–140
Atmospheric N deposition	4.2	5–14	8–14
BNF	16–44	12–28	6–35
N gas emissions ($NO+N_2O$)	0.4	0.5–4.0	1.1–3.4
Biomass burning N losses	25.26[g]	8–27	
Other			
% Woody cover	20–50%		42% (s.d. = 12%)
% Herbaceous cover			85% (s.d. = 15%)
Woody plant foliar N (%)	0.7–1.8[e]		3.2–4.1
Herbaceous plant foliar N (%)	0.5–0.6[d]		1.1–1.5
Woody plant root N (%) (0–100 cm)	0.65–0.36 (< 2 mm)[h]		
	0.75–0.48 (2–5 mm)		
	0.49–0.20 (> 20 mm)		
Herbaceous plant root N (%)			

[a]Cerrado sensu stricto located in IBGE Reserve, Brasilia, Brazil; [b]*Trachypogon* savannas in Central Venezuela (Medina 1982); [c]Mesquite savannas located throughout Texas, USA (Archer 1989; Asner et al. 2003); [d]Siqueira (2002); [e]Nardoto et al. (2006); [f]Kauffmann et al. (1994) – all living and dead vegetation less than 2 m in height above the soil surface excluding trunks of large trees and shrubs. Grasses comprised 17% of the total N; [g]Kauffmann et al. (1994) – 20.54 kg ha^{-1} lost as particulates, 4.72 kg ha^{-1} lost through volatilization; [h]Nardoto, Bustamante et al., unpublished data; [i] *Axonopus purpusii-Leptocoryphium lanatum* (Barinas Venezuela) (Sarmiento 1984).

Only a small part (~5 kg ha^{-1} yr^{-1}) of the nitrogen content of the grass layer in an African savanna enters through mineralization of the soil organic matter. Most of the N comes from the recycling of the nitrogen stock in dead roots, before humification (~40 kg ha^{-1} yr^{-1}) as the annual bush fire destroys the major part of stems and leaves (Abbadie et al. 1992). This indicates that nitrogen is mostly internally recycled within the vegetation–soil system.

In the cerrado sites the rates of net N mineralization increased during the rainy season while reductions in soil microbial biomass were observed at burned and unburned sites. This suggested a peak of microbial activity with the onset of the rainy season and initial net immobilization followed by net mineralization (Nardoto and Bustamante 2003). This microbial biomass dynamic leads to the accumulation and conservation of nutrients in a biologically active form during the dry period when the activity of the plants is low. At the beginning of the rainy period, nutrients are released and taken up by plants (Singh et al. 1989). In the Cerrado, although net N mineralization and net nitrification rates were highest during the rainy season, inorganic-N concentrations decreased during the rainy season, indicating that plant uptake draws down inorganic-N stocks and consumes the N released from net mineralization (Nardoto and Bustamante 2003).

Despite the transient nature of nitrate in soils, the low content of NO_3-N in soils of the Cerrado sites could be related to low nitrification rates, even though significant production of NH_4-N occurs. Competition between plants and microorganisms for ammonium is probably intense. Hence, the population of nitrifying bacteria may be low in these soils. In soils of the Mediterranean zone and under eucalyptus forests of Australia, similar results were also explained by the low density of nitrifiers found in those soils, and no increase in nitrification after fire was detected (Prieto-Fernandez et al. 1993). Verchot et al. (1999) reported low rates for potential nitrification in a Cerrado area but Poth et al. (1995) detected chemoautotrophic nitrifier populations in Cerrado soils. Nitrifying bacteria could be present but rates of net nitrification and nitrate accumulation are modest. Plants and microorganisms could rapidly take up small increases in NO_3^- production in soil. This is a further indication of N limitation in ecosystems (Davidson et al. 1992).

Similar to the Cerrado, inorganic-N in semi-arid regions of Argentina (Chaco) and the United States (Texas) accumulated in the soil during the dry season, while net N mineralization rates decreased in the same period and increased in the following rainy season (Oliva et al. 1993; Hibbard et al. 2003). Higher microbial immobilization also occurred at the onset of the wet season. In the selective logging sites studied by Oliva et al. (1993), higher N mineralization rates and consequently higher N availability occur as a consequence of the accumulation of dead plant material and increase of light and water availability (Oliva et al. 1993). Increases in N mineralization rates after selective logging are due to increases in soil temperature (high radiation), input of new leaves with lower C:N ratios and reduction in competition for available nutrients between plants and microorganisms.

The effects of different fire frequencies on carbon and nitrogen dynamics of a savanna of the 'Chaco Semiárido Occidental' were compared for a single point in time by Gonzalez et al. (2001). Soil was described as a Torriorthentic Haplustoll. Fire frequencies were characterized as: high (one fire every year), medium or normal (one fire every 3–4 years), and low (one fire every 10 years). Organic carbon was lower under the high frequency than under the medium and low frequency areas. Total soil organic carbon, total nitrogen, soil nitrate and soil microbial biomass nitrogen were lower under high frequency fires, although medium and low frequency fires did not differ. Particulate organic carbon and nitrogen were also lower under the high frequency, and were more sensitive than total carbon and nitrogen to the effects of fire history. There were no differences in soil respiration among fire histories. On the other hand, soil respiration and microorganisms were strongly affected in soil samples taken immediately after fire and after periods of 30, 180, 360, and 720 days in a native forest and a grassland in the Chaco region (Chancani Forest Reserve of Cordoba Province) with burned and unburned patches. The conditions observed were stable over the 2 years of study, nevertheless, there was a tendency to recover the original values (Gonzalez et al. 1999). These results showed that repeated burns reduce soil organic matter as well as soil biological activity and may increase soil susceptibility to erosion processes. However, areas under medium or normal fire frequency maintained organic nitrogen and carbon pools and showed higher nitrogen availability.

Beyond the effects of selective logging and fires, the N cycling in the Chaco is markedly influenced by vegetation composition. Soil respiration and soil nitrogen dynamics were measured for 1 year underneath five vegetation types in the Chaco region: a leguminous tree (*Prosopis flexuosa*), a non-leguminous tree (*Aspidosperma quebracho-blanco*), a non-leguminous shrub (*Larrea* spp.), open interspaces, and a pure grassland (Mazzarino et al. 1991a, b). As stated earlier, during the dry season, microbial biomass N and net N mineralization were low, while accretion of easily mineralizable C occurred, but this was reversed with the onset of rain. The highest values of total N, N mineralization, inorganic N, microbial biomass N, density of nitrifiers, N content in litter, total organic C and easily mineralizable C were found under *Prosopis* and the lowest values under shrubs and the interspaces. The main differences between tree species were in N mineralization at the beginning of the wet season, in total and inorganic N pools, and in nitrifier densities; all of which were significantly lower under *Aspidosperma* than under *Prosopis*. N mineralization in the pure grassland was very low despite high values of total N and C sources.

Emissions of NO and N_2O

Nitrification and denitrification are key processes in the production of nitrogen oxides, NO and N_2O (Firestone and Davidson 1989), and low emissions are generally correlated with the dominance of NH_4^+ over NO_3^- (Davidson et al. 2000).

In a recent review, Davidson et al. (2001) concluded that soil emissions of NO from tropical savannas such as the Brazilian cerrado remain a large and important uncertainty. Poth et al. (1995) suggested that Cerrado, burned or unburned, is an important source of NO to the troposphere. However, the high NO production observed by these authors was reported after an artificial water addition in a short-term experiment. Pinto et al. (2002) measured monthly soil fluxes of NO and N_2O in cerrado areas of central Brazil during a year. The study focused on two vegetation types, cerrado *stricto sensu* (20–50% canopy cover) and *campo sujo* (open, grass-dominated), which were either burned every 2 years or protected from fire. N_2O fluxes were very low and below the detection limit in all of the vegetation-fire treatments. Soil moisture and vegetation type were more important in controlling NO fluxes than fire regime (early-dry season, middle-dry season or late-dry season burning). NO emissions increased after burning, but flux returned quickly to pre-fire levels and even lower. In comparison, NO emissions increased 100-fold (to 10.5 ng NO-N cm^{-2} h^{-1}) during a water addition experiment in unburned campo sujo. After the first rains NO fluxes increased to 1.0 ng NO-N cm^{-2} h^{-1} in unburned cerrado and to 1.9 ng NO-N cm^{-2} h^{-1} in burned cerrado. These results demonstrate that wetting of dry soil in the Cerrado causes an increase in NO emissions of a factor of 10 or more, but that the pulse is short lived and does not persist during the rainy season. The annual emissions of N via NO fluxes were similar in the burned campo sujo (0.5 kg N ha^{-1} yr^{-1}) and in the unburned and burned cerrado areas (0.4 kg N ha^{-1} yr^{-1}) while the unburned campo sujo fluxes were the lowest (0.1 kg N ha^{-1} yr^{-1}). A positive relationship between NO emission and soil nitrogen status had been shown for South America savanna soils (Johansson and Sanhueza 1988; Sanhueza et al. 1990; Rondón et al. 1993; Cárdenas et al. 1993) and for savannas of South Africa (Parsons et al. 1996; Levine et al. 1996). However positive correlations between NO emissions and mineralization and nitrification were not found for the Cerrado, although both NO emissions and nitrification rates were low (Pinto et al. 2002).

Similar to Cerrado systems, mesquite savannas of Texas, USA have very low N_2O emissions, but NO fluxes can be relatively high (Table 5), with values ranging from 0.3 kg N ha^{-1} yr^{-1} on dystrophic clayey soils to 2.4 kg N ha^{-1} yr^{-1} on loamy soils (Martin et al. 2003). Many of these savanna-grassland systems in Texas have undergone woody encroachment in the past century or more (Archer 1994), and this has caused substantial increases in NO emissions at both the local and regional scales. Locally, NO fluxes increase by about 200% following increases in woody mesquite cover (Martin et al. 2003). Martin and Asner (2004) used field and remote sensing data to show that the contemporary N cycle of North Texas savannas has a regional NO flux of about 1.6 kg NO-N ha^{-1} yr^{-1}, a value that has probably increased more than threefold in the past 100 years with woody encroachment.

Interactions between N and other elements controlling the N cycle

The N/P ratio in leaves can be used as an indicator of nutrient limitation where N/P < 16 indicates N limitation, values $14 < NP < 16$ co-limitation by N and P and N/P > 16 P limitation (Aerts and Chapin 2000). The mean N/P in leaves of 10 Cerrado woody species (evergreen and deciduous) was 18 indicating a strong P limitation, and although the N resorption rates were similar to the values (30%) presented by Aerts and Chapin (2000), the P resorption rates were much higher ($> 50\%$) Nardoto et al. (2006). The P supply for these woody species seems to be more dependent on biochemical recycling (resorption), whereas N supply is more dependent on N uptake (biogeochemical cycling). Although N and P uptake by plants is low due to the low availability, P limitation is stronger. For example, in a Cerrado area, the flux of N that reaches the soil (bulk precipitation, throughfall and litter leachate) was 9.3 kg ha^{-1} yr^{-1}, whereas for P it was only 0.03 kg ha^{-1} yr^{-1} (Resende 2001).

BNF can, in the long-term, adjust the supply of N close to the availability of other resources, such as P (Schimel et al. 1997). The importance of P availability controlling the N fixation rates is discussed in several studies. In the bacteria, P seems to activate the genes for the nitrogenase synthesis (Stock et al. 1990), and in many cases, the N fixation rates seem to be controlled by soil P availability. As P availability is generally low in the highly weathered soils of the Cerrado (Goedert 1983; Le Mare et al. 1987), this deficiency might be one of the factors limiting the rate of nitrogen fixation in tropical legumes. Data on soybeans indicated that the symbiotic N_2 fixation process has a higher requirement for P than host plant growth (Israel 1987). It is also suggested that symbiotic N_2 fixation has a higher P requirement than the nitrate-assimilation process. These results are consistent with the interpretation that plants solely dependent upon BNF have a higher internal P requirement for optimal growth than plants supplied with nitrate. It can be concluded that P has specific roles in nodule initiation, growth and functioning in addition to its role in host plant growth. Indeed, the whole plant N concentration was significantly greater for plants solely dependent upon BNF than for nitrate-supplied plants when high P concentrations were present in the nutrient solution.

Phosphorus shows low mobility in soil, and the diffusion rate strongly limits its availability to the roots. Mycorrhizal associations can act as important mediators for the transfer of nutrients, and 12 species sampled in a Venezuelan savanna on nutrient-poor soil and dominated by Gramineae and Cyperaceae had arbuscular mycorrhizae (Cuenca and Lovera 1992). Andrade et al. (1996) found in the Venezuelan Gran Sabana colonization by arbuscular mycorrhizae in 18 native species distributed in 7 families. Even in Cyperaceae, generally considered non-mycotrophic, mycorrhizal colonization was observed, demonstrating the importance of these associations in dystrophic soils. Surveys in Cerrado soils indicated that infections of arbuscular mycorrhizae also occur in a large number of native plants (Thomazini 1974; Bononi and Trufem 1983; Siqueira et al. 1989; Miranda and Miranda 1996).

Disturbances/land use changes

Conversion to pastures

The major land-use changes in the Cerrado are conversion to pasture and soybean croplands, but the ecological consequences of these land uses are still poorly understood in Cerrado ecosystems. The introduction of African grasses of the genus *Brachiaria* in the 1970's led to widespread land conversion (Kichel et al. 1996). The region hosts approximately 44% of the national herd, with approximately 50 million hectares of the Brazilian Cerrado planted with African grasses (mainly *Brachiaria* and *Andropogon* spp). Most of the pastures, however, are in an advanced stage of degradation. Generally pure *Brachiaria* or *Andropogon* pastures in the Cerrado suffer a decline in productivity after 4–10 years of grazing (Macedo 1995). The nutrient-poor organic matter produced (6–10 Mg C ha^{-1} yr^{-1} in well-managed pastures; Cadisch et al. 1994a, b) may decrease the sustainability of these pastures, because the scarce nutrients are immobilized and not available for plant growth. The existence of a dense and permanent rooting system within the pasture limits nutrient losses by leaching. The main loss pathways are urine and dung patches in which nutrient concentrations can be very high over a limited area (Boddey et al. 1996).

Studies indicated that BNF could introduce 30–45 kg N ha^{-1} yr^{-1} in *Brachiaria* pastures (Boddey and Victoria 1986; Loureiro and Boddey 1988). This N comes mainly from non-symbiotic associations with endophytic bacteria or bacteria in the rhizophere. Several bacteria such as *Azospirillum*, *Herbaspirillum* and *Gluconacetobacter* can be associated with grasses.

Modifications of soil organic matter and nutrient cycling after land conversion can change the magnitude and direction of NO and N_2O fluxes. The few existing data indicate that the annual emissions of nitrogen oxides are low. Saminêz (1999) measuring soil N_2O fluxes estimated the annual emission in a native cerrado area of 0.52 and of 0.51 kg N ha^{-1} yr^{-1} in a 5-year old pasture (*Andropogon gayannus*). In a 10-year old pasture (*Paspalum* sp.), Nobre (1994) measured fluxes of 0.2 ng N-N_2O cm^{-2} h^{-1}. These authors did not measure NO fluxes, but Varella et al. (2004) compared fluxes of NO and N_2O, soil microbial biomass, and N-mineralization rates in a 20-year old *Brachiaria* pasture of low productivity and in a native cerrado area. Net N-mineralization in the pasture was low, and net N-immobilization was observed in the dry season. Artificial water addition in pastures during the dry season resulted in short-lived pulses of NO (1.3 ng N-NO cm^{-2} h^{-1}). N_2O fluxes were below the detection limit at both sites. Recently, Pinto et al. (2006) compared the fluxes of N oxides in degraded pastures subjected to recovery treatments (fertilization and consortium with the legume *Stylosanthes guianensis*) and a young pasture

(2 years old – formerly a dense cerrado). Considering the measurements between January and April (rainy season), the young pasture emitted 0.03 kg N-NO ha^{-1} versus 0.01 kg N-NO ha^{-1} for the other treatments. In the same period, a cerrado emitted 0.11 kg N-NO ha^{-1}, while a campo sujo showed similar values to the old pastures. Higher fluxes of N_2O (9.5 ng N cm^{-2} h^{-1}) were only measured in the young pasture during the transition from dry to wet seasons.

Conversion to croplands

Soy plantations in Brazil began to expand more vigorously in the second half of the seventies, propelled by expansion in international demand. Expansion affected mainly the states located in the southern region of the country. In 1980, soy had not significantly penetrated the savannas, only about 15% of the soy planted areas in Brazil were outside the southern states boundaries. Ten years later, however, soy plantations formed a continuous zone in central Brazil, largely associated with the expansion of soy in the savannas. The impacts on the regional and global environment of this massive change in land use in only two decades have not been fully assessed.

The expansion of soy was strongly influenced by the existing natural conditions in the savannas, by investments in transport infrastructure and, especially by the development of soybean varieties adapted to the region and highly efficient in N fixation. Today soy is cultivated in the Cerrado (ca. 7 million hectares are used for grain production) without the addition of N fertilizers. It is estimated that 70–85% of the nitrogen in the soy plants comes from BNF (Boddey et al. 1984, 1990). At least 80 kg N (grain + vegetative parts) are necessary for the production of one ton of grain with 6.5% of N. Considering an average production of 2571 kg ha^{-1}, approximately 200 kg N ha^{-1} are needed, and consequently BNF must contribute ~170 kg N ha^{-1}, which is equivalent to 378 kg of urea ha^{-1} (Reis et al. 2003)!

Information on the effects of this input of N in the savanna ecosystems is still very limited. The data now available indicate that the conversion of natural systems to agricultural uses may significantly increase the N_2O emissions. Nobre (1994) has shown that the N_2O emissions have increased during the first 100 days of soy implementation, reaching 0.5 ng N cm^{-2} h^{-1} presumably as a result of the increased N fixation by the cultures. In another study on cultivation of soy in rotation with corn, the emissions during the rainy season varied from 0.8 to 2.5 ng N cm^{-2} hr^{-1} (Saminêz 1999). The relatively dry weather of the Cerrado does not favor high emissions of N_2O, but the increasing use of irrigation can significantly enhance the N_2O emissions.

Another important aspect of soy cultivation in the Cerrado is the link between organic matter management and N cycling. Currently, no-tillage systems are used in about 70% of the soybean cultivated area versus 30% that still use conventional tillage systems. Studies of N cycling impacts associated to soy cultivation should include a survey of the different forms of agricultural management, particularly under no-tillage systems, and their regional distribution and representation.

Over-grazing and fire suppression

In North and South America, Africa, Australia and elsewhere, woody vegetation cover has increased significantly in grazed savannas during the past few decades. Cited causes of woody encroachment include over-grazing of herbaceous cover that reduces competition for woody seedlings, fire suppression that enhances woody plant survival, atmospheric CO_2 enrichment that favors C3 (woody) plant growth, and nitrogen pollution which also favors woody encroachment (Archer et al. 1995). It is noted in most encroachment studies that the woody plants were present somewhere on the landscape prior to the installment of managed grazing. For example, in a South Texas rangeland containing a diverse array of trees, shrubs and sub-shrubs, heavy grazing caused increases in the cover of the nitrogen-fixing tree *Prosopis glandulosa* (mesquite). The same species of mesquite has increased dramatically in cover in a North Texas rangeland during the past century (Asner et al. 2003), but there are very few other woody species established in this region. Most other species are confined to riparian zones, thus few if any woody plants can be found in association with the mesquite cover. Precipitation conditions are similar between the North (650 mm) and South (680 mm) Texas sites, but temperatures are substantially lower in the North, with temperatures below freezing in many months (Asner et al. 2003). Low temperatures in the North likely preclude the presence of many warm-climate woody plants found in the South (*Acacia*, *Diospyros* spp.), and thus the biological and ecological dynamics of woody encroachment are very different between the sites.

Although woody encroachment reduces the quality of land for animal production, in some cases, it enriches total ecosystem C and N stocks (Asner et al. 2004). With the shift to woody vegetation comes a large increase in aboveground NPP and C storage. Increases in aboveground NPP of up to 1400 kg C ha^{-1} have been observed when the dominant woody species is a nitrogen fixer (Geesing et al. 2000). Increases in the aboveground C pool can range from 300 to 44,000 kg C ha^{-1} in less than 100 years of woody encroachment (Asner et al. 2003). When the dominant woody species is a N-fixer, nitrogen accumulation can be 9–40 kg N ha^{-1} yr^{-1} greater in the woody areas than the grasslands (Geesing et al. 2000), with aboveground nitrogen increasing 39–468 kg N ha^{-1} following woody encroachment (Hughes et al. in review).

Conclusions

American tropical and temperate savannas are limited by nitrogen availability. However, land use changes are affecting the savanna N cycles differently in these regions. Grazing and conversion to pasture has been the predominant land use change in most savannas including the Cerrados, although an intensification of mechanized agriculture and irrigation has been observed in the last decade. In the case of the Cerrado and the Llanos, the intensification of fire for pasture management is causing the reduction of woody plant density. Oppositely, in the Chaco and North American savannas, exclusion of fire and grazing are leading to an increase in woody density. In addition, the higher soil P availability in the Gran Chaco and the higher N deposition in North American savannas are resulting in an increase of N cycling and net productivity rates. Besides the impacts of land use changes, some aspects of the N budget for savannas of the American continent are still unclear and require increased analysis of the N fixation rates (both free-living and symbiotic fixers), the spatial and temporal soil heterogeneity controlling N fluxes through soil solution, and the role of riparian zones on N fluxes through streams. Changes in vegetation structure may also have impacts on climate. Hoffmann and Jackson (2000) proposed that the conversion of tropical savannas to grasslands would decrease precipitation but also change the seasonal patterns of rainfall. Biogeochemical cycles in savannas are regulated by seasonal distribution of precipitation and pulses of nutrients resulting from wetting of dry soil which are relevant to maintain the dynamics between microorganism and vegetation. The alteration of seasonal patterns would decouple "supply and demand" processes resulting in higher N losses.

Acknowledgements

We would like to thank Dr. Luiz Martinelli (CENA, University of São Paulo) and Dr. Bob Howarth (Cornell University) for the organization of the workshop "N fluxes and processes in tropical and temperate systems" (Ubatuba, 15–18th March 2003) that motivated the preparation of this manuscript. We are grateful to Dr. Eric Davidson (Woods Hole Research Center), Dr. Daniel Markewitz (University of Georgia) and the reviewers for their very helpful comments and suggestions. G. Asner and M. Bustamante were supported by NASA LBA grant NCC5–675 (LC-21).

References

Abbadie L., Mariotti A. and Menaut J. 1992. Independence of savanna grasses from soil organic matter for their nitrogen supply. Ecology 73(2): 608–613.

Abril A. and Bucher E.H. 1999. The effects of overgrazing on soil microbial community and fertility in the Chaco dry savannas of Argentina. Appl. Soil Ecol. 12: 159–167.

Aerts R. and Chapin F.S. 2000. The mineral nutrition of wild plants revisited: a re-evaluation of processes and patterns. Adv. Ecol. Res. 30: 1–67.

Andrade Z., Cuenca G. and Escalante G. 1996. Preliminary studies on mycorrhizal status and morphology of arbuscular mycorrhizae of some native plant species from La Gran Sabana,

Venezuela. In: Azcon-Aguilar C. and J.M. Barea (eds), Proceedings of the Fourth European Symposium on Mycorrhizas, European Commission, Brussel, pp. 67–70.

Archer S. 1989. Have Southern Texas savannas been converted to woodlands n recent history? Am. Naturalist 134: 545–561.

Archer S. 1994. Woody plant encroachment into southwestern grasslands and savannas: rates, patterns, and proximate causes. In: Vavra M., Laycock W.A. and Pieper R.D. (eds), Ecological Implications of Livestock Herbivory in the West, Society for Range Management, Denver, USA, pp. 13–68.

Archer S. 1995. Tree-grass dynamics in a *Prosopis*-thornscrub savanna parkland: reconstructing the past and predicting the future. Ecoscience 2: 83–99.

Archer S., Boutton T.W. and Hibbard K.A. 2001. Trees in grasslands: biogeochemical consequences of woody plant expansion. In: Schimel D.S. (ed.), Global Biogeochemical Cycles in the Climate System, Academic Press, San Diego.

Asner G.P. and Martin R.E. 2004. Biogeochemistry of desertification and woody encroachment in grazing systems. In: Defries R., Asner G.P. and Houghton R.A. (eds), Ecosystem Interactions with Land Use Change, American Geophysical Union, Washington, DC.

Asner G.P., Townsend A.R., Riley W., Matson P.A., Neff J.C. and Cleveland C.C. 2001. Physical and biogeochemical controls of terrestrial ecosystems responses to nitrogen deposition. Biogeochemistry 54: 1–39.

Asner G.P., Archer S.A., Hughes R.F., Ansley J.N. and Wessman C.A. 2003. Net changes in regional woody vegetation cover and carbon storage in North Texas rangelands, 1937–1999. Glob. Change Biol. 9: 316–335.

Asner G.P., Elmore A.J., Olander L.P., Martin R.E. and Harris A.T. 2004. Grazing systems, ecosystem responses, and global change. Ann. Rev. Environ. Res. 29: 261–299.

Bailey R.G. 1996. Ecoregions: The Ecosystem Geography of the Oceans and Continents. Springer Verlag, New York, USA.

Barrios S. and Gonzalez V. 1971. Rhizobial symbiosis in Venezuelan savannas. Plant Soil 34: 707–719.

Batmanian G.J. (1983) Efeitos do fogo sobre a produção primária e a acumulação de nutrientes do estrato rasteiro de um Cerrado. MSc. Thesis. Universidade de Brasília, Brasília, Brazil.

van Berkum P. and Day J.M. 1980. Nitrogenase activity associated with soil cores of grasses in Brazil. Soil Biol. Biochem. 12: 137–140.

Boddey R.M., Chalk P.M., Victoria R.L. and Matsui E. 1984. Nitrogen-fixation by nodulated soybean ander tropical field conditions estimated by the N-15 isotope dilution technique. Soil Biol. Biochem. 16(6): 583–588.

Boddey R.M., Urquiaga S., Neves M.C.P., Suhet A.R. and Peres J.R. 1990. Quantification of the contribution of N_2-fixation to field-grown grain legumes – a strategy for the practical application of the N15 isotope dilution technique. Soil Biol. Biochem. 22: 649–655.

Boddey R.M. and Döbereiner J 1995. Nitrogen fixation associated with grasses and cereals: recent progress and perspectives for the future. Fert. Res. 42: 241–250.

Boddey R.M. and Victoria R.L. 1986. Estimation of biological nitrogen fixation associated with *Brachiaria* and *Paspalum* grasses using ^{15}N labeled organic matter and fertilizer. Plant Soil 90: 256–292.

Boddey R.M., Alves B.J.R. and Urquiaga S. 1996. Nutrient cycling and sustainability of improved pastures in the Brazilian Cerrados. In: Pereira R.C. and Nasser L.C.B. (eds), Proceedings of the 1st International Symposium on Tropical Savannas 24–29th March 1996. EMBRAPA-CPAC, Brasília, Brazil, pp. 33–38.

Bononi V.L.R. and Trufem S.F.B. 1983. Endomicorrizas vesículo arbusculares do cerrado da Reserva de Moji-Guaçú, São Paulo, Brasil. Rickia 10: 55–84.

Bourlière F. and Hadley M. 1983. Present-day savannas: an overview. In: Bourlière F. (ed.), Tropical Savannas. Ecosystems of the World 13, Elsevier Scientific Publishing Company, Amsterdam, pp. 1–17.

Bucher E.H. and Schofield J. 1981. Economis assault on chagas disease. New Sci. 92: 321–324.

Bustamante M.M.C., Martinelli L.A., Silva D.A., Camargo P.B., Klink C.A., Domingues T.F. and Santos R.V. 2004a. ^{15}N natural abundance in woody plants and soils of the savannas of Central Brazil (Cerrado). Ecol. Appl. 14(4): S200–S213.
Bustamante M.M.C., Nardoto G.B. and Martinelli L.A. 2004b. Aspectos comparativos de ciclaje de nutrientes entre bosques amazónicos de terra-firme y sabanas tropicales (Cerrado brasileiro). In: Cabrera H.M. (ed.), Fisiologia Ecologica en Plantas: mecanismos e respuestas a Estrés en los ecosistemas, EUV Valparaíso, Chile, pp. 189–205.
Cadisch G., Carvalho E.F., Suhet A.R., Vilela L., Soares W., Spain J.M., Urquiaga S., Giller K.E., Boddey R.M. 1994a. The importance of legume N_2-fixation in sustainability of pastures in the Cerrados of Brazil. In: Proceedings of the International Grassland Congress. Palmerston North, New Zealand, pp. 1915–1918.
Cadisch G., Schunke R.M. and Giller K.E. 1994b. Nitrogen cycling in a pure grass pasture and a grass-legume mixture on a red latosol in Brazil. Trop. Grass. 28: 43–52.
Campelo A.B. 1976. Caracterização e especificidade de *Rhizobium* ssp. de leguminosas florestais, M.Sc. Thesis, Universidade Federal do Rio de Janeiro, Brazil.
Cárdenas L., Rondón A., Johansson C. and Sanhueza E. 1993. Effects of soil moisture, temperature, and inorganic nitrogen on nitric oxide emissions from acidic tropical savannah soils. J. Geophys. Res. 98: 14783–14790.
Chacón P., López-Hernández I.D. and Lamotte M. 1991. Le cycle de l'azote dans une savane à Trachypogon au centre du Venezuela. Rev. Ecol. Biol. Sol 28: 67–75.
Cleveland C.C., Townsend A.R., Schimel D.S., Fisher H., Howarth R.W., Hedin L.O., Perakis S.S., Latty E.F., Von Fischer J.C., Elseboard A. and Wasson M.F. 1999. Global patterns of terrestrial biological nitrogen (N_2) fixation in natural ecosystems. Glob. Biogeochem. Cycles 13: 623–645.
Cuenca G. and Lovera M. 1992. Vesicular-arbuscular mycorrhizae in disturbed and revegetated sites from La Gran Sabana, Venezuela. Can. J. Bot 70: 73–79.
Crutzen P. and Andreae M.O. 1990. Biomass burning in the tropics: impact on atmospheric chemistry and biogeochemical cycles. Science 250: 1669–1678.
D'Antonio C.M. and Vitousek P.M. 1992. Biological invasions by exotic grasses, the grass/fire cycle, and global change. Ann. Rev. Ecol. Syst. 23: 63–87.
Davidson E.A., Hart S.C. and Firestone M.K. 1992. Internal cycling of nitrate in soils of a mature coniferous forest. Ecology 73: 1148–1156.
Davidson E.A., Keller M., Erickson H.E., Verchot L.V. and Veldkamp E. 2000. Testing a conceptual model of soil emissions of nitrous and nitric oxide. BioScience 50: 667–680.
Davidson E.A., Bustamante M.M.C. and Pinto A.S. 2001. Emissions of nitrous oxide and nitric oxide from soils of native and exotic ecosystems of the Amazon and Cerrado regions of Brazil. Sci. World 1(S2): 312–319.
Döbereiner J. and Campelo A.B. 1977. Importance of legumes and their contribution to tropical agriculture. In: Hardy R.W.F. and Gibson A.H. (eds), A Treatise of Dinitrogen Fixation. Section IV: Agronomy and Ecology, Wiley & Sons, New York, pp. 191–220.
Eiten G. 1972. The Cerrado vegetation of Brazil. Bot. Rev. 38: 201–341.
Faria S.M., Franco A.A., Jesus R.J., Menandro S., Baitello J.B., Mucci E.S.F., Döbereiner J. and Sprent J.I. 1984. New nodutaling legume trees from south-east Brazil. New Phytol. 98: 317–328.
Felfili J.M., Silva M.C. Jr., Rezende A.V., Machado J.W.B., Walter B.M.T., Silva P.E.N. and Hay J.D. 1992. Análise comparativa da florística e fitossociologia da vegetação arbórea do cerrado sensu stricto na Chapada Pratinha, D.F. Brasil. Acta Bot. Brasilica 6: 27–84.
Firestone M.K. and Davidson E.A. 1989. Microbiological basis of NO and N_2O production and consumption in soil. In: Andreae M.O. and Schimel D.S. (eds), Exchange of Trace Gases between Terrestrial Ecosystems and the Atmosphere, Wiley, New York, pp. 7–21.
Frost P., Medina E., Menaut J.C., Solbrig O., Swift M. and Walker B. 1986. Responses of Savannas to Stress and Disturbance. Biology International, Special Issue 10. IUBS, Paris, France.
Galloway J.N., Dentener F.J., Capone D.G., Boyer E.W., Howarth R.W., Seitzinger S.P., Asner G.P., Cleveland C.C., Green P., Holland E., Karlo D.M., Michaels A.F., Porter J.H., Townsend

A.R. and Vörösmarty C. 2004. Nitrogen cycles: past, present and future. Biogeochemistry 70: 153–226.

Geesing D., Felker P. and Bingham R.L. 2000. Influence of mesquite (Prosopis glandulosa) on soil nitrogen and carbon development: implications for global carbon sequestration. J. Arid Environ. 46: 157–180.

Giller K.E. and Day J.M. 1985. Nitrogen fixation in the rhizosphere: significance in natural and agricultural systems. In: Fitter A.H. (ed.), Ecological Interactions in Soils, Blackwell Scientific Publishers, London.

Goedert W.J. 1983. Management of Cerrado soils: a review. J. Soil Sci. 34: 405–428.

Gonzalez C., Abril A. and Acosta M. 1999. Efecto del fuego sobre la fertilidad edafica y las comunidades microbianas en el Chaco occidental argentino. Ecol. Austral 9(12): 3–10.

Gonzalez C.C., Albanesi A., Studdert G.A. and Knust C. 2001. Comportamiento de algunas propriedades del suelo en una sabana del Chaco semiarido Occidental bajo distintas frecuencias de fuego. Ciência Del Suelo 19(2): 92–100.

Hibbard K.A., Schimel D.S., Archer S., Ojima D.S. and Parton W. 2003. Grassland to woodland transitions: integrating changes in landscape structure and biogeochemistry. Ecol. Appl. 13: 911–926.

Hibbard K.A., Archer S., Schimel D.S. and Valentine D.W. 2001. Biogeochemical changes accompanying woody plant encroachment in a subtropical savanna. Ecology 82: 1999–2011.

Hoffmann W.A. and Jackson R.B. 2000. Vegetation-climate feedbacks in the conversion of tropical savanna to grassland. J. Climate 13(9): 1593–1602.

Huntley B. and Walker B.H. 1982. Ecology of Tropical Savannas. Springer, Berlin, 671 pp.

Israel D.W. 1987. Investigation of the role of phosphorus in symbiotic dinitrogen fixation. Plant Physiol. 84(3): 835–840.

Izaguirre-Mayoral M.L., Carballo O., Flores S., Sicardi de Mallorca M. and Oropeza T. 1992. Quantitative analysis of the symbiotic N_2-fixation, non-structural carbohydrates and chlorophyll content in sixteen native legume species collected in different savanna sites. Symbiosis 12: 293–312.

Johansson C. and Sanhueza E. 1988. Emission of NO from savanna soils during rainy season. J. Geophys. Res. 93: 14193–14198.

Johnson H.B. and Mayeux H.S. 1990. *Prosopis glandulosa* and the nitrogen balance of rangelands: extent and occurrence of nodulation. Oecologia 84: 176–185.

Kauffmann J.B., Cummings D.L. and Ward D.E. 1994. Relationships of fire, biomass and nutrient dynamics along a vegetation gradient in the Brazilian Cerrado. J. Ecol. 82: 519–531.

Kichel A.N., Miranda C.H.B. and Macedo M.C. 1996. Conventional and multiple cropping systems of upland rice for reclamation of degraded *Brachiaria decumbens* pastures. In: Pereira R.C. and Nasser L.C.B. (eds), Proceedings of the 1st International Symposium on Tropical Savannas, EMBRAPA-CPAC, Brasília, pp. 443–445.

Le Mare P.H., Pereira J. and Goedert W.J. 1987. Effects of green manure on isotopically exchangeable phosphate in a dark-red latosol in Brazil. J. Soil Sci. 38: 199–209.

Leitão M.R.S.M.M. 1997. Fixação biológica do nitrogênio por espécies arbóreas. In: Vargas M.A.T. and Hungria M. (eds), Biologia dos solos dos Cerrados, EMBRAPA-CPAC, Brasília, pp. 155–186.

Levine J.S., Winstead E.L., Parsons D.A.B., Scholes M.C., Scholes R.J., Cofer W.R. III, Cahoon D.R. and Sebacher D.I. 1996. Biogenic soil emissions of nitric oxide (NO) and nitrous oxide (N_2O) from savannas in South Africa: the impact of wetting and burning. J. Geophys. Res. 101: 23689–23697.

Loureiro M.F. and Boddey R.M. 1988. Balanço de nitrogênio em quarto gramíneas do gênero *Brachiaria*. Pesq. Agropec. Bras. 23: 1343–1353.

Macedo M.C.M. 1995. Pastagens no ecossistema Cerrados: pesquisa para o desenvolvimento sustentável. In: Andrade R.P.de, Barcellos A.O. and da Rocha C.M.C. (eds), Proceedings of the Simpósio sobre pastagens nos ecossistemas brasileiros: pesquisas para o desenvolvimento sustentável, Sociedade Brasileira de Zootecnia, Universidade Federal de Viçosa, Viçosa, MG, Brazil, pp. 28–62.

Magalhães F.M.M., Magalhães L.M.S., Oliveira L.A. and Döbereiner J. 1982. Ocorrência de de nodulação em leguminosas florestais de terra firme nativas da região de Manaus. Acta Bot. 12: 509–514.
Markewitz D., Davidson E.A., Figueiredo R.O., Victoria R.L. and Krusche A.V. 2001. Control of cation concentrations in stream waters by surface soil processes in an Amazonian watershed. Nature 410: 802–805.
Martin R.E., Asner G.P., Ansley R.J. and Mosier A.R. 2003. Effects of woody vegetation encroachment on soil nitrogen oxide emissions in a temperate savanna. Ecol. Appl. 13(4): 897–910.
Mazzarino M.J., Buffa E., Oliva L., Nuñez A. and Nunez G. 1991a. Nitrogen mineralization and soil fertility in the dry Chaco ecosystem (Argentina). Soil Sci. Soc. Am. 55(2): 515–522.
Mazzarino M.J., Oliva L., Abril A. and Acosta M. 1991b. Factors affecting nitrogen dynamics in a semiarid woodland (Dry Chaco, Argentina). Plant Soil 138: 85–98.
McKeon G.M., Day K.A., Howden S.M., Mott J.J., Orr D.M., Scattini W.J. and Weston E.J. 1991. Northern Australian savannas: management for pastoral production. In: Werner P.A. (ed.), Savanna Ecology and Management, Blackwell Scientific Publications, Oxford, UK, pp. 11–28.
Medina E. 1993. Mineral nutrition: tropical savannas. Prog. Bot. 54: 237–253.
Medina E. 1982. Nitrogen balance in the *Trachypogon* grasslands of Central Venezuela. Plant Soil 67: 305–314.
Medina E. and Bilbao B. 1991. Significance of nutrient relations and symbiosis for the competitive interactions between grasses and legumes in a tropical savanna. In: Esser G. and Overdieck E. eds, Modern Ecology Basic and Applied Aspects. Elsevier Publ. Co., Amsterdam Netherlands, pp. 295–320.
Medina E. and Izaguirre M.L. 2004. N_2-fixation in tropical American savannas evaluated by the natural abundance of ^{15}N in plant tissues and soil organic matter. Trop. Ecol. 45: 87–95.
Medina E. 1987. Nutrients; requirements, conservation and cycles of nutrients in the herbaceous layer. In: Walker B.H. (ed.), Determinants of Tropical Savannas, IR Press, Oxford, UK, pp. 39–65.
Miranda J.C.C. and Miranda L.N. 1996. Efeito de fungos micorrízicos arbusculare nativos de cerrado no crescimento da soja adubada com nitrogênio ou inoculada com *Rhizobium*. In: Pereira R.C. and Nasser L.C.B. (eds), Proceedings of the 1st International Symposium on Tropical Savannas, EMBRAPA-CPAC, Brasília, Brazil, pp. 393–395.
Miranda H.S., Bustamante M.M.C. and Miranda A.C. 2003. The fire factor. In: Oliveira O.S. and Marquis R.J. (eds), The Cerrados of Brazil: Ecology and Natural History of a Neotropical Savanna, Columbia University Press, USA, pp. 55–68.
Miranda H.S., Rocha e Silva E.P. and Miranda A.C. 1996. Comportamento do fogo em queimadas de campo sujo. In: Miranda H.S., Saito C.H. and Dias B.F.S. (Orgs) Impactos de Queimadas em áreas de cerrado e restinga. ECL/UnB, Brasília DF Brazil, pp. 1–10.
Mistry J. 2000. World Savannas: Ecology and Human Use. Pearson Education Limited, Great Britain.
Montes R. and San Jose J.J. 1989. Chemical composition and nutrient loading by precipitation in the Trachypogon savannas of the Orinoco llanos, Venezuela. Biogeochemistry 7: 241–256.
Morello J. and Saraiva-Toledo C. 1959. El bosque chaqueño II. Ganadería y el bosque en el oriente d Salta. Revista Agronômica Del Noroeste Argentino 3: 209–258.
NADP 2002. National Atmospheric Deposition Program (NRSP-3)/National Trends Network. Natural Resource Ecology Laboratory, Colorado State University, Fort Collins, Colorado.
Nardoto G.B. and Bustamante M.M.C. 2003. Effects of fire on soil nitrogen dynamics and microbial biomass in savannas of Central Brazil. Pesq. Agropec. Bras. 38(8): 955–962.
Nardoto G.B., Bustamante M.M.C., Pinto A.S. and Klink C. A. 2006. Nutrient use efficiency at ecosystem and species level in savanna areas of Central Brazil and impacts of fire. J. Trop. Ecol. 22(2): 191–201.
Neff J.C., Keller M., Holland E.A., Weitz A.W. and Veldkamp E. 1995. Fluxes of nitric oxide from soils following the clearing and burning of a secondary tropical rain forest. J. Geophys. Res. 100: 25913–25922.

Nobre A.D. 1994. Nitrous oxide emissions from tropical soils. Ph.D thesis, University of New Hampshire, New Hampshire, USA.

Oliva L., Mazzarino M.J., Nuñez G., Abril A. and Acosta M. 1993. Dinamica del nitrogeno y del agua del suelo en un desmonte selectivo en el Chaco arido argentine. Pesq. Agropec. Bras. 28(6): 709–718.

Olson J.S., Watts J.A. and Allison L.J. 1983. Carbon in Live Vegetation of Major World Ecosystems. Report ONRL-5862. Oak Ridge National Laboratory, Oak Ridge Tenn.

Pacheco M., Donoso L. and Sanhueza E. 2004. Soluble organic nitrogen in Venezuelan rains. Tellus 56: 393–395.

Parron L.M. 2004. Aspectos da ciclagem de nutrients em função do gradiente topográfico em uma mata de galeria no Distrito Federal. Ph.D. thesis, University of Brasília, Brasília, 187 pp.

Parsons D.A.B., Scholes M.C., Scholes R.J. and Levine J.S. 1996. Biogenic NO emissions from savanna soils as a function of fire regime, soil type, soil nitrogen, and water status. J. Geophys. Res. 101: 23683–23688.

Pinto A.S., Bustamante M.M.C., Kisselle K., Burke R., Zepp R., Viana L.T., Varella R.F. and Molina M. 2002. Soil emissions of N_2O, NO and CO_2 in Brazilian Savannas: effects of vegetation type, seasonality and prescribed fires. J. Geophys. Res. 107: 8089–8096.

Pinto A.S. 2003. Contribuição dos solos do cerrado do Brasil Central para as emissões de gases traço (CO_2, N_2O e NO): sazonalidade, queimadas prescritas e manejo de pastagens degradadas. Ph.D. thesis. University of Brasília, Brasília.

Pinto A.S., Bustamante M.M.C., Silva M.R.S.S., Kisselle K.W., Brossard M., Kruger R., Zepp R.G. and Burke R.A. 2006. Effects of different treatments of pasture restoration on soil trace gas emissions in the Cerrados of Central Brazil. Earth Interactions 10(1): 1–26.

Pivello V.R. and Coutinho L.M. 1992. Transfer of macro-nutrients to the atmosphere during experimental burnings in an open cerrado (Brazilian savanna). J. Trop. Ecol. 8: 487–497.

Poth M.A., Anderson I.C., Miranda H.S., Miranda A.C. and Riggan P.J. 1995. The magnitude and persistence of soil NO, N_2O, CH_4, and CO_2 fluxes from burned tropical savanna in Brazil. Glob. Biogeochem. Cycles 9: 503–513.

Prieto-Fernandez A., Villar M.C., Carballas M. and Carballas T.T. 1993. Short-term effects in a wildfire on the nitrogen status and its mineralization kinetics in an Atlantic forest soil. Soil Biol. Biochem. 25(12): 1657–1664.

Reatto A., Correia J.R. and Spera S.T. 1998. Solos do bioma Cerrado: aspectos pedológicos. In: Sano S.M. and Almeida S.P. (eds), Cerrado: ambiente e flora, EMBRAPA-Cerrados, Brasília Brazil, pp. 47–86.

Reis F.B., Jr., Mendes I.C., Hungria M. 2003. Contribuição agronômica da FBN. www.revistasafra.com.br/artigo, 14th Nov 2003.

Resende J.C.F. 2001. A ciclagem de nutrientes em áreas de Cerrado e a influência de queimadas controladas. PhD Thesis, Universidade de Brasília, Brasília, Brazil.

Robertson G.P. and Rosswal T. 1986. Nitrogen in West Africa: the regional cycle. Ecol. Monog. 56: 43–72.

Rondón A., Johanson C. and Sanhueza E. 1993. Emission of nitric oxide from soils and termite nests in a Trachypogon savanna of the Orinoco basin. J. Atmos. Chem. 17: 293–306.

Rundel P.W., Nilsen E.T., Sharifi M.R., Virginia R.A., Jarrel W.M., Kohl D.H. and Shearer G.B. 1982. Seasonal dynamics of nitrogen cycling for *Prosopis* woodland in the Sonoran desert. Plant and Soil 67: 343–353.

Saminêz T.C.O. 1999. Efeito do sistema de cultivo, tensão de água, biomassa microbiana e temperatura do solo nos fluxos de CH_4 e N_2O em solos de Cerrados. MSc. Thesis, Universidade de Brasília, Brasília, Brazil.

Sanhueza E. and Crutzen P.J. 1998. Budgets of fixed nitrogen in the Orinoco savannah region: role of pyrodenitrification. Glob. Biogeochem. Cycles 12: 653–666.

Sanhueza E., Hao W.M., Scharffe D., Donoso L. and Crutzen P.J. 1990. N_2O and NO emissions from soils of the Northern Part of the Guayana Shield, Venezuela. J. Geophys. Res. 95: 22481–22488.

Santaella S.E. 1985. Fijación biológica de nitrógeno en sabanas de *Trachypogon* spp. BSc Thesis, Universidad Central de Venezuela, Caracas, Venezuela.

Sarmiento G. 1983. Patterns of specific and phonological diversity in the grass community of the Venezuelan tropical savannas. J. Biogeogr. 10: 373–391.

Sarmiento G. 1984. The Ecology of Neotropical Savannas. Harvard Press, Cambridge, USA.

Sarmiento G. 1996. Ecologia de pastizales y sabanas en America Latina. In: Sarmiento G. and Cabido M. (eds), Biodiversidad y Funcionamento de Pastizales y Sabanas em América Latina, CYTED y CIELAT, Venzeula.

Sato M.N., Garda A.A. and Miranda H.S. 1998. Effects of fire on the mortality of woody vegetation in Central Brazil. In: Viegas D.X. (ed.), Proceedings of the 14th Conference of Fire and Forest Meteorology II 16–20th Nov. 1998. University of Coimbra, Portugal, pp 1785–1792.

Scholes R.J and Walker B.H. 1993. An African savanna: Synthesis of the Nylsvley study. Cambridge University Press, 318 pp.

Schimel D.S. et al. 1997. Continental scale variability in ecosystems processes: models, data, and the role of disturbance. Ecol. Monogr. 67(2): 251–271.

Scholes R.J. and Archer S.R. 1997. Tree–grass interactions in savannas. Ann. Rev. Ecol. Syst. 28: 517–544.

Scholes R.J. and Hall D.O. 1996. The Carbon Budget of Tropical Savannas, Woodlands and Grasslands. In: Breymeyer A.L., Hall D.O., Melillo J.M. and Agren G.I. (eds), Global Change: Effects on Coniferous Forests and Grasslands. SCOPE 56. Wiley & Sons.

SCS 1962. Soil Survey of Wilbarger County, Texas. 18, United States Department of Agriculture Soil Conservation Service, Fort Worth, TX.

Sicardi de Malloorca M. and Izaguirre-Mayoral M.L. 1993. A comparative evaluation of the symbiotic N_2-fixation and physiological performance of thirty-six native legume species collected in a tropical savanna during the rainy and dry season. Symbiosis 16: 225–247.

Silva F. 1990. Compartilhamento de nutrientes em diferentes componentes da biomassa aérea em espécies arbóreas de um cerrado. MSc. Thesis, Departamento de Ecologia, Universidade de Brasília, Brasília.

Singh J.S., Raghubanshi A.S., Singh R.S. and Srivastava S.C. 1989. Microbial biomass acts as a source of plant nutrients in dry tropical forest and savanna. Nature 3388: 499–500.

Siqueira K.P. 2001. Mineralização de nitrogênio em área de campo sujo:efeito do fogo. BSc. Thesis. Universidade de Brasília, Brasília, Brazil.

Siqueira J.O., Colozzi Filho A., Oliveira E. and Schenk N.C. 1989. Ocorrência de micorrizas vesículo-arbusculares em agro e ecossistemas naturais do Estado de Minas Gerais. Pesq. Agropec. Bras. 24: 1499–1506.

Solbrig O.T. 1991. Savanna Modeling for Global Change. Biology International Special Issue 24. IUBS, Paris, France.

Solbrig O.T. 1996. The diversity of the savanna ecosystem. In: Solbrig O.T., Medina E. and Silva J.A. (eds), Biodiversity and Savanna Ecosystem Processes, Springer Verlag, Heidelberg, pp. 1–27.

Sprent J.I., Geoghegan I.E., Whitty P.W. and James E.K. 1996. Natural abundance of ^{15}N and ^{13}C in nodulated legumes and other plants in the cerrado and neighboring regions of Brazil. Oecologia 105: 440–446.

Stock J.B., Stock A.M. and Mottonen J.M. 1990. Signal transduction in bacteria. Nature 344: 395–400.

Thomazini L.I. 1974. Mycorrhiza in plants of the "Cerrado". Plant Soil 41: 707–711.

Varella R.F., Bustamante M.M.C., Pinto A.S., Kisselle K.W., Santos R.V., Burke R.A., Zepp R.G. and Viana L.T. 2004. Soil fluxes of CO_2, CO, NO, and N_2O from an old pasture and from native savanna in Brazil. Ecol. Appl. 14(4): S221–S231.

Verchot L.V., Davidson E.A., Cattânio J.H., Ackerman I.L., Erickson H.E. and Keller M. 1999. Land use change and biogeochemical controls of nitrogen oxide emissions from soils in eastern Amazonia. Glob. Biogeochem. Cycles 13: 31–46.

Viana L.T. 2002. Comparação das dinâmicas de mineralização de nitrogênio, biomassa e estrutura das comunidades microbianas do solo emáreas de cerrado nativo e pastagem. MSc. Thesis. Universidade de Brasília, Brasília, Brazil.

Vicentini K.R.C.F. 1999. História do fogo no Cerrado: uma análise palinológica. Ph.D. thesis, Departamento de Ecologia, Universidade de Brasília, Brasil.

Villecourt P. and Roose E. 1978. Charge en azote et en elements minéraux divers des eaux de pluie, de pluviolessivage et de drainage dans la savane de Lamto (Côte d'Ivoire). Rev. Ecol. Biol. Sol 15: 1–20.

Vitousek P.M. 1982. Nutrient cycling and nutrient use efficiency. Am. Nat. 119: 553–572.

Wahlen S.C., Phillips R.L. and Fischer E.N. 2000. Nitrous oxide emission from an agricultural field fertilized with liquid lagoon swine effluent. Glob. Biogeochem. Cycles 14: 545–558.

Ward D.E., Sussott R.A., Kauffman J.B., Babbitt R.E., Cummings D.L., Dias B., Holben B.N., Kauffman Y.J. and Setzer A.W. 1992. Smoke and fire characteristics for Cerrado and deforestation burns in Brazil: BASE-B experiment. J. Geophys. Res. 97: 14601–14619.

Weitz A.M., Veldkamp E., Keller M., Neff J. and Crill P.M. 1998. Nitrous oxide, nitric oxide, and methane fluxes from soils following clearing and burning of tropical secondary forest. J. Geophys. Res. 103: 28047–28058.

Winter W.H., Mott J.J. and McLean R.W. 1989. Evaluation of management options for increasing the productivity of tropical savanna pastures 2. Legume species. Aust. J. Exp. Agricult. 29: 623–630.

Zitzer S.F., Archer S.R. and Boutton T.W. 1996. Spatial variability in the potential for symbiotic N_2 fixation by woody plants in a subtropical savanna ecosystem. J. Appl. Ecol. 33: 1125–1136.

Biogeochemistry (2006) 79: 239–250
DOI 10.1007/s10533-006-9007-9

Nutrient–chlorophyll relationships in tropical–subtropical lakes: do temperate models fit?

VERA L.M. HUSZAR[1,*], NINA F. CARACO[2], FABIO ROLAND[3] and JONATHAN COLE[2]

[1]*Depto. de Botânica, Museu Nacional, Universidade Federal do Rio de Janeiro, Rio de Janeiro, RJ, 20940-040, Brasil;* [2]*Institute of Ecosystem Studies, Box ABMillbrook, NY, 12545, USA;* [3]*Laboratório de Ecologia Aquática, Instituto de Biologia, Universidade Federal de Juiz de Fora, Juiz de Fora, MG, 36036-330, Brasil; *Author for correspondence (e-mail: vhuszar@gbl.com.br; phone: + (5521)-2529-6946; fax: +47-55236379)*

Key words: Eutrophication, Inland waters, Nutrient–chlorophyll relationship, Tropical and subtropical lakes

Abstract. In tropical lakes relatively little is known about the general relationship between nutrient concentration and phytoplankton biomass. Using data from 192 lakes from tropical and subtropical regions we examine the relationship between total P (TP) and chlorophyll (Chl). The lakes are all located between 30° S to 31° N include systems in Asia, Africa, and North and South America but are dominated by Brazilian ($n = 79$) and subtropical N. American ($n = 67$) systems. The systems vary in morphometry (mean depth and lake area), trophic state as well total N (TN) to total P (TP) ratios and light extinction. Despite a nearly 500-fold range in TP concentrations (2–970 μg P l^{-1}), there was a poorer relationship between log TP and log Chl ($r^2 = 0.42$) than is generally observed for temperate systems from either narrow or broad geographic regions. N limitation is not a likely explanation for the relatively weak TP–Chl relationship in the tropical–subtropical systems. Systems had high average TN:TP ratios and neither a multiple regression with log TP and log TN nor separating systems with high TN:TP (> 17 by weight) improved the predictive power of the log TP–log Chl relationship.

Introduction

The relationship between nutrients and phytoplankton biomass in aquatic systems has yielded insights on nutrient limitation and is a valuable management tool (Peters 1986). For freshwater lakes the great majority of studies on the relationship between nutrients and phytoplankton biomass are for temperate systems (Peters 1986; Kalff 2002). These studies have advanced from early studies (Sakamoto 1966; Vollenweider 1968), which focused on simple, linear relationships between log total P (log TP) and log chlorophyll (log Chl), to include the secondary impacts of N, lake depth, color, inorganic turbidity and food web relationships (Smith 1982; Quirós 1990; Mazumder 1994; Nürnberg 1996). However, compared to even what was known nearly half a century ago for temperate systems, little is known about the general relationship between nutrients and chlorophyll in tropical systems or the primary limiting nutrient in these systems (Canfield and Hodgson 1983; Lewis 1990; Salas and Martino 1991; Fisher et al. 1995; Sarnelle et al. 1998; Kalff 2002).

One prevailing view is that nitrogen limitation may be more common in tropical than in temperate systems (Talling and Lemoalle 1998; Lewis 2000) possibly due to greater P supply by chemical weathering of rocks and greater internal N loss at higher temperatures (Lewis 2000, 2002). This possible N limitation suggests that tropical systems may have a poorer relationship between TP and Chl than temperate systems but Chl may be more closely related to total N (TN) concentrations. To date, however, there are relatively few studies examining the relationship between Chl and nutrients in either tropical (Walker and Tyler 1983 *apud* Ferris and Tyler 1985; Salas and Martino 1991; Lewis 1990; Sarnelle et al. 1998; Jones et al. 2000) or subtropical regions (e.g., Canfield 1983; Brown et al. 2000). Further, many of the existing studies are based on relatively narrow geographic regions (e.g., Florida, Thailand), or relatively small data sets. Thus, it is not yet clear if there is a general difference in nutrient limitation or nutrient–chlorophyll relationships between temperate and tropical areas (Fisher et al. 1995; Kalff 2002). The question of nutrient limitation and the relationship between nutrient concentration and Chl is increasingly important as human activity in tropical systems is accelerating rapidly and tropical systems are increasingly confronted with water quality problems associated with eutrophication (Meybeck et al. 1990; Downing et al. 1999).

In this study we examine nutrient status of lakes from tropical to subtropical regions. We develop empirical models of the relationship between Chl and nutrients and compare these relationships to those developed based largely on temperate data sets.

Methods

Database. Our data set of tropical and subtropical inland waters are from both N and S. Hemisphere between latitude 31° N and 30° S. This working definition is based on a minimum water temperature of 10 °C under normal conditions with a minimum annual average of 15 °C as used by Salas and Martino (1991). High altitude tropical lakes (>3000 m above the sea level) and saline lakes were not included in the analysis.

The data set includes 192 aquatic systems (136 lakes, 56 reservoirs) sampled on seasonal bases at least during 1 year and we use average annual values in our analysis. Samples are from surface mixed waters. The data base is primarily drawn from Florida, United States ($n = 67$, Beaver and Crisman 1991) and from 3° N to 30° S in Brasil ($n = 79$). Brazilian data were taken from the following publications: (FEEMA 1982; Pontes 1980; Andrade et al. 1988; Branco 1991; Salas and Martino 1991; Thomaz 1991; Huszar 1994; Rodrigues 1994; Silva 1995; Konrath 1995; Debertd 1997; Mitamura and Hino 1997;

Salomoni 1997; Oliveira 1998; Rodrigues 1998; Sant'Anna et al. 1998; Laudares-Silva 1999; Loverde-Oliveira 1999; Barroso 2000; Bressan 2001; Melo 2001; Szawka 2001; Ferreira 2002; Marinho and Huszar 2002; Cardoso and Motta Marques 2004) but also included unpublished data (D.C. Bicudo, personal communication and L.H.S. Silva, personal communication). The remaining 46 lakes are from Africa ($n = 18$, Sarnelle et al. 1998; Lewis 1990; Kalff and Watson 1986), China ($n = 18$, Jin 1994), Phillippines ($n = 1$, Lewis 1990), and from Mexico ($n = 5$), Puerto Rico ($n = 1$), Texas, United States ($n = 1$) and Venezuela ($n = 2$) (Salas and Martino 1991). All data are from tables from the above references except for Sarnelle et al. (1998), which were taken by measurement from Figure 3b. The complete data set is available on request. For all 192 systems we obtained data for chlorophyll (Chl) and total phosphorus (TP). For many of them we were able to also obtain total N (TN), Secchi depth (SD), lake area and mean depth (Table 1).

Analysis. Statistical analysis of data was done in StatView® (version 5.01). Least square regressions were done on logarithmic (base 10) transformed data. Differences between data sets were calculated by non-paired *t*-tests. To test if the slopes were significantly different ($p<0.05$), a covariance analysis was used (Zar 1996). With the exception of light extinction all analysis were performed on raw data.

Total light extinction (K) was calculated from SD as: $K = -1 \times \ln(\text{SDlight})/\text{SD}$. SDlight is the fraction of surface light penetration at the SD and is generally reported as 0.1 (e.g., 10%, Wetzel and Likens 1991). Light extinction (K) is the sum of extinction from phytoplankton themselves (K_p) as well as extinction from colored organic matter and inorganic turbidity (residual K, K_r). $K_r = K - K_p$. K_p was calculated as: $K_p = \text{EP} \times \text{Chl-}a$, where EP, is the biomass specific extinction (in m^2 mg Chl^{-1}) of phytoplankton, which generally varies between 0.01 and 0.02 for different phytoplankton (McBride et al. 1993) here we use a value of 0.016.

For some analyses we grouped the systems by N status. We followed the procedure of Smith (1998) in considering high N:P systems to have TN:TP ratios (by weight) greater than 17.

Table 1. Summary statistics of available average limnological data from 192 tropical–subtropical lakes.

Entire data set	Median	Mean	SD	Minimum	Maximum	n
Lake area (km^2)	6.6	770.6	6079.7	0.1	66250.0	148
Mean depth (m)	4.1	12.7	60.9	0.7	700.0	132
Secchi depth (m)	0.9	1.4	1.4	0.1	6.6	132
Extinction coefficient (m^{-1})	2.5	3.2	2.6	0.4	16.5	132
Chlorophyll-*a* ($\mu g\ l^{-1}$)	11.0	34.2	67.0	0.9	556.0	192
Total phosphorus ($\mu g\ l^{-1}$)	59.0	131.9	181.4	2.0	970.0	192
Total nitrogen ($\mu g\ l^{-1}$)	925.1	1506.4	2064.6	42.4	21000.0	156
TN:TP (by weight)	19.8	26.0	30.9	0.7	221.5	156

Results

Lakes used in this study represent a wide range of limnological conditions. For example mean annual TP and TN ranged nearly 3-orders of magnitude across systems (Table 1). Phytoplankton biomass and light extinction (*K*) also varied greatly between systems but was generally high (Table 1). On average, 19% and 10%, respectively, of the light extinction is due to phytoplankton (K_{p},) the remainder being due to inorganic turbidity or dissolved color.

Combining all systems, there was a highly significant relationship between log TP and log Chl ($p<0.0001$) but log TP explained only 42% of the variance in log Chl (Figure 1). Further the slope of the log TP–Chl relationship was significantly lower ($p<0.0001$) than that from temperate relationships (Figure 2). Thus predicted chlorophyll at a given TP level was, on average, slightly but significantly ($p<0.001$) lower than that expected from published relationships for temperate systems. In addition the prediction of Chl did not improve separately considering systems with high TN:TP (<17; $r^2=0.37$; Figure 1). The relationship between log TN and log Chl was also significant and had significantly similar ($p<0.0001$) explanatory power ($r^2=0.39$) to TP. The residuals of the TP–Chl relationship were significantly related to TN concentration ($p=0.03$) but TN explained only 3% of the variance in these residuals (Figure 3). Similarly, a multiple regression which included both TN and TP, improved only modestly ($r^2=0.47$) the prediction of Chl with TP alone (Table 2). Lastly, a regression of predictions from a non-linear regression that considers both TP and TN (McCauley et al. 1989) did not perform substantially better than linear regressions with or without TN ($r^2=0.49$).

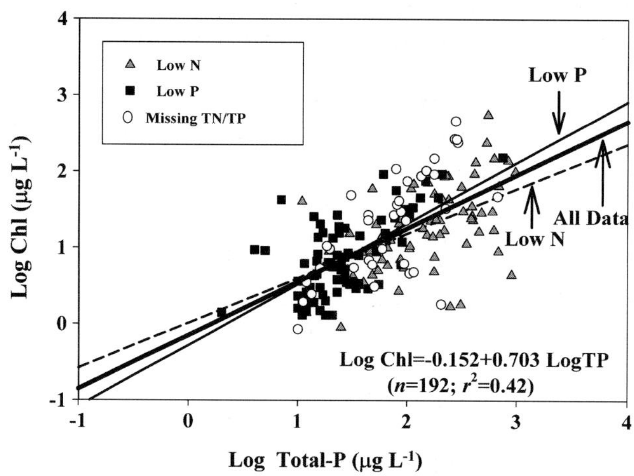

Figure 1. Relationship between annual mean of log-total phosphorus (µg l^{-1}) vs. log chlorophyll (µg l^{-1}) for surface waters of 192 tropical and subtropical lakes from Africa, Asia, and South and North America. The data are coded as P-limited lakes (TN:TP>17, by weight) and N- or N and P-limited lakes (TN:TP<17, by weight).

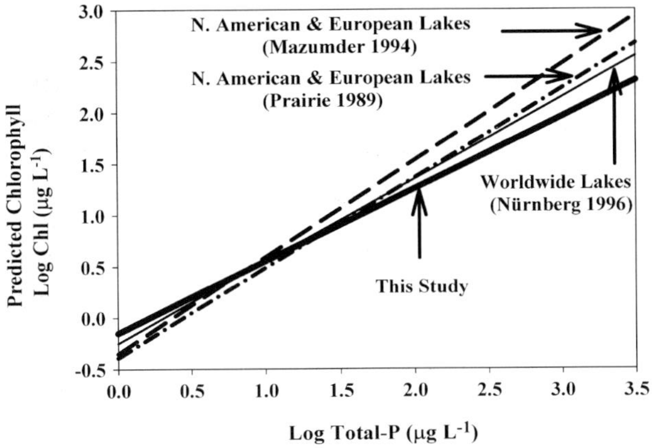

Figure 2. Comparison of the log-total phosphorus (µg l^{-1}) vs. log chlorophyll (µg l^{-1}) relationship among our tropical and subtropical lakes with selected temperate regressions (see Table 3).

The generally high light extinction (low SD, Table 1) and the high proportion of light extinction by non-phytoplankton particulates or dissolved organic matter suggest that light limitation could account for the high variability between TP and Chl and the generally low Chl yield at a given P concentration. However, indicators of light limitation were either significantly positively related to the residual of the log TP–log Chl relationship or were not significantly related to these residuals. There was no significant relationship between residuals and K_r ($p=0.69$) and residuals were positively related to total K ($p<0.05$) and negatively related to both log SD and log SD/average depth ($p<0.01$).

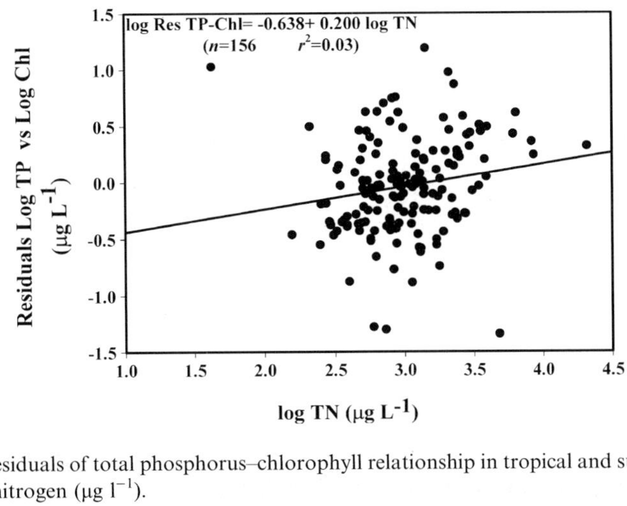

Figure 3. Residuals of total phosphorus–chlorophyll relationship in tropical and subtropical lakes vs. log total nitrogen (µg l^{-1}).

Table 2. Nutrient–chlorophyll and chlorophyll–Secchi depth regressions for the 192 tropical–subtropical lakes.

Regression models	n	r^2	p
log Chl = −0.152 + 0.703 log TP	192	0.42	<0.0001
log Chl = −1.753 + 0.936 log TN	156	0.39	<0.0001
log Chl = −0.059 + 0.615 log TP (low TN:TP)	71	0.26	<0.0001
log Chl = −0.262 + 0.789 log TP (high TN:TP)	85	0.37	<0.0001
log Chl = −1.188 + 0.413 log TP + 0.504 log TN	156	0.47	<0.0001

Discussion

This paper represents the largest compilation of nutrients and Chl data from a broad range of tropical and subtropical systems and analysis of their relationship. Although conclusions may be changed by future analyses that include more data particularly from Asia, Oceania and Africa our analyses does suggest several similarities and some differences to temperate studies to the relationship between nutrients and Chl. First our data show that, tropical lakes, like temperate ones, vary dramatically not only in morphometry, light extinction and trophic status. Using Chl concentrations, as a criterion (Wetzel 2001), 15% of systems were oligotrophic (<3 μ g l^{-1} Chl-*a*), 35% were mesotrophic, 50% were eutrophic (>11 μ g l^{-1} Chl-*a*). Further, a full 25% have average annual chlorophyll concentrations greater than 50 μg l^{-1} Chl-*a*, suggesting hypereutrophic conditions. The, high chlorophyll concentrations are related to both high nitrogen and phosphorus concentrations in tropical as for temperate systems. Like temperate systems N:P ratios are generally greater than the 7:1 ratio (by weight) required for balanced phytoplankton growth (Wetzel 2001) and phosphorus relates somewhat better to chlorophyll concentration than does nitrogen. There are, however, substantial differences in the quantitative relationship between chlorophyll and nutrients in comparison to temperate regions and to some studies from relatively narrow geographic regions in tropical areas (e.g., Sarnelle et al. 1998). Our analysis suggests both a more variable relationship between log TP and log Chl and a somewhat lower chlorophyll yield per unit TP than do regressions from the temperate zone (Table 3).

The differences we found between tropical and temperate TP–Chl relationships could have several causes. It is possible that the lower Chl yield is related to differences in sampling. Many studies in temperate systems are based on only summer growing season sampling, when Chl is likely higher than the annual average, while the tropical samples are based on year round sampling (Lewis 1990). We do not believe, however, that this would explain the higher between system variance between TP and Chl in the tropical systems. This higher variation could occur, if TP had a low dynamic range as compared to temperate studies. Our TP data, however, range by nearly 500-fold between systems, a range as large as seen in the better-fitting temperate data sets.

Table 3. Selected regression equations for selected temperate lakes used in comparisons with our tropical–subtropical data.

	n	r^2	p	Distribution
TP vs. Chl				
(a) log Chl = −0.35 + 0.94 log TP	367	0.71	<0.0001	North American and European lakes[a]
(b) log Chl = −0.25 + 0.799 log TP	180	0.64	<0.0001	Worldwide lakes[b]
(c) log Chl = −0.390 + 0.874 log TP	133	0.69	<0.0001	North American and European lakes[c]
TP and TN vs. Chl				
(d) log Chl = −0.83 log TP + 0.34 log TP^2 − 0.12 log TP^3 + 0.43 (log TP × log TN) + 0.03	875	0.71	<0.01	USA lakes[d]
(e) log Chl = −2.213 + 0.517 log TP + 0.838 log TN	133	0.81	<0.0001	North American and European lakes[c]
(f) log Chl = −1.517 + 0.653 log TP + 0.548 log TN	127	0.76	<0.01	North temperate lakes[e]

[a]Mazumder (1994).
[b]Nürnberg (1996).
[c]Prairie et al. (1989).
[d]McCauley et al. (1989).
[e]Smith (1982).

Another possible cause for the relatively poor linear relationship between log TP and log Chl is that systems are outside the range of linearity between TP and Chl relationships (McCauley et al. 1989), however, 85% of the data have TP between 10 and 300 μg l^{-1}, a range where log TP relates nearly linearly to log Chl (McCauley et al. 1989).

Widespread N limitation in tropical–subtropical systems could explain both the lower Chl yield and higher variance in TP–Chl relationship. It has also been suggested that while N limitation occurs in temperate freshwater systems (Smith 1982) and P limitation occurs in tropical systems (Fisher et al. 1995; Sarnelle et al. 1998; Kalff 2002) for tropical systems N limitation may be more prevalent (Lewis 1996, 2002). Our studies do not suggest, however, widespread N limitation. For our tropical–subtropical lakes, TN:TP ratios were generally high and exclusion of those systems with relatively low TN:TP ratios (Figure 1) did not improve the predictability of Chl from TP. Similarly, TN did not explain well residuals of the TP–Chl relationship and a multiple regression with TN and TP did not explain substantially better the variance in Chl to TP. Further, inorganic nutrient concentrations, which may be a better indication of this limitation that are TN:TP ratios, were also on average far above 16:1 (Redfield et al. 1963) in a subset of Brazilian lakes.

In agreement with our results based on nutrient ratios and nutrient–Chl relationships, nutrient addition experiments in tropical systems as for temperate systems (Elser et al. 1990) do not show uniform N limitation but rather suggest that systems can vary between N limitation, P limitation and co-limitation of N and P. Reviewing experimental nutrient enrichment studies in 10 Brazilian lakes and reservoirs, Arcifa et al. (1995) did not find clear cut N limitation, rather limitation varied both between systems and seasonally within single systems. Similarly, nutrient limitation in other tropical areas as inferred from nutrient additions, physiological indicators, or dissolved N:P ratios do not show uniform N or P limitation but rather show seasonal and between system variance (Fisher et al. 1995). Thus, the total weight of evidence does not support the view that N limitation is necessarily the norm in tropical to subtropical systems or even that it is more prevalent than in temperate systems (Elser et al. 1990). Other factors may explain the somewhat lower Chl yield per unit TP and more variable Chl–TP relationsip that we observed.

In temperate systems, in addition to TN or TN:TP ratios, it has been suggested that variable light limitation could explain variance in the TP to Chl relationships (Canfield and Bachman 1981). While the tropical–subtropical database, as a whole, had high light extinction (K) that was dominated by color or turbidity (K_r), neither K nor K_r were negatively related to the residuals of the Chl–TP relationship. It is possible that a better indicator of light availability (e.g., the amount of light in the mixed layer) would relate well to residuals of the TP to Chl relationship, however, we did not have information on thermal stratification for most of the 192 systems in the data set.

There is also evidence that in temperate systems grazing, especially by large zooplankton, can regulate the empirical relationships between nutrients and phytoplankton biomass (Pace 1984; Quirós 1990; Mazumder 1994). However, tropical and subtropical lakes are frequently dominated by small cladocerans, rotifers, and by juveniles and small copepodites (Lewis 1996; Branco et al. 2002; Jeppesen et al. 2005). While the classic control of phytoplankton by large zooplankton may not hold in tropical systems, more complex food web interactions could be responsible for the tropical–subtropical vs. temperate differences in the predictive power of the log TP vs. log Chl relationship (Lazzaro 1997).

In summary, our analysis showed some differences to expectations based on N limitation in tropical systems and to similar empirical studies from both broad temperate regions and narrow tropical areas. A number of possible mechanisms could explain these differences and different mechanisms could be operating across different regions of the tropics and subtropics. Further empirical as well as experimental studies on a diverse set of tropical systems are needed to elucidate these differences.

Acknowledgements

This research was supported by grants from Coordenação de Aperfeiçoamento de Pessoal de Nível Superior (CAPES) and A.W. Mellon Foundation, USA, and by scholarship to Vera Huszar from Conselho Nacional de Desenvolvimento Científico e Tecnológico (CNPq), Brasil. The authors want to thank Dr Lucia H.S. Silva and Dr Denise Bicudo for unpublished data, Eduardo Welp for entering data, Dr W.M. Lewis Jr and an anonymous reviewer for providing criticism of the manuscript. This is a contribution to the Institute of Ecosystem Studies.

References

Andrade L.F., Brunkow R.F., Xavier C.F. and Domingues L.L. 1988. Fitoplâncton e características fisico-químicas do reservatório de Itaipu, Paraná, BR. In: Tundisi J.G. (ed.), Limnologia e manejo de represas., EESC-USP/CRHEA/ACIESP, São Carlos, pp. 205–268.

Arcifa M.S., Starling F.R.L., Sipaúba-Tavares L.H. and Lazzaro X. 1995. Experimental limnology. In: Tundisi J.G., Bicudo C.E.M. and Tundisi T.M. (eds.), Limnology in Brazil, ABC/SBL, Rio de Janeiro, pp. 257–281.

Barroso A.D. 2000. Comunidade fitoplanctônica do reservatório Duas Bocas (Reserva, Biológica de Duas Bocas) – ES: variação vertical e temporal em duas épocas distintas. M.Sc. Dissertation, Univ. Fed. São Carlos, São Carlos, SP.

Beaver J.R. and Crisman T. 1991. Importance of latitude and organic color on phytoplankton primary productivity in Florida lakes. Can. J. Fish. Aquat. Sci. 48: 1145–1150.

Branco C.W.C. 1991. A comunidade planctônica e a qualidade da água no Lago Paranoá. M.Sc. Dissertation, Univ Brasília, Brasília, DF.

Branco C.W.C., Rocha M.I., Pinto G.F.S., Gômara G. and Filippo R.D. 2002. Limnological features of Funil reservoir (RJ, Brazil) and indicator properties of rotifers and cladocerans of the zooplankton community. Lakes Res.: Res. Manag. 7: 87–92.

Bressan F.A. 2001. Fatores reguladores da dominância de *Cylindrospermopsis raciborskii* (Woloszynska) Seenayya & Subba-Raju no Reservatório Tabocas. M.Sc. Dissertation, Univ. Fed. Rio de Janeiro, Rio de Janeiro, RJ.

Brown C.D., Hoyer M.V., Bachmann R.W. and Canfield D.E. 2000. Nutrient–chlorophyll relationships: an evaluation of empirical nutrient–chlorophyll models using Florida and north-temperate lake data. Can. J. Fish. Aquat. Sci. 57: 1574–1583.

Canfield D.E. and Bachmann R.W. 1981. Prediction of total phosphorus concentrations, chlorophyll-*a* and Secchi depths in natural and artificial lakes. Can. J. Fish. Aquat. Sci. 38: 414–423.

Canfield D.E. 1983. Prediction of chlorophyll-*a* concentrations in Florida lakes: the importance of phosphorus and nitrogen. Water Res. Bull. 19: 255–262.

Canfield D.E. and Hodgson L.M. 1983. Prediction of Secchi disc depths in Florida lakes: impact of algal biomass and organic color. Hydrobiologia 99: 51–60.

Cardoso L.S. and Motta Marques D. 2004. The influence of hydrodynamics on the spatial and temporal variation of phytoplankton pigments in a large, sub-tropical coastal lake (Brazil). Braz. Arch. Biol. Tech. 47: 587–600.

Debertd G.L.B. 1997. Produção primária e caracterização da comunidade fitoplanctônica no reservatório de Salto Grande (Americana SP) em duas épocas do ano. M.Sc. Dissertation, Univ. São Paulo, São Paulo, SP.

Downing J.A., McClain M., Twilley R., Melack J.M., Elser J., Rabalais N.N., Lewis W.M. Jr., Turner R.E., Corredor J., Soto D., Yanez-Arancibia A., Kopaska J.A. and Howarth R.W. 1999. The impact of accelerating land-use change on the N-cycle of tropical aquatic ecosystems: current conditions and projected changes. Biogeochemistry 46: 109–148.

Elser J.J., Marzolf E.R. and Goldman C.R. 1990. Phosphorus and nitrogen limitation of phytoplankton growth in the freshwaters of North America: a review and critique of experimental enrichment. Can. J. Fish. Aquat. Sci. 47: 1468–1477.

FEEMA-Fundação Estadual de Engenharia do Meio Ambiente. 1982. Estudo preliminar do reservatório de Funil. Relatório Técnico, Rio de Janeiro, RJ.

Ferreira A.C. 2002. Dinâmica do fitoplâncton de um reservatório hipereutrófico (Reservatório Tapacurá, Recife, PE), com ênfase em *Cylindrospermopsis raciborskii* e seus morfotipos. M.Sc. Dissertation, Univ. Fed. Rio de Janeiro, Rio de Janeiro, RJ.

Ferris J.M. and Tyler P.A. 1985. Chlorophyll–total phosphorus relationships in lake Burragorang, New South Wales and some other Southern Hemisphere lakes. Mar. Freshwater Res. 36: 157–168.

Fisher T.R., Melack J.M., Grobbelaar J.U. and Howarth R.W. 1995. Nutrient limitation of phytoplankton and eutrophication of inland, estuarine, and marine waters. In: Tiessen H. (ed.), Phosphorus in the Global Environment, John Wiley & Sons Ltd., New York, NY, pp. 301–322.

Huszar V.L.M. 1994. Fitoplâncton de um lago amazônico impactado por rejeito de bauxita (Lago Batata, Pará, Brasil): estrutura da comunidade, flutuações espaciais e temporais. Ph.D. Thesis, Univ. Fed. São Carlos, São Carlos, SP.

Jeppesen E., Søndergaard M., Mazzeo N., Meerhoff M., Branco C.W.C., Huszar V.L.M. and Scasso F. 2005. Lake restoration and biomanipulation in temperate lakes: relevance for subtropical and tropical lakes. In: Reddy V. (ed.), Tropical Eutrophic Lakes: Their Restoration and Management, Oxford & I.B.H Publ., New Hampshire, pp. 331–359.

Jin X.C. 1994. An analysis of lake eutrophication in China. Verh. Int. Verein. Theor. Angew. Limnol. 24: 207–211.

Jones J.R.B., Perkins D., Witt A., Kaiser M.S., Thamasara S., Siriworakul M. and Benyasut P. 2000. Limnological characteristics of some reservoirs in Thailand. Verh. Int. Verein. Theor. Angew. Limnol. 27: 2158–2166.

Kalff J. 2002. Limnology; Inland Water Ecosystems. Prentice Hall, New Jersey.

Kalff J. and Watson S.B. 1986. Phytoplankton and its dynamics in two tropical lakes: a tropical and temperate zone comparison. Hydrobiologia 138: 161–176.

Konrath J. 1995. Flutuação temporal da taxa de fixação de carbono particulado, estrutura do fitoplâncton e fatores ambientais em uma lagoa costeira eutrófica (Lagoa Caconde, Osório, RS). M.Sc. Dissertation, Univ. Fed. Rio Grande do Sul, Porto Alegre, RS.

Laudares-Silva R.L. 1999. Aspectos limnológicos, variabilidade espacial e temporal na estrutura da comunidade fitoplanctônica da Lagoa do Peri, Santa Catarina, Brasil. Ph.D. Thesis, Univ. Fed. São Carlos, São Carlos, SP.

Lazzaro X. 1997. Do the trophic cascade hypothesis and classical biomanipulation approaches apply to tropical lakes and reservoirs?. Verh. Int. Verein. Theor. Angew. Limnol. 26: 719–730.

Lewis W.M. Jr. 1990. Comparisons of phytoplankton biomass in temperate and tropical lakes. Limnol. Oceanogr. 35: 1838–1845.

Lewis W.M. Jr. 1996. Tropical lakes: how latitude makes a difference. In: Schiemer F. and Boland K.T. (eds.), Perspectives in Tropical Limnology, SPB Academic Publ., Amsterdam, pp. 43–64.

Lewis W.M. Jr. 2000. Basis for the protection and management of tropical lakes. Lake Reserv. Res. Manag. 5: 35–48.

Lewis W.M. Jr. 2002. Causes for the high frequency of nitrogen limitation in tropical lakes. Verh. Int. Verein. Theor. Angew. Limnol. 28: 210–213.

Loverde-Oliveira S.M. 1999. Variáveis limnológicas e análise temporal da comunidade fito planctônica da Baía Sá Mariana, Pantanal de Barão de Melgaço, Cuiabá, MT. M.Sc. Dissertation, Univ. Fed. Mato Grosso, Cuiabá, MT.

Marinho M.M. and Huszar V.L.M. 2002. Nitrogen availability and physical conditions as controlling factors of phytoplankton composition and biomass in a tropical reservoir (Southern Brasil). Arch. Hydrobiol. 153: 443–468.

Mazumder A. 1994. Phosphorus–chlorophyll relationships under contrasting herbivory and thermal stratification: predictions and patterns. Can. J. Fish. Aquat. Sci. 51: 390–400.

McBride G.B., Vant W.N., Cloern J.E. and Liley J.B. 1993. Development of a model of phytoplankton blooms in Manukau Harbour. NIWA Ecosyst. 3: 1–52.

McCauley E., Downing J.A. and Watson S.B. 1989. Sigmoid relationships between nutrients and chlorophyll among lakes. Can. J. Fish. Aquat. Sci. 46: 1171–1175.

Melo S. 2001. Fitoplâncton da lagoa Imboassica (Macaé, RJ, Brasil): flora, estrutura de comunidade e variações espaciais e temporais. Ph.D. Thesis, Univ. Fed. do Rio de Janeiro, Rio de Janeiro, RJ.

Meybeck M., Chapman D.V. and Helmer R. 1990. Global Environment Monitoring System. Global Freshwater Quality; A First Assessment. Oxford Univ. Press, New York.

Mitamura O. and Hino K. 1997. Distribution of biogeochemical constituents in the waters. In: Tundisi J.G. and Saijo Y. (eds.), Limnological Studies on the Rio Doce Valley Lakes, Brazil, Brazilian Academy of Sciences, Rio de Janeiro, RJ, pp. 97–108.

Nürnberg G.K. 1996. Trophic state of clear and colored, soft- and hardwater lakes with special consideration of nutrients, anoxia, phytoplankton and fish. Lake Reserv. Manag. 12: 432–447.

Oliveira H.T. 1998. Avaliação das condições limnológicas de um compartimento (Braço da Capivara) e sua interação com o reservatório de Barra Bonita, SP, com ênfase na comunidade fitoplanctônica. Ph.D. Thesis, Univ. São Paulo, São Carlos, SP.

Pace M.L. 1984. Zooplankton community structure, but not biomass, influences the phosphorus–chlorophyll relationship. Can. J. Fish. Aquat. Sci. 41: 1089–1096.

Peters R.H. 1986. The role of prediction in limnology. Limnol. Oceanogr. 31: 1143–1159.

Pontes M.C.F. (1980) Produção primária, fitoplâncton e fatores ambientais no Lago Dom Helvécio, Parque Florestal do Rio Doce. M.Sc. Dissertation, Univ. Fed. São Carlos.

Prairie Y.T., Duarte C.M. and Kalff J. 1989. Unifying nutrient–chlorophyll relationships in lakes. Can. J. Fish. Aquat. Sci. 46: 1176–1182.

Quirós R. 1990. Factors related to variance of residuals in chlorophyll–total phosphorus regressions in lakes and reservoirs of Argentina. Hydrobiologia 200/201: 343–355.

Redfield A.C., Ketchum B.H. and Richards F.A. 1963. The influence of organisms on the composition of sea-water. In: Hill M.N. (ed.), The Sea, Wiley-Interscience, New York, pp. 26–77.

Rodrigues L.C. 1998. Estrutura da comunidade fitoplanctônica de uma lagoa marginal do rio Invinhema (lagoa dos Patos, planície de inundação do alto rio Paraná) em diferentes períodos do ciclo hidrológico. M.Sc. Dissertation, Univ. Est. Maringá, Maringá, PR.

Rodrigues M.S. 1994. Biomassa e produção fitoplanctônica do Lago Camaleão (Ilha de Marchantaria, Amazonas). Ph.D. Thesis, Fund. Univ. Amazonas/INPA, Mnáus, AM.
Sakamoto M. 1966. Primary production by phytoplankton community in some Japanese lakes and its dependence upon lake depth. Arch. Hydrobiol. 62: 1–28.
Salas H.J. and Martino P. 1991. A simplified phosphorus trophic state model for warm tropical lakes. Water Res. 25: 341–350.
Salomoni E.S. 1997. Aspectos da limnologia e poluição das lagoas costeiras Marcelino, Peixoto e Pinguela (Osório, RS): uma abordagem baseada no fitoplâncton. M.Sc. Dissertation, Univ. Fed. Rio Grande do Sul, Porto Alegre, RS.
Sant'Anna C.L., Sormus L., Tucci A. and Azevedo M.T. 1998. Variação sazonal do fitoplâncton do lago das Graças, São Paulo, SP. Hoehnea 24: 67–86.
Sarnelle O., Cooper S.D., Weisman S. and Mavuti K. 1998. The relationship between nutrients and trophic-level biomass in turbid tropical ponds. Freshwater Biol. 40: 65–75.
Silva L.H.S. 1995. Variabilidade temporal na estrutura da comunidade fitoplanctônica de um reservatório eutrófico (Lago Monte Alegre, Ribeirão Preto, São Paulo, Brasil). M.Sc. Dissertation, Univ. Fed. Rio de Janeiro, Rio de Janeiro, RJ.
Smith V.H. 1982. The nitrogen and phosphorus dependence of algal biomass in lakes: an empirical and theoretical analysis. Limnol. Oceanogr. 27: 1101–1112.
Smith V.H. 1998. Cultural eutrophication of inland, estuarine, and coastal waters. In: Pace M.L. and Groffmann P.M. (eds.), Successes, Limitations and Frontiers in Ecosystem Science, Springer, New York, pp. 7–49.
Szawka C. 2001. Estrutura e dinâmica espaço-temporal do fitoplâncton no reservatório da Usina Hidrelétrica de Salto Caxias, Paraná. M.Sc. Dissertation, Univ. Fed. Paraná, Curitiba, PR.
Talling J.F. and Lemoalle J. 1998. Ecological Dynamics of Tropical Inland Waters. Cambridge University Press, United Kingdom.
Thomaz S.M. 1991. Influência do regime hidrológico (pulsos) sobre algumas variáveis limnológicas de diferentes ambientes aquáticos da planície de inundação do alto Rio Paraná. M.Sc. Dissertation, Univ. Fed. São Carlos, São Carlos, SP.
Vollenweider R.A. 1968. Scientific Fundamentals of the Eutrophication of Lakes and Flowing Waters, with Particular Reference to Nitrogen and Phosphorus as Factors in Eutrophication. Technical Repport DAS/CSI/68, Paris.
Wetzel R.G. and Likens G.E. 1991. Limnological Analyses. Springer, New York.
Wetzel R.G. 2001. Limnology: Lake and River Ecosystems. Academic Press, San Diego.
Zar J.H. 1996. Biostatistical Analysis3. Prentice Hall, New Jersey.

Biogeochemistry (2006) 79: 251–274
DOI 10.1007/s10533-006-9008-8

The stable carbon and nitrogen isotopic composition of vegetation in tropical forests of the Amazon Basin, Brazil

JEAN P. H. B. OMETTO[1,*], JAMES R. EHLERINGER[2], TOMAS F. DOMINGUES[2], JOSEPH A. BERRY[3], FRANÇOISE Y. ISHIDA[1], EDMAR MAZZI[1], NIRO HIGUCHI[4], LAWRENCE B. FLANAGAN[5], GABRIELA B. NARDOTO[1] and LUIZ A. MARTINELLI[1]

*[1]Centro de Energia Nuclear na Agricultura, Av.Centenário 303, 13416-000, Piraciaba-SP, Brazil; [2]Department of Biology, University of Utah, Salt Lake City, UT, USA; [3]Carnegie Institute, Washington, USA; [4]Instituto de Pesquisas da Amazonia, Manaus, Brazil; [5]University of Lethbridge, Lethbridge, Canada; *Author for correspondence (e-mail: jpometo@cena.usp.br)*

Key words: Amazon, Carbon, Nitrogen, Stable isotope, Primary forest

Abstract. Here we present the within-site, seasonal, and interannual variations of the carbon ($\delta^{13}C$) and nitrogen ($\delta^{15}N$) isotope ratios of leaves, wood, bark and litter from four sites in the Amazon region, Brazil. Samples were collected in Manaus (3° 06′07″ S; 60°01′30″ W), Ji-Paraná (10°53′07″ S; 61°57′06″ W), and Santarém (2°26′35″ S; 54°42′30″ W) with mean annual precipitation of 2207, 2040 and 1909 mm respectively. The overall average for all leaf samples was-32.3 ± 2.5‰ for $\delta^{13}C$ and$+5.8 \pm 1.6$‰ for $\delta^{15}N$ ($n=756$). The leaf δ values at these sites were often but not always statistically distinct from each other. The $\delta^{13}C$ values varied from-37.8‰ to-25.9‰. Pronounced differences in $\delta^{13}C$ values occurred with height associated with differences in forest structure. The $\delta^{13}C$ of leaf dry matter showed seasonal variations associated with the length of the dry season, despite the fact that total annual precipitation was similar among the studied sites. Leaf $\delta^{15}N$ values ranged from$+0.9$‰ to a maximum value of$+10.9$‰, and the Santarém sites showed more enriched values than Manaus and Ji-Paraná sites. No seasonal variation was detected in the $\delta^{15}N$ of leaves, but significant differences were observed among sites and with changes in canopy height. The isotope ratio data are consistent with our current understanding of the roles of light, water availability, and recycling of soil-respired CO_2 influences on $\delta^{13}C$ and consistent with our understanding that an open nitrogen cycle can lead to high $\delta^{15}N$ values despite a significant number of legumes in the vegetation.

Introduction

Stable isotope analyses are among the approaches used to characterize components of the carbon and nitrogen cycles in the Amazon Basin (Buchmann et al. 1997; Martinelli et al. 1999; Ehleringer et al. 2002; Ometto et al. 2002). Given that tropical forests play a central role in the global carbon cycle, understanding the factors that influence net carbon gain or loss in these regions can help improve our understanding of the constraints on the global carbon cycle (Schimel 1995; Field et al. 1998; Gash et al. 2004). Recent rainforest studies within the Amazon Basin have suggested that some regions might act as a substantial carbon sink (Grace et al. 1995a, b, 1996; Malhi et al. 1999; Araújo et al. 2002), whereas other sites might have been either neutral or even a small source of carbon to the atmosphere during the past two to three years (Saleska et al. 2003; Miller et al. 2004). Part of the difference of these estimates may be attributable to the complexity and diversity of tropical forests (Cuevas and Medina 1988; Martinelli et al. 1999; Luizão et al. 2004; Vieira et al. 2004). Additionally, the Amazon region does not have a homogeneous climate with similar precipitation and temperature (Obregon and Nobre 1990; Marengo 1992; Marengo et al. 1993; Fisch et al. 1998) or forests of equivalent age and structure (Vieira et al. 2004). Thus, structural and climatic features may contribute to the different carbon sink conclusions of different studies.

Carbon isotope ratio ($\delta^{13}C$) analyses of plant materials can provide carbon-cycle insights by providing as quantitative assessement of the stomatal limitations to photosynthetic activity (Farquhar et al. 1989; Ehleringer et al. 1993). The $\delta^{13}C$ value of a leaf is in part determined by its photosynthetic pathway and in part by the gas exchange constraints associated with environmental conditions, nutrient availability, and to some extent life history. Farquhar et al. (1982) first showed the $\delta^{13}C$ values in C_3 photosynthetic pathway plants is basically controlled by the ratio between the CO_2 concentration inside the leaf intercellular space (c_i) and in the adjacent atmosphere (c_a). As c_i is affected by the photosynthetic demand for CO_2 relative to the stomatal control over CO_2 supply, environmental factors affecting carbon gain, such as water supply, light availability, nitrogen content, just to mention some, are recorded in the $\delta^{13}C$ values of plant tissues (Farquhar et al. 1989).

Variations in the $\delta^{13}C$ of leaves in tropical forests are strongly correlated with position in the canopy profile, with lighter foliar $\delta^{13}C$ (more negative) values observed in the understory vegetation as compared to the upper canopy (Medina and Minchin 1980; Ehleringer et al. 1986). This clear effect of variation in c_i/c_a ratio between the lower- and the upper-canopy leaves has been reported in rainforests in Brazil (Merwe and Medina 1989; Sternberg et al. 1989; Medina et al. 1991; Kapos et al. 1993; Kruijt et al. 1996; Martinelli et al. 1998; Ometto et al. 2002), China (Ehleringer et al. 1986), French Guiana (Buchmann et al. 1997), Panamá (Sternberg et al. 1989), Trinidad (Broadmeadow et al. 1992), and Costa Rica (Leffler and Enquist 2002). The drivers for variations in $\delta^{13}C$ values among sites are not as clear but it is hypothesized to be related to the forest stand structure affecting for instance, light availability and vapor pressure deficit (VPD) within the canopy profile.

Nitrogen isotope ratios ($\delta^{15}N$) provide information related to nitrogen cycling within ecosystems (Högberg 1990; Evans and Ehleringer 1993; Högberg and Johannisson 1993; Högberg 1997; Austin and Vitousek 1998; Roggy et al. 1999a; Martinelli et al. 1999; Amundson et al. 2003). Martinelli et al. (1999) showed evidence that foliar and soil $\delta^{15}N$ patterns in tropical forests were on average higher than values in leaves from temperate forests. One possible explanation is that gaseous nitrogen losses associated with microbial activities in tropical forests leave behind a substrate enriched in ^{15}N (Austin and Vitousek 1998). Enriched soil $\delta^{15}N$ was also observed in agricultural fields (Meints et al. 1975), temperate forests under high N fertilization (Högberg 1990; Högberg and Johannisson 1993), and in tropical regions of Hawaii, where the drier sites, with high N losses were ^{15}N-enriched in comparison with wetter sites (Austin and Vitousek 1998).

On the other hand, the abundance of legumes trees and their capability of fixing N from the atmosphere are other important characteristics of tropical forests (Moreira et al. 1992; McKey, 1994). The $\delta^{15}N$ value of leguminous trees is often near 0‰ because of symbiotic nitrogen fixation (Evans 2001; Robinson 2001). Several studies have shown that species with symbiotic N fixation occur in sites with reduced N availability (McKey 1994), often associated with disturbance (Sylvester-Bradley et al. 1980). However, Roggy et al. (1999a) found that legumes trees (comprising 7.5% of all trees) were fixing N in a primary rainforest of French Guiana without any apparent disturbance and that this amounted to approximately 7 kg N ha^{-1} $year^{-1}$ (Roggy et al. 1999b). These activities would be expected to drive $\delta^{15}N$ values toward 0‰ in tropical ecosystems. Yet Martinelli et al. (1999) suggested that $\delta^{15}N$ of tropical species should be above 0‰ because the nitrogen cycle was open in tropical forests, with microbial activities resulting in the loss of isotopically light nitrogen gases which left isotopically heavy nitrogen available to internal cycling with tropical ecosystems.

Our objective in this paper was to investigate the variability of the carbon and nitrogen isotope ratios as a proxy for interpreting aspects of the carbon and nitrogen cycles in Amazonian rainforests. The samples were conducted to analyze the temporal variations of the isotope ratios and their relationships to the input precipitation in three locations of the Brazilian Amazônia.

Material and methods

The original and summarized datasets presented in this study were collected as part of the Large-Scale Biosphere Atmosphere Project (LBA), an international effort to better understand ecosystem processes at regional scales in the Amazon Basin. All data are available at http://www.lba.cptec.inpe.br/beija-flor/.

Study area

The study sites were official sites of LBA program and located near the cities of Manaus (AM), Ji-Paraná (RO), and Santarém (PA) in Brazil. The mean annual precipitation for these sites are 2285, 2040, 1909 mm $year^{-1}$ respectively. The forests consisted of dense evergreen terra-firme tropical vegetation in Manaus and Santarém, and open evergreen terra-firme forest in Ji-Paraná

(Higuchi et al. 1998; Alder and Silva 2000). The Manaus forest site was located 70 km north of the city (2.59° S; 60.11° W), in a reserve controlled by the Instituto Nacional de Pesquisas da Amazônia. The designation for this site is ZF2, referring to the road allowing access to the sites C14 and K34 where eddy covariance towers are installed (Araújo et al. 2002). The "Rebio" primary forest site is located 80 km north of Ji-Paraná in a Federal Biological Reserve known as Reserva Biológica do Jaru (10.08° S; 61.92° W), and is controlled by the Brazilian Environment Protection Agency (IBAMA). In Santarém there were two sites located at the Floresta National do Tapajós (Flona Tapajós) 67 km south of the city, which was established as a federal reserve in 1974. One site, "Flona-1", at 2.86° S; 54.96° W, and the second site, "Flona-2", located approximately 3 km from the Flona-1 site. "Flona 2" was the control plot of a rain exclusion experiment (Seca Floresta) conducted by the Instituto de Pesquisas Amazônicas (IPAM) and the Woods Hole Research Institute (Davidson et al. 2004).

The three locations contrasted in the length of the dry season, which is defined as the number of months with total precipitation less than 100 mm. Manaus had a 3-month dry season from July to September. At Ji-Paraná the dry season, extended for 4 months from May to August, and Santarém had the longest dry season (5 months) extending from July through November.

The soils at the Manaus and Santarém sites were deeply weathered oxisols (Hapludox) with high clay content (60–80%), low pH (4.0–4.3) and low nutrient content (Telles et al. 2003). The soil in Ji-Paraná site have been classified by Hodnett et al. (1996) as an orthic acrisol, with 85%, or more, of sand at the surface layer.

Sampling

Field samples were collected from May 1999 to June 2004. The sampling at the ZF-2 sites (C14 and K34) were from May 1999 to November 2002; at Flona-1 from June 2000 to June 2004, at Flona-2 from May 1999 to June 2000, February and September 2001, and March 2002; at Rebio in February and September 2000, and February 2001. Leaf samples were collected from plants along a height profile through the canopy, using a scaffold tower in all sites as well from catwalks in the Flona-2. Each sample from a single individual consisted of mature, healthy leaves combined to form a simple sample. A total of 756 tree leaves samples were collected. At ZF2 site, from a total of 176 leaf samples 83 were collected from trees with botanical identification. In Santarém area, 284 and 165 tree leaf samples were collected in the Flona-1 and in the Flona-2, respectively. From these totals, 207 and 125 samples were collected from trees classified to species level, respectively. Finally, at the Rebio site 48 tree leaf samples were collected without botanical classification. Litter, bark and wood samples were randomly collected from some study sites. At total 129 samples of litter were collected, mostly from Flona-1 (80 samples) and at ZF2

(42 samples) sites. Only 8 litter samples were collected at Rebio. Twenty-three samples of living wood and thirteen of bark were collected, at ZF2 (14, 7) and at Flona-1 (10, 6), for wood and bark respectively. Additionally, 11 samples of decaying wood were collected at Flona-1 and 16 were collected at Flona-2.

Analyzes

Soil, litter and plant material were dried at 65° C until a constant weight was obtained and then ground to a fine powder. A 1–2 mg sub sample was placed in a tin capsule. Samples were combusted in an elemental analyzer (Carlo Erba) coupled with an isotope ratio mass spectrometer (IRMS Delta Plus, Finnigan Mat, San Jose, CA, USA) operating in a continuous flow mode. From these analyses, we obtained both isotope ratio (δ^{13} C; δ^{15} N) and elemental content (%C; %N) for carbon and nitrogen.

Data are expressed in "delta" notation (δ^{13} C and δ^{15} N) as:

$$\delta = (R_{sample}/R_{standard} - 1) \times 1000 \qquad (1)$$

where, R_{sample} and $R_{standard}$ are the ratio ^{13}C:^{12}C or the ratio ^{15}N:^{14}N of the sample and standard, respectively. and the standards for carbon and nitrogen are PDB and AIR, respectively.

Statistical analysis

An unequal "n" HSD post hoc test and a Tukey HSD test were used to determine significant differences between the group means of the performed analysis of variance of the sites. The "Statistica 5.0" package was used for the analysis.

Results

Overall tendencies

With 756 separate leaf observations, the overall average $\delta^{13}C$ and $\delta^{15}N$ leaf values were$-32.2 \pm 2.5‰$ and $+5.8 \pm 1.6‰$, respectively (Table 1). We will come back to these values in the discussion, comparing the range presented in this study with ranges of leaf values in different terrestrial tropic ecosystems. There were no significant relationships between $\delta^{13}C$ and $\delta^{15}N$ leaf values (Figure 1). The $\delta^{13}C$ values of leaf materials were depleted in ^{13}C relative to litter, bark, wood, and decaying wood and with statistically significant differences among some of these components (see Table 1). By comparison, the

Table 1. The average and standard deviations for the carbon isotope ratios ($\delta^{13}C$), nitrogen isotope ratios ($\delta^{15}N$), % tissue carbon, % tissue nitrogen, and C:N ratios of terra-firme forests studied in the Amazon Basin.

	$\delta^{13}C$(‰)	$\delta^{15}N$ (‰)	%C	%N	C:N
Leaf (756)	−32.2 ± 2.5a	5.8 ± 1.6a	46.9 ± 4.6a	2.29 ± 0.7a	22.6 ± 7.7a
Litter (129)	−30.3 ± 1.6b	6.0 ± 1.6a	45.8 ± 5.9b	1.93 ± 0.7b	25.4 ± 6.0a
Decaying Wood (27)	−29.5 ± 1.6b	5.8 ± 1.4b	43.6 ± 4.0c	1.58 ± 0.6c	31.1 ± 11.2b
Wood (24)	−28.3 ± 1.5c	4.6 ± 2.0c	47.4 ± 2.6a	0.60 ± 0.6d	145.4 ± 82.1c
Bark (13)	−29.4 ± 2.4b	3.5 ± 2.1d	51.5 ± 7.9d	0.78 ± 0.3d	70.4 ± 26.0d

The number in parentheses indicates the total number of samples). Different letters indicate significant statistical difference at the $p < 0.05$ or less level.

leaf $\delta^{15}N$ values were statistically similar to those observed for both litter and decaying wood. Yet leaf $\delta^{15}N$ values were statistically heavier than observed in either undecayed wood or bark, implying that when plant materials were intact on standing trees there were pronounced spatial $\delta^{15}N$ different among components that disappeared once plant materials hit the ground and began to decay (Table 1).

Undecayed wood, in its $\delta^{13}C$ values, was statistically heavier than decaying wood, bark, or litter. Consistent with the expectation that would had not been impacted by soil microbes and that these other decay components had been, we

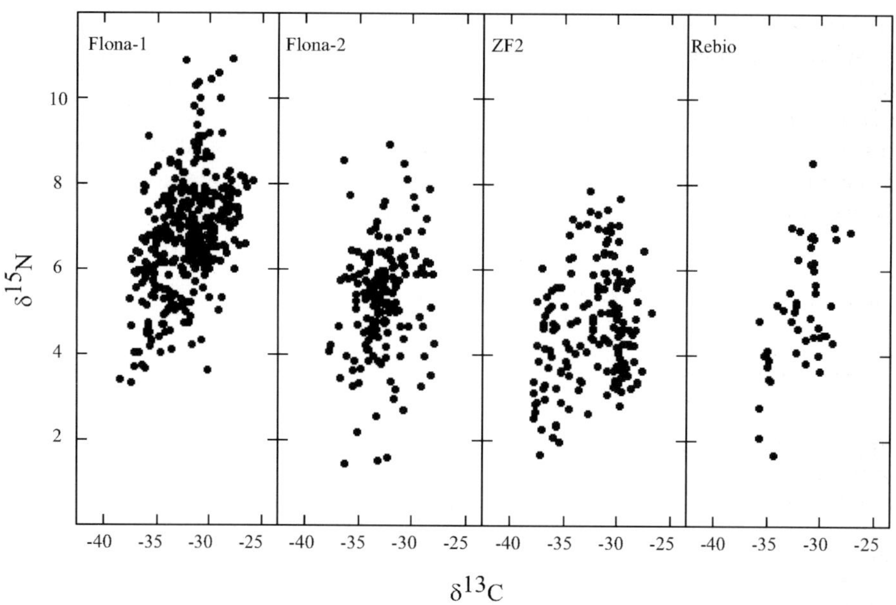

Figure 1. Plots of the nitrogen isotope ratio ($\delta^{15}N$) versus carbon isotope ratio ($\delta^{13}C$) for leaves from all study sites.

detected the highest %C, lowest %N, and highest C:N ratio in undecayed wood.

As expected, the highest %N contents were observed in leaves and these concentrations were statistically higher than in all other tissue components analyzed (Table 1). Undecayed wood and bark had the lowest %N contents, roughly 3–4 fold lower than in leaves. Yet surprisingly the litter and decaying wood components was roughly 2–3 folds higher in %N than undecayed wood.

Difference among study sites

The Flona-1 site stood out as distinctive from the other three forest site in two ways (Figure 1). First, although $\delta^{15}N$ leaf values for all locations were generally clustered between +3 to +8‰, the very highest $\delta^{15}N$ leaf values were measured at Flona-1 with many species having values exceeding +9‰. Second, while the lowest range of $\delta^{15}N$ leaf values at other sites were + 1 to +3‰, these lowest values were absent at Flona-1. As a consequence, the $\delta^{15}N$ average value was significantly higher at Flona-1 in relation to all other sites (Table 2). Interestingly, Flona-1 and Flona-2 are only 3 km apart and yet had statistically different $\delta^{15}N$ values. The basis of these differences was not clear once not all species were represented in both sites. Looking at differences between trees families occurring in Flona-1 and Flona-2, *Caesalpinaceae* (legume), *Burseraceae, Rubiaceae, Sapotaceae, Lecythidaceae*, showed the same pattern discussed above. Only *Lecythidaceae* had the $\delta^{15}N$ values higher at Flona-2, however the species within this family were not the same.

When a comparison is made of the $\delta^{13}C$ values of leaves at all canopy heights across these study sites, there was no significant difference across sites (Table 2). The very isotopically light ^{13}C leaf values reflected a strong contribution from partly-to-fully shaded vegetation elements of the canopy. For an alternative comparison, we considered vegetation samples up to a height of 26 m, which is where each of the canopies opened up. If we considered these

Table 2. The average and standard deviations of the carbon isotope ratios ($\delta^{13}C$), nitrogen isotope ratios ($\delta^{15}N$), % tissue carbon, % tissue nitrogen, and C:N ratios of terra-firme forests studied in the Amazon Basin.

	$\delta^{13}C$(‰)	$\delta^{15}N$ (‰)	%C	%N	C:N
ZF2 C14 – 133 (133)	−32.1 ± 2.9a (−32.2 ± 3.0)a	4.6 ± 1.3a	49.7 ± 3.7a	2.07 ± 0.7a	26.8 ± 8.3a
ZF2 K34 (42)	−32.6 ± 2.7a	4.33 ± 1.7a	50.5 ± 3.3a	2.04 ± 0.6a	27.5 ± 9.5a
Flona-2 – 165 (150)	−32.8 ± 2.2a (−33.1 ± 1.9)b	5.3 ± 1.3a	45.9 ± 4.5b	2.31 ± 0.79b	22.0 ± 7.6b
Flona-1 – 366 (280)	−32.1 ± 2.6a (−32.9 ± 2.1)b	6.8 ± 1.3b	46.5 ± 5.3b	2.45 ± 0.76b	20.9 ± 7.6b
Rebio – 48 (40)	−31.9 ± 2.2a (−32.3 ± 2.0)ab	5.0 ± 1.4a	45.9 ± 3.5b	2.48 ± 0.65b	19.8 ± 5.6b

Different letters indicate significant statistical difference at the $p < 0.05$ or less level. Numbers following the sites names indicate the number of samples. Average $\delta^{13}C$ values between brackets were obtained considering a maximum canopy height of 26 m (see text for further explanation).

"shaded" conditions, then the average $\delta^{13}C$ values at Flona-1 and Flona-2 were significantly lower than the ZF2 site (Table 2). As no statistical differences in carbon or nitrogen isotope ratios were found between the two tower sites in Manaus, they will be considered hereafter only as ZF2. Latter in the discussions we will consider the implication of the foliar isotopic signature in these sites in relation to different carbon assimilation by the vegetation proposed by Araujo et al. (2002).

The average N concentration was significantly lower and the C average concentration was significantly greater at the ZF2 site when compared with other sites (Table 2). As a consequence, the C:N ratio was significantly higher in the ZF2 site than in the other sites (Table 2).

Nitrogen isotope ratios and nitrogen contents of leguminous species

From the 756 tree leaf samples, 353 had botanical identifications, allowing enough samples to evaluate distinctions between legume and non-legume plants. We expected lower leaf $\delta^{15}N$ values if these plants were fixing N from the atmospheric air; generally legume leaves have been found richer in N than non-legume plants (McKey 1994; Evans 2001). The Rebio site had no species-level identifications at all and could not be considered in these evaluations. Considering the other three forest sites, the average foliar N concentration was statistically higher in legume leaves than in non-legume leaves at each of these forest sites ($p < 0.01$. Table 3). Considering all samples, the total average foliar N concentration among 95 legume trees was $2.77 \pm 0.68\%$, which was statistically higher ($p < 0.01$) than the total average foliar concentration of 258 non-legume trees ($2.06 \pm 0.66\%$.). The foliar N concentration of non-legume trees was not significantly different among sites; nor was the foliar N concentration of legume trees significantly different between sites (Table 3).

The foliar $\delta^{15}N$ values of a fixing legume tree should reflect the atmosphere N source (ca. 0‰) and therefore should contrast with the $\delta^{15}N$ values of non-fixing species. However, at the ZF2 and Flona-1 sites there were no significant differences in the foliar $\delta^{15}N$ among legume and non-legume trees (Table 3).

Table 3. Foliar average and standard deviations of the N concentration and $\delta^{15}N$ among the study sites.

	Legume (%N)	Non-legume (%N)	Legume $\delta^{15}N$(‰)	Non-legume $\delta^{15}N$(‰)
ZF2	2.45 ± 0.74Ab (19)	1.94 ± 0.62Ba (61)	5.5 ± 1.5Ab (19)	4.8 ± 1.3Ab (61)
Flona-2	3.27 ± 0.67Aa (20)	2.07 ± 0.67Ba (105)	6.3 ± 1.3Aa (20)	5.1 ± 1.1Bb (105)
Flona-1	2.67 ± 0.54Ab (37)	2.16 ± 0.67Ba (66)	7.0 ± 0.9Aa (37)	6.9 ± 1.0Aa (66)

The number in parentheses indicates the number of values. Different upper case letters indicate difference between columns, and different lower case letters indicate significant difference between rows at the $p < 0.05$ or less level.

The total percentage of legumes samples to the total leaf samples at Flona-1 was 21%, while at Flona-2 and ZF2 were 15 to 16%, respectively.

Soil organic matter

The $\delta^{15}N$ of soil organic matter at Flona 1 and 2 ranged from +7.5 to+11.5‰ in 50-cm soil profiles (Figure 2). In a broader survey (Telles et al. 2003) have shown $\delta^{15}N$ values from +8 to +10‰ at surface and from +10 to +12‰ at depth within the Manaus and Santarem regions. Soil organic nitrogen content mirrored $\delta^{15}N$ values, with $\delta^{15}N$ becoming increasingly more positive with soil depth.

At the same time, the soil $\delta^{13}C$ values increased with depth, ranging almost 2‰ from−28‰ to −26‰ (Figure 2), and the organic matter content decrease in depth, with most of the carbon located in the upper soil layer (0–10 cm). Interestingly the soil $\delta^{13}C$ values spanned a much smaller range than did the canopy leaf $\delta^{13}C$ values, suggesting that the leaves with very negative $\delta^{13}C$ values were likely not contributing significantly to the soil carbon pool.

Temporal variability

The sampling at ZF2 (Manaus) and Flona-1 (Santarém) allowed us to investigate the temporal variability of the $\delta^{13}C$ and $\delta^{15}N$ values of tree leaves over a 5-year period. For ZF2 there was slight response in $\delta^{13}C$ values of ca. 2‰ in the understory vegetation to variations in precipitation, with higher $\delta^{13}C$ values occurring drier periods and lower values occurring during the wet seasons (Figure 3). A statistical analysis indicated a significant, negative correlation between understory $\delta^{13}C$ values of vegetation and the monthly mean precipitation ($p < 0.05$). In the mid-canopy heights ($h = 5$–20 m), the changes in $\delta^{13}C$

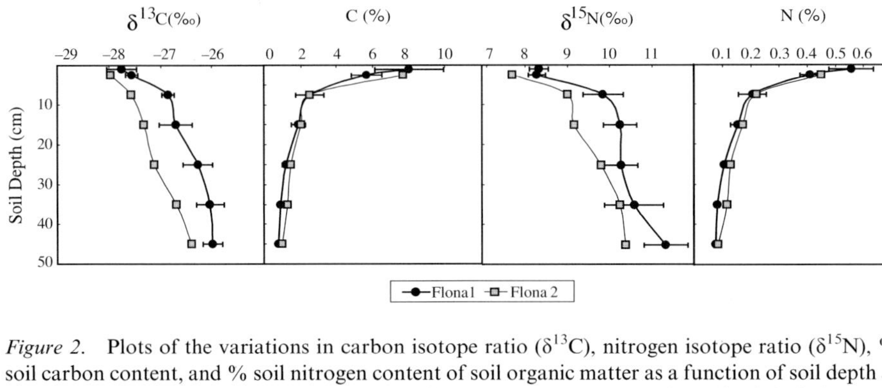

Figure 2. Plots of the variations in carbon isotope ratio ($\delta^{13}C$), nitrogen isotope ratio ($\delta^{15}N$), % soil carbon content, and % soil nitrogen content of soil organic matter as a function of soil depth at the Flona 1 and Flona 2 study sites in Santarém. Data are averages and standard deviations.

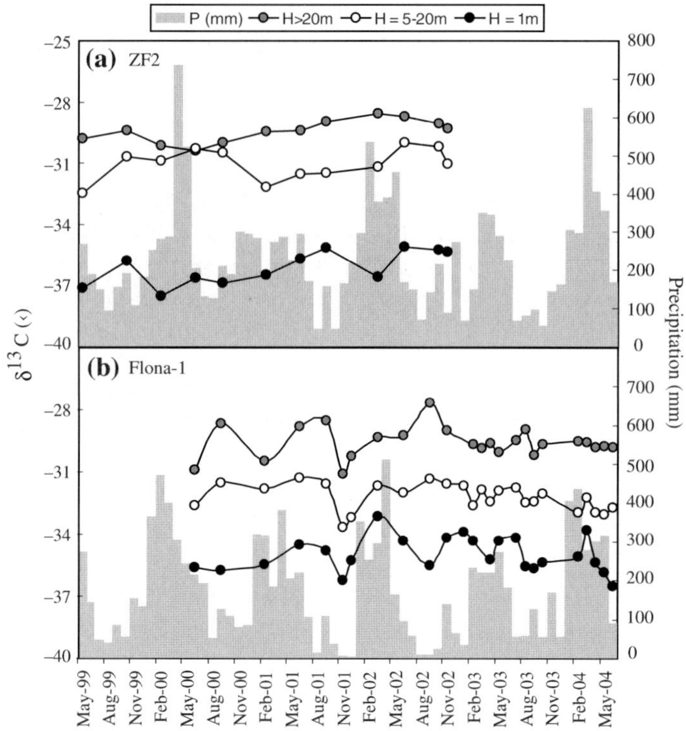

Figure 3. Temporal variations in the leaf carbon isotope ratios ($\delta^{13}C$) of tree leaves at three different heights and average monthly precipitation at (a) Manaus (ZF2) and (b) Santarém (Flona-1) sites. $H = 1.0$ m represents the average $\delta^{13}C$ of all leaves collected in the understory at a height of 1.0 m. $H = 15 - 20$ m represents the average $\delta^{13}C$ of all leaves collected between a height of 15 and 20 m in the canopy. $H > 20$ m represents the average $\delta^{13}C$ of all leaves collected in the canopy at a height higher than 20 m.

values were muted showing no significant temporal variability, suggesting a damped response of the vegetation to changes in the seasonal moisture inputs (Figure 3a). At Flona-1 in Santarém, the $\delta^{13}C$ values showed a positive correlation with monthly precipitation although with a lag of 5 months, for both upper canopy and understory leaves (Figure 3b). The increase of water availability in the system would change the $\delta^{13}C$ of the leaves by changes in the c_i/c_a ratio, although the patterns observed indicated a difference between the expected isotopic signature of a recent carbon assimilated and the bulk leaf carbon. An important aspect of Figure 3a and 3b), is the different species specific responses to seasonal-scale environment variability, in this case precipitation. Two trees (*Copaifera multijuga* and *Manilkara huberi*) at the top of the canopy had distinct foliar $\delta^{13}C$ values and distinct response to precipitation that seemed to follow a multi-year cycle (Figure 4a). Similarly, the lianas at the top of the canopy (*Priomostemma aff. aspera and Tetrapterrys* sp.) at Flona 1 site also showed long-term $\delta^{13}C$ changes that were not annual (Figure 4b).

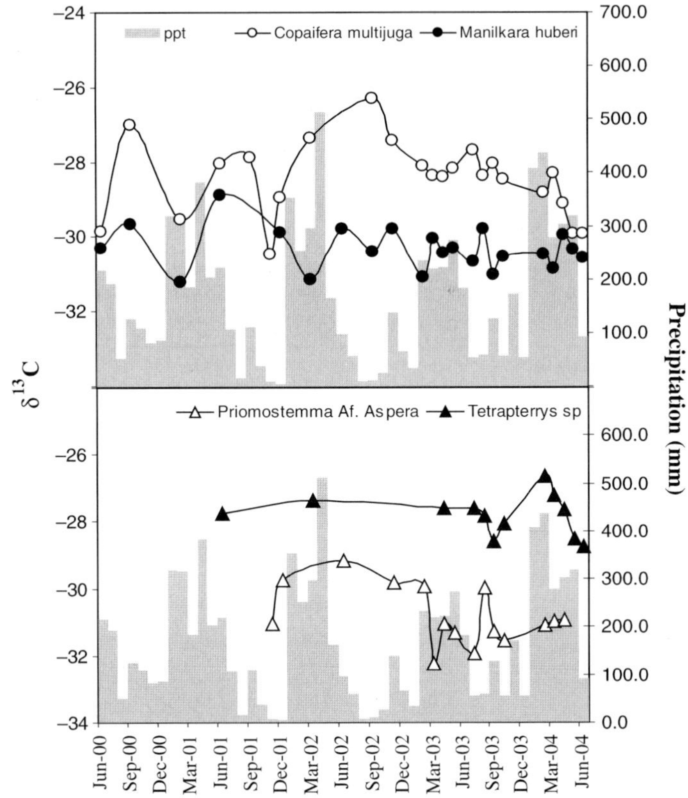

Figure 4. Temporal variations in the leaf carbon isotope ratios ($\delta^{13}C$) of two canopy trees occurring at the highest canopy strata at Flona 1: *Copaifera Multijuga* (open circles) *and Manilkara huberi* (close circles), and lianas: *Priomostemma Af. Aspera* (open triangles), *and Tretrapterrys* sp. (close triangles).

A 2‰ temporal variation of the foliar $\delta^{15}N$ values was seen in the leaves of the understory vegetation at the ZF2 site; this pattern was positively correlated to variations in precipitation in the understory vegetation, but not significantly correlated with precipitation in the mid-canopy and top-canopy layers (Figure 5a).The seasonal variation of the foliar $\delta^{15}N$ at the Flona-1 showed significant negative correlations between $\delta^{15}N$ of both the understory and mid-canopy vegetation and the average monthly precipitation ($p < 0.01$); for the mid canopy vegetation there was a 5-month lag correlation with the monthly precipitation (Figure 5b).

Height variability of the foliar $\delta^{13}C$ values

The foliar $\delta^{13}C$ values were significantly correlated with canopy height at all forest sites (Figure 6). Linear equations between canopy height and $\delta^{13}C$ were

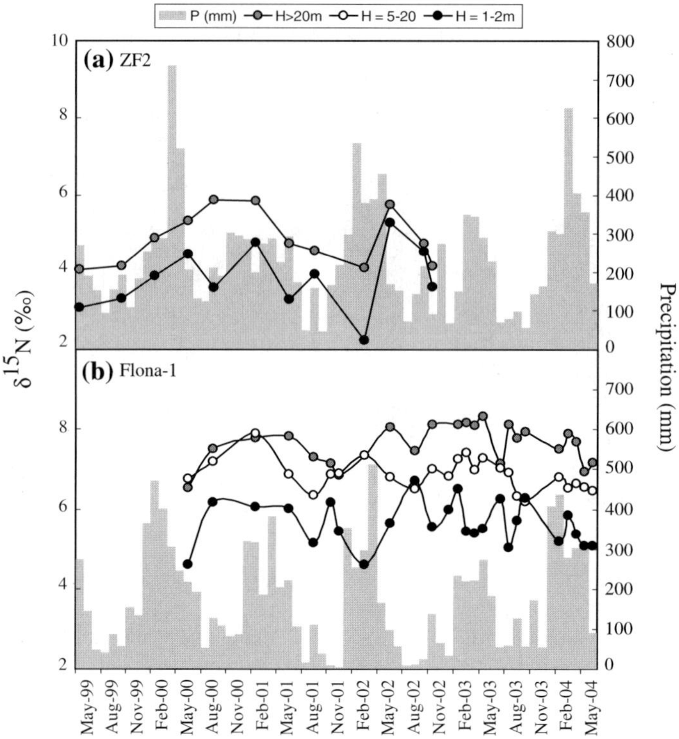

Figure 5. Temporal variations in the nitrogen isotope ratios ($\delta^{15}N$) of tree leaves from different canopy heights at (a) Manaus (ZF2) and (b) Santarém (Flona-1) sites. $H = 1.0$ m represents the average $\delta^{15}N$ of all leaves collected in the understory at a height of 1.0 m. $H = 15$–20 m represents the average $\delta^{15}N$ of all leaves collected between a height of 15 and 20 m in the canopy. $H > 20$ m represents the average $\delta^{13}C$ of all leaves collected in the canopy at a height higher than 20 m.

determined and the intercepts and slopes of these height relationships at Flona-1 and Flona-2 were not distinguishable statistically (Table 4). The equation describing $\delta^{13}C$ variation with canopy height at the Rebio site had a similar slope to the Flona observations, but a higher intercept. Finally, the equation describing $\delta^{13}C$ variation with canopy height at Manaus had a smaller intercept but a higher slope than the equations for the other sites (Table 4).

Height variability of the foliar $\delta^{15}N$ values

We evaluated correlations between $\delta^{15}N$ and canopy height for each of these forest sites. At the ZF-2 and Flona-1 sites, there were significant, positive correlations between $\delta^{15}N$ and canopy height, however the proportion of the variance in the $\delta^{15}N$ values explained by canopy height was small at both sites (Table 5, Figure 7). We separated the $\delta^{15}N$ values of tree leaves into four

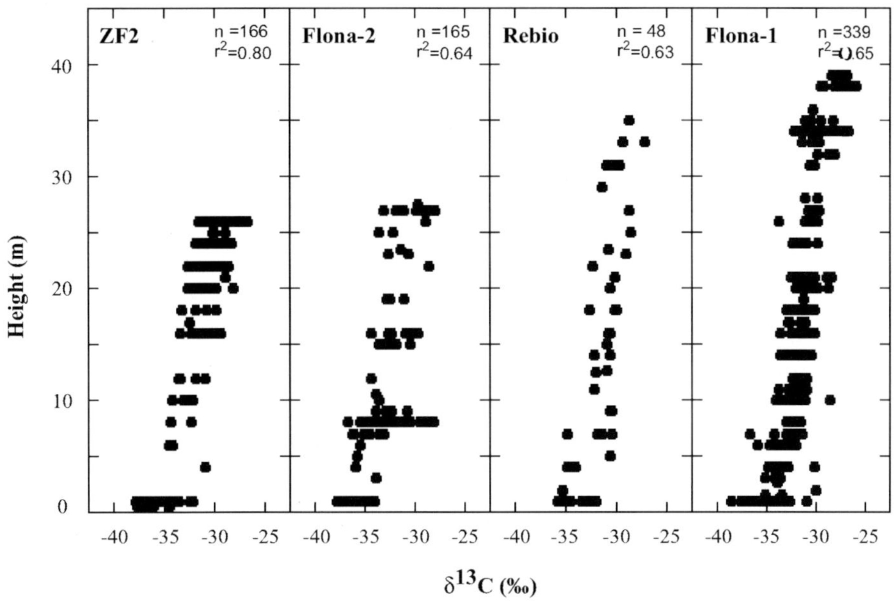

Figure 6. Variations in the carbon isotope ratios ($\delta^{13}C$) of leaves with height for each of the four study sites (regression lines found in Table 4).

height categories at all sites (Table 6). Significant differences in $\delta^{15}N$ values were detected among understory vegetation and the upper canopy layers for ZF2 and Flona-1, but none of these relationships were statistically significant at Rebio or Flona-2. For all regions no statistical difference on the $\delta^{15}N$ of tree leaves were detected among the upper canopy vegetation.

Discussion

The use of stable isotope analysis of plant material can provide important information about the ecology and dynamic of the vegetation and its interaction with the surrounding environment. In the Amazon region of Brazil,

Table 4. Equations describing the statistical relationships between $\delta^{13}C$ and height (H), the R^2 value for that relationship, and number of data (N) for the four forest sites.

Site	Equation	R^2	N
Manaus-ZF2	$\delta^{13}C = -35.9 + 0.25^{*}H$	0.80**	166
Santarém-Flona-2	$\delta^{13}C = -34.8 + 0.19^{*}H$	0.64**	165
Santarém-Flona-1	$\delta^{13}C = -34.9 + 0.17^{*}H$	0.75**	339
Ji-Paraná-Rebio	$\delta^{13}C = -34.0 + 0.16^{*}H$	0.63**	48

All regressions were statistically significant at the $p < 0.05$ or less level.

Table 5. Equations describing the relationships between foliar $\delta^{15}N$ and height (H), the R^2 value for that relationship, and number of data (N) for the four forest sites.

Site	Equation	R^2	N
Manaus-ZF2	$\delta^{15}N = 4.3 + 0.030^{*}H$	0.09**	166
Santarém-Flona-2	$\delta^{15}N = 5.05 + 0.025^{*}H$	0.01ns	165
Santarém-Flona-1	$\delta^{15}N = 5.95 + 0.053^{*}H$	0.24**	340
Ji-Paraná-Rebio	$\delta^{15}N = 4.63 + 0.030^{*}H$	0.03ns	48

All regressions were statistically significant at the $p < 0.05$ or less level.

despite being the largest contiguous tropical forest of the world, few studies have been carried on the past looking at the carbon and nitrogen isotopic variation of the vegetation (Martinelli et al. 1998).

The data present here indicate a variability of the $\delta^{13}C$ among species and sites, but with most of the variation associated with canopy height. Similar patterns have been noted in other tropical forest sites (Kruijt et al. 1996; Buchmann et al. 1997; Martinelli et al. 1998; Guehl et al. 1998a, b, and Bonal et al. 2000a, b). The overall average $\delta^{13}C$ value for all our samples ($-32.2 \pm 2.5‰$; $n = 756$) was similar to average values reported for other tropical forests by Buchmann et al. (1997) and consistent with an expected high photosynthetic carbon isotopic discrimination (Farquhar et al. 1989). Between 63% and 80% of the variance of $\delta^{13}C$ values was explained by the canopy

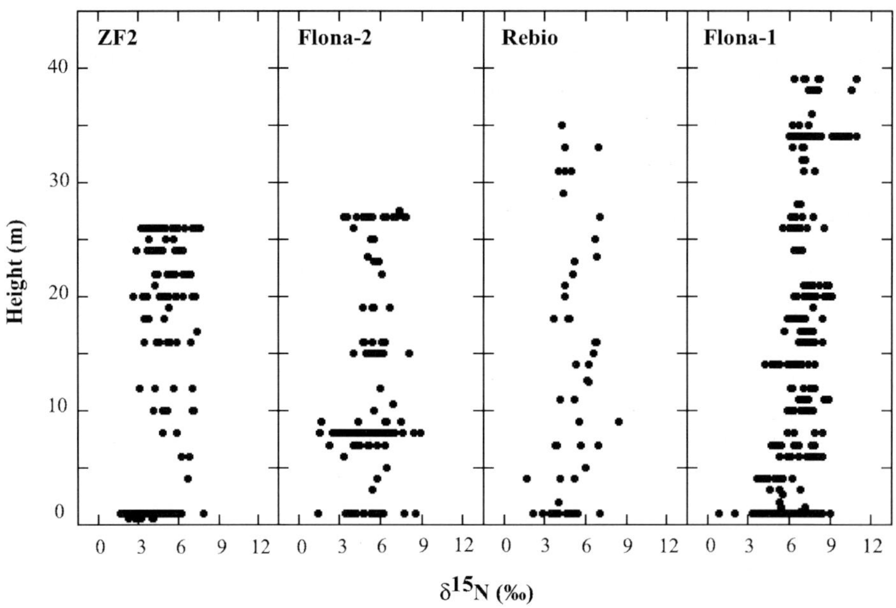

Figure 7. Variations in the nitrogen isotope ratios ($\delta^{15}N$) of leaves with height for each of the four study sites study sites (regression lines found in Table 5).

Table 6. Foliar average $\delta^{15}N$ and standard-deviation at different canopy heights at the different study sites.

Canopy height	ZF2	Flona-1	Flona-2	Rebio
≤ 1 m	4.1 ± 1.3 (57)a	5.7 ± 1.3 (100)a	5.0 ± 1.3 (18)a	4.2 ± 1.4 (12)a
5 to 10 m	5.6 ± 1.4 (13)b	7.1 ± 1.0 (62)b	5.2 ± 1.3 (101)a	5.1 ± 1.5 (10)a
15 to 20 m	4.9 ± 1.2 (33)b	7.0 ± 1.1 (78)b	5.7 ± 1.2 (24)a	5.5 ± 1.4 (12)a
≥ 20 m	4.9 ± 1.2 (63)b	7.4 ± 1.2 (100)b	5.6 ± 1.3 (22)a	5.2 ± 1.3 (14)a

Numbers in parentheses are the number of samples for each canopy height. Different letters indicate statistical significant differences between averages of the different heights at the $p < 0.05$ or less level.

height (Table 4). These data agree with findings by Martinelli et al. (1998) in a *terra-firme* forest at the western Brazilian Amazon. Difference in the foliar carbon isotope ratios of forest canopies profiles have also been detected in the past (Ehleringer et al. 1986; Medina et al. 1986, 1991, Jackson et al. 1993; Lloyd et al. 1996; Bassow and Bazzaz 1997; Berry et al. 1997; Buchmann et al. 1997; Bonal et al. 2000b; Fessenden and Ehleringer 2002; Ometto et al. 2002). Upon further examination, the observations in Table 4 suggest a canopy that is more closed in the Manaus forest when compared to tropical forests in Santarém and Ji-Paraná. These patterns are consistent with known differences in the lengths of the dry season among these sites, and are predicted given that the Manuas forests had the shortest dry season. The $\delta^{13}C$ data can also be interpreted as indicating that some portion of recycled respired CO_2 as source to the photosynthesis of understory leaves, contributing to the leaf $\delta^{13}C$ values (Sternberg et al. 1989). At an extremely high c_i/c_a ratio of 0.9, we would expect that leaf $\delta^{13}C$ values would approach −33‰ under typical atmospheric [CO_2] values today. Yet many of the leaf $\delta^{13}C$ observations were less than −33‰ (e.g., Figure 1), strongly inferring a more negative atmospheric $\delta^{13}C$ value than −8‰. Ometto et al. (2002) did provide evidence that the $\delta^{13}C$ values of atmospheric CO_2 in the bottom of these forest canopies were more negative than at the top of the canopy. In the past, variations in $\delta^{13}C$ values have been correlated with other parameters not immediately related to canopy height, including water-use efficiency (Guehl et al. 1998a, b; Bonal et al. 2000a), branch morphology and hydraulic conductivity (Walcroft et al. 1996; Panek 1996), structural and compositional differences between leaves (Broadmeadow and Giffiths 1993; Bonal et al. 2000a, b; Leffler and Enquist 2002), and leaf age and position in the crown (Gebauer and Schulze 1991; Donovan and Ehleringer 1992). Just how much additional these factors contributed to variations in $\delta^{13}C$ values is unknown.

Ehleringer et al. (2000) noted that soils tend to become ^{13}C enriched with depth as soil carbon became older and more processed by soil microbes. The $\delta^{13}C$ of the soil organic matter for Flona-1 and Flona-2 reflected this decay process being ca.2‰ heavier than the ground surface litter component (Table 1 and Figure 7), and becoming ^{13}C enriched in the deeper soil layers as soil carbon decreased. These observations support previous observations by Camargo et al. (1999) for other rainforest regions within the Amazon basin. The soil-depth profiles in tropical regions are distinct from temperate regions because of the more rapid decline in soil organic carbon with soil depth. Similar to our observations, Natelhoffer and Fry (1988) had shown that soil $\delta^{15}N$ values increased with soil depth. The results from these tropical forest profiles are again distinct in that most of the changes occur in the initial 10-cm depth into the soil.

The $\delta^{15}N$ of leaves can reflect the nutrient pools that the vegetation is exploiting, nitrogen fixation, soil pools, re-translocation within the plant, among others. The average $\delta^{15}N$ value for non-leguminous trees in our data ($5.3 \pm 1.2‰$, Table 3) was similar to the value presented by Martinelli et al. (1999) for several tropical forests around the world. These high $\delta^{15}N$ values have been interpreted as an indication of relatively high nitrogen abundance in these ecosystems. However, our dataset contrasts with the Roggy et al. (1999a) observations for $\delta^{15}N$ foliar values for French Guiana. In that study, they found 82% of the foliar $\delta^{15}N$ values falling between$-1‰$ to$+3‰$. In our study only ca. 5% of the values were lower than3‰, which agrees with global model predictions by Amundson et al. (2003). Another aspect to point out was the slight, but significantly lower foliar $\delta^{15}N$ values observed in the understory vegetation when compared with the upper canopy strata at the ZF2 and Flona-1 forests (Table 5). The contrasting $\delta^{15}N$ values within a site might be interpreted as indicating that these plants were obtaining different nitrogen sources in the soil. Here the form of nitrogen uptake could differ among species or perhaps nitrogen losses by volatilization at upper canopy leaves might leave behind enriched ^{15}N in relation to understory leaves (J.Berry, personal communication). Finally, a preferential uptake of NH_4 by larger trees and a preferential uptake of NO_3 by plants of the understory would also contribute to increase the $\delta^{15}N$ in the upper canopy, since in the soil the $\delta^{15}N–NH_4$ is generally higher than $\delta^{15}N–NO_3$ (Högberg 1997). The high average $\delta^{15}N$ value found in this study confirms that tropical forests are more ^{15}N- enriched than temperate forests (Martinelli et al. 1999). The $\delta^{15}N$ values were also more enriched relative to the vegetation in savannas to the south of Brazil (Cerrado, Bustamante et al. 2004).

Precipitation gradient

The response of photosynthesis and respiration to precipitation or water availability will be reflected in the carbon isotope ratios of the vegetation (Farquhar et al. 1989). The carbon isotope ratios of foliar tissue represent a balance between stomatal conductance to supply CO_2 and photosynthetic demand consuming CO_2, integrated in the c_i/c_a ratio. A recent study of ecosystem gas flux in Santarém showed that the gross ecosystem production (GEP) responded weakly to seasonal changes in precipitation while respiration had a much stronger response (Saleska et al. 2003). Our data for Santarém Flona-1 site (Figure 3b) showed that the seasonal distribution of precipitation was correlated with the $\delta^{13}C$ variability of tree leaves. Even though an expected trend would indicate smaller $\delta^{13}C$ signal with increase of water availability and c_i/c_a ratio (e.g. Walcroft et al. 1997), our observations suggested a time lag between the carbon structurally allocated on the leaves tissues and the carbon fixed by photosynthesis, being correlated, therefore, to previous rain events. The eddy-covariance studies at the Flona-1 site predict carbon fixation that is more ^{13}C enriched associated with less woody growth in the dry season and with a significant increase at the beginning of the rainy season (Saleska et al. 2003; Goulden et al. 2004; Miller et al. 2004). The mechanism suggested by Saleska et al. (2003) for the increase of wood increment "just before" the returning of the rain could be related to the specific carbon re-allocation by the plant and then reflected in the lag of the $\delta^{13}C$ values as shown in our temporal data.

In Manaus, the seasonal variation of foliar $\delta^{13}C$ values were small (Figure 3a), with the exception of changes in the $\delta^{13}C$ values of understory vegetation. In this strata there were ca. 2‰ variations between seasons. Consistent with the smaller interseasonal variations in ^{13}C values, NEE (Net Ecosystem Exchange) observations using eddy-covariance approaches at ZF2 in Manaus have also indicated a smaller seasonal variation (Grace et al. 1996; Araújo et al. 2002), especially when compared to Flona-1 in Santarém (Saleska et al. 2003). The basis for these differences is likely to be the differences in the lengths of the dry season, with Manaus experiencing a much shorter dry season even though the total annual precipitation values are similar.

The $\delta^{15}N$ of leaves were quite positive and generally did not show significant variations within sites. A possible explanation for this pattern would be a more open nitrogen cycle in which N gas losses by microbes resulted in the preferential loss of light N and the residual accumulation of ^{15}N components in the soil. Here observations and the predictions of $\delta^{15}N$ with the Amundsen et al. (2003) global model are in disagreement. In their model soils should become ^{15}N-depleted with the high precipitation amounts that characterize rainforests. In fact, differences in the annual amount of rainfall are not big between Manaus and Santarém regions, but the length of the dry season is quite different among these sites (see also Figure 2 a and b). The biological processing of the soil organic nitrogen might present local peculiarity between the tropical regions, causing changes in the soil nitrogen mineral form available to plants. Nardoto et al. (in preparation) suggest that in Santarém the major nitrogen form available to plants in the soil is ^{15}N-enriched ammonium (NH_4^+). In Figure 1 one can observe the spatial variation of the leaf nitrogen isotope ratios at our study sites, especially at Flona-1. The basis for this difference remains unclear.

Leguminous trees in a non limiting nitrogen environment

The foliar N content was especially high in legume trees if compared with non-legumes (Table 3), in agreement with earlier findings by McKey (1994). The N enrichment in leaves of legume trees was already noted for tropical forests (Roggy et al. 1999a; Vitousek et al. 2002). One of the possible explanations is that a fraction of the N present in the legumes is derived from the biological fixation from the atmosphere. Although we observed an average foliar $\delta^{15}N$ for legume trees close to +5‰ at ZF2 and Flona-2 and close to +7‰ at Flona-1 (Table 3), suggesting that most of the legume trees were not fixing nitrogen in the terra-firme forests of the study regions of the Amazon. A similar conclusion was proposed by others authors looking at nitrogen fixation in tropical forests (Sylvester-Bradley 1980; Yoneyama et al. 1993; Vitousek et al. 2002; Gehring 2003). Based on these studies Vitousek et al. 2002 suggested that symbiotic N fixation in tropical forests might occur only during temporary N shortage. However, in contrast, Roggy et al. (1999a and 1999b) found a significant contribution of legume atmospheric fixed-N_2 to the nitrogen budget of a tropical forest on Oxisol soils in French Guiana.

Most of the legume tree species in the Flona-1 and Flona-2 sites belong to the sub-family Caesalpinoideae, which is known by having few N-fixing species (Souza et al. 1994). N-fixing species are much more common in the sub-families Mimosoideae and Papilionoideae (Sprent 1995). At Flona-1, from 37 samples of Caesalpinoideae, 35% belongs to the genus *Sclerolobium* (a nitrogen fixing genus). However, none of them had $\delta^{15}N$ values significantly lower than the average of non-legume trees. Only one species (*Ormosia* sp.) belonged to the sub-family Papilinoideae. This specie had a $\delta^{15}N$ value of +4.5‰, which is lower than +6.9‰, the average found for non-legume trees in this site. Five species of the sub-family Mimosoideae were found at Flona-2 and most of the Caesalpinoideae trees belong to the genus *Tachigali* and *Sclerolobium* (both nodulating trees with potential for fixing nitrogen, Sprent 1995). However, from this total, only two species of *Inga* (Mimosoideae) had $\delta^{15}N$ values significantly lower than the average $\delta^{15}N$ value found for non-legume trees. Finally, at ZF-2 site most samples belonged to the Mimosoideae family, but one tree (*Inga* sp.) had a $\delta^{15}N$ value (+3.0‰) significantly lower than the average $\delta^{15}N$ value of non-legume trees (+4.8‰) (Table 3).

A significant difference on the $\delta^{15}N$ of tree leaves was found between Flona-1 and Flona-2 sites, despite the proximity of the sites (Table 6). A similar result was not detected in the understory vegetation, although in all other canopy heights the $\delta^{15}N$ values at Flona-1 were significantly higher than at Flona-2 (Table 6 and Figure 1). We also compared the foliar $\delta^{15}N$ of legume and non-legume trees at both sites. For non-legume trees the difference between these two sites was still significant, but there was no statistical difference for legume trees (Table 6). Therefore, it seems that non-legume trees were responsible for the observed difference in $\delta^{15}N$ values between these two sites. Based on this fact, we selected two species of non-legume trees that occurred in the two sites.

This procedure rules-out any potential difference in the foliar $\delta^{15}N$ values caused by physiological differences among species. Both plants, *Coussarea racemosa* and *Manilkara huberi* had higher foliar $\delta^{15}N$ values at Flona-1 (6.10 ± 0.77a, n = 14; 6.60 ± 0.48a, n = 16) than at Flona-2 (4.76 ± 0.46b, n = 9; 4.73 ± 0.70b, n = 6) for each species at the sites (same letters indicate statistical significance). As there is a small but significant correlation between foliar $\delta^{15}N$ and tree height, we tested for differences between the two tree species using an ANOVA analysis with a co-variance for tree height and the different sampling heights in the two sites did not produce the differences the $\delta^{15}N$ of both species.

It is not easy to explain the differences in $\delta^{15}N$ values observed in non-legume trees of Flona-1 and Flona-2, since the sites are quite close to each other, with similar vegetation, and the same precipitation and temperature regimes. In a general sense the $\delta^{15}N$ value may vary with the type of vegetation (Handley et al. 1999), dominant nitrogen form assimilated by the different plant species (Stewart et al. 1992; Roggy et al. 1999a), or climatic patterns (Handley et al. 1999; Amundson et al. 2003). Martinelli et al. (1999) working with white sand-soils ("campinarana" forest) and savannas within the Amazon forest observed that the foliar $\delta^{15}N$ of tree species have much lower values of $\delta^{15}N$ than the more extensive terra-firme forests in similar climatic pattern. The $\delta^{15}N$ value of total-N in the soil is not always similar to the nitrogen available-N to plants, especially when nitrate is formed by nitrification (Hogberg 1997). Our results show a $\delta^{15}N$ of soil organic matter that was on average 0.7‰ heavier at the Flona-1 forest site when compared to the Flona-2 forest site (Figure 2). The foliar $\delta^{15}N$ values of Flona-1 were ca. 1.5‰ heavier in this site when compared to Flona-2 (Table 8). Thus, it is reasonable to hypothesize that part of the difference between the two sites may be explained by differences in the $\delta^{15}N$ of the soils. Although, further investigations on the dominant form of inorganic nitrogen available to the plants (NH_4^+ or NO_3^-), on the organic N pool and on the isotopic fractionation during the uptake and assimilation process (Hogberg 1997) would be very helpful for further understanding of the nitrogen isotope variation between these two sites.

The differences in forest structure are important in the ecological understanding of the system and the $\delta^{13}C$ values obtained in this study showed a consistent response changing it value with the canopy height, reflecting the forest structure. Changes in canopy heights also observed for the nitrogen probably indicate different pools been explored in the soil, once the nitrogen availability in the Amazon primary forest is high. Isotopic variations associated with the length of the dry season, observed in carbon indicate an interesting patterns and the importance of recent fixed carbon to the integrated isotopic signature of the leaf organic carbon. Our current ecophysiological knowledge concerning light and water availability, nutrients and recycling of soil-respired CO_2 show the isotopic data here presented consistent with an open carbon and nitrogen cycle by an highly diverse vegetation dealing with high energy and water input and nitrogen availability.

Acknowledgements

This study was supported by a grant from NASA to project CD-02 in the LBA Terrestrial Ecology Program. We would like to thank the technical support from Dr. Plínio B. de Camargo and Dr. Marcelo Z. Moreira (CENA/USP).

References

Alder D. and Silva J.N.M. 2000. An empirical cohort model for management of Terra Firme forests in Brazilian Amazon. Forest Ecol. Manage. 130: 141–157.

Amundson R., Austin A.T., Schuur E.A.G., Yoo K., Matzek V., Kendall C., Uebersax A., Brenner D. and Baisden W.T. 2003. Global patterns of the isotopic composition of soil and plant nitrogen. Global Biogeochem. Cycles 17(1): 1031Doi: 10.1029/2002GB001903.

Araujo A.C., Nobre A.D., Kruijt B., Culf A.D., Stefani P., Elbers J., Dallarosa R., Randow C., Manzi A.O., Valentini R., Gash J.H.C. and Kabat P. 2002. Dual tower longterm study of carbon dioxide fluxes for a central Amazonian rain forest: The Manaus LBA site. J. Geophys. Res. Atmosph. 107(D20): 8090, doi: 10.1029/2001JD000676.

Austin A. and Vitousek P.M. 1998. Nutrient dynamics on a precipitation gradient. Oecologia 113: 519–529.

Bassow S.L. and Bazzaz F.A. 1997. Intra- and inter-specific variation in canopy photosynthesis in a mixed deciduous forest. Oecologia 13: 507–515.

Berry S.C., Varney G.T. and Flanagan L.B. 1997. Leaf $\delta^{13}C$ in *Pinus resinosa* trees and understory plants: variation associated with light and CO_2 gradients. Oecologia 13: 499–506.

Bonal D., Barigah T.S., Graniers A. and Guehl J.M. 2000a. Late stage canopy tree species with extremely low δ^{13} C and hight stomatal sensitivity to seasonal soil drought in the tropical rain forest of French Guiana. Plant Cell Environ. 23: 445–459.

Bonal D., Sabatier D., Montpied P., Tremeaux D. and Guehl J.M. 2000b. Interspecific variability of $\delta^{13}C$ among trees in rainforests of French Guiana: functional groups and canopy integration. Oecologia 124: 454–468.

Broadmeadow M.S.J. and Griffiths H. 1993. Carbon isotope discrimination and the coupling of CO2 fluxes within forest canopies. In: Ehleringer J.R., Hall A.E. and Farquhar G.D. (eds), Stable Isotopes and Plant Carbon-Water Relations, Academic Press, San Diego, pp. 109–130.

Broadmeadow M.S.J., Griffiths H., Maxwell C. and Borland A.M. 1992. The carbon isotope ratio of plant organic material reflects temporal and spatial variation in CO_2 within tropical forest formations in Trinidad. Oecologia 89: 435–441.

Buchmann N., Guehl J.M., Barigah T.S. and Ehleringer J.R. 1997. Interseasonal comparison of CO_2 concentrations, isotopic composition and carbon dynamics in an Amazonian rainforest (French Guiana). Oecologia 110: 120–131.

Bustamante M.M.C., Martinelli L.A., Silva D.A., Camargo P.B., Klink C.A., Domingues T.F. and Santos R.V. 2004. $\delta^{13}N$ natural abundance in woody plants and soils of central Brazilian savannas (cerrado). Ecol. Appl. 14(4): S200–S213Suppl. S, AUG.

Camargo P.B., Trumbore S., Martinelli L.A., Davidson E., Nepstad D. and Victoria R.L. 1999. Soil carbon dynamic in regroying forest in Eastern Amazonia. Global Change Biol. 5: 693–702.

Cuevas E. and Medina E. 1988. Nutrient dynamics within Amazonian forests. II fine root growth, nutrient availability and leaf litter decomposition. Oecologia 76: 222–235.

Davidson E., Ishida F.Y. and Nepstad D.C. 2004. Effects of an Experimental Drought on Soil Emissions of Carbon Dioxide, Methane, Nitrous Oxide, and Nitric Oxide in a Moist Tropical Forest. Global Change Biol. v. 10.

Donovan L.A. and Ehleringer J.R. 1992. Contrasting water-use patterns among size and life-history classes of a semi-arid shrub. Funct. Ecol. 6: 482–488.

Ehleringer J.R. and Monson R.K. 1993. Evolutionary and ecological aspects of photosynthetic pathway variation. Ann. Rev. Ecol. Systemat. 24: 411–439.

Ehleringer J.R., Bowling D.R., Flanagan L., Fessender J., Helliker B., Martinelli L.A. and Ometto J.P.H.B. 2002. Stable isotopes and carbon cycle in forests and grasslands. Plant Biol. 4: 181–189.

Ehleringer J.R., Field C.B., Lin Z.F. and Kuo C.Y. 1986. Leaf carbon isotope and mineral composition in subtropical plants along an irradiance cline. Oecologia 70: 520–526.

Ehleringer J.R., Buchmann N. and Flanagan L.B. 2000. Carbon isotope ratios in belowground carbon cycle processes. Ecol. Appl. 10: 412–422.

Evans R.D. and Ehleringer J.R. 1993. A break in the nitrogen cycle in aridlands? Evidence from $\delta^{13}N$ of soils. Oecologia 94: 314–317.

Evans R.D. 2001. Physiological mechanisms influencing plant nitrogen isotope composition. Trends plant science 6(3): 121–126.

Farquhar G.D., Ehleringer J.R. and Hubick K.T. 1989. Carbon isotope discrimination and photosynthesis. Ann. Rev. Plant Physiol. Plant Mol. Biol. 40: 503–537.

Farquhar G.D., O'Leary M.H. and Berry J.A. 1982. On the relationship between carbon isotope discrimination and the intercellular carbon dioxide concentration in leaves. Aust. J. Plant Physiol. 9: 121–137.

Fessenden J.E. and Ehleringer J.R. 2002. Age dependent variations in the $\delta^{13}C$ of ecosystem respiration across a coniferous forest chronosequence in the Pacific Northwest. Tree Physiol. 22: 159–167.

Field C.B., Behrenfeld M.J., Randerson J. and Falkowski 1998. Primary production of the biosphere: integrating terrestrial and oceanic components. Science 281: 237–240.

Fisch G., Marengo J.M. and Nobre C.A. 1998. Uma revisão geral sobre o clima na Amazônia. Acta Amazonica 28(2): 101–126.

Gash J.H.C., Huntingford C., Marengo J.A., Betts R.A., Cox P.M., Fisch G., Fu R., Gandu A.W., Harris P.P., Machado L.A.T., von Randow C. and Silva Dias M.A. 2004. Amazonian climate: results and future research. Theor. Appl. Climatol. 78: 187–193, LBA Special Issue.

Gebauer G. and Schulze E.D. 1991. Carbon and nitrogen isotope ratios in different compartments of a healthy and a declining Picea abies forest in the Fichtelgebirge, NE Bavaria. Oecologia 87: 198–207.

Gehring C. 2003. The Role of Biological Nitrogen Fixation in Secondary and Primary Forests of Central Amazonia Doctoral Dissertation. Faculty of Agriculture Rheinische Friedrich-Wilhelm-Universität Bonn, Germany.

Goulden M.L., Miller S.D., Rocha H.R., Menton M.C., Freitas H.C., Figueira A.M.S. and Souza C.A.D. 2004. Diel and seasonal patterns of tropical forest CO_2 exchange. Ecol. Appl. 14: S42–S54.

Grace J., Lloyd J., McIntyre J., Miranda A.C., Meir P., Miranda H.S., Moncrieff J., Massheder J., Wright I. and Gash J. 1995b. Fluxes of carbon dioxide and water vapor over an undisturbed tropical forest in south-west Amazonia. Global Change Biol. 1: 1–12.

Grace J., Lloyd J., McIntyre J., Miranda A.C., Meir P., Miranda H.S., Nobre C., Moncrieff J., Massheder J., Malhi Y., Wright I. and Gash J. 1995a. Carbon dioxide uptake by an undisturbed tropical rain forest in Southwest Amazonia, 1992–1993. Science 270: 778–780.

Grace J., Malhi Y., Lloyd J., McIntyre J., Miranda A.C., Meir P. and Miranda H.S. 1996. The use of eddy covariance to infer the net carbon dioxide uptake of Brazilian rain forest. Global Change Biol. 2: 208–217.

Guehl J.M., Domenach A.M., Bereau M., Barigah T.S., Casabianca H., Ferhi A. and Garbaye J. 1998a. Functional diversity in an Amazonian rainforest of French Guyana: a dual isotope approach ($\delta^{13}N$ and $\delta^{13}C$). Oecologia 116(3): 316–330, SEP.

Guehl J.M., Domenach A.M., Bereau M., Barigah T.S., Casabianca H., Ferhi A. and Garbaye J. 1998b. Functional diversity in an Amazonian rainforest of French Guyana. A dual isotope approach ($\delta^{13}N$ and $\delta^{13}C$). Oecologia 116: 316–330.

Handley L., Austin A., Robinson D., Scrimgeour C., Raven J., Heaton T., Schmidt S. and Stewart G. 1999. The 15-N natural abundance ($\delta^{13}N$) of ecosystem samples reflects measures of water availability. Aust. J. Plant Phys. 26: 185–199.

Higuchi N., Santos J., Ribeiro R.J., Minette L. and Biot Y. 1998. Biomassa da parte aérea da vegetação da floresta tropical úmida de terra-firme da Amazônia Brasileira. Acta Amazônica 28: 153–166.

Hodnett M.G., Oyama M.D., Tomazella J. and Marques Filho A.O. 1996. Comparisons of long term soil water storage behavior under pasture and forest in three areas of Amazonia. In: Gash J.H.C., Nobre C.A., Roberts J.M. and Victoria R.L. (eds), Amazonian Deflorestation and Climate, John Wiley & Sons, Chichester, U.K, pp. 79–100.

Högberg P. and Johannisson C. 1993. $\delta^{13}N$ abundance of forests is correlated with losses of nitrogen. Plant Soil 157: 147–150.

Högberg P. 1990. 15N natural abundance as a possible marker of the ectomycorrhizal habit of trees in mixed African woodlands. New Phytol. 115: 483–486.

Hogberg P. 1997. ^{15}N natural abundance in soil-plant systems. New Phytol 137: 179–203.

Jackson P.C., Meinzer F.C., Goldstein G., Holbrook N.M., Cavelier J. and Rada F. 1993. Environmental and physiological influences on carbon isotope composition of gap and understorey plants in a lowland tropical forest. In: Ehleringer J.R., Hall A.E. and Farquhar G.D. (eds), Stable Isotopes and Plant Carbon–Water Relations, Academic Press, San Diego, pp. 131–140.

Kapos V., Ganade G., Matsui E. and Victoria R.L. 1993. $\delta^{13}C$ as an indicator of edge effects in tropical rain forest reserves. J. Ecol. 81: 425–432.

Kruijt B., Lloyd J., Grace J., McIntyre J., Farquhar G.D., Miranda A.C. and McCracken P. 1996. Sources and sinks of CO_2 in Rondonian tropical forest, inferred from concentrations and turbulence along a vertical gradient. In: Gash J.H.C., Nobre C.A., Roberts J.M. and Victoria R.L. (eds), Amazonian Deflorestation and Climate, John Wiley & Sons, Chichester, U.K, pp. 331–351.

Leffler Aj. and Enquist Bj 2002. Carbon isotope composition of tree leaves from Guanacaste, Costa Rica: comparison Across tropical forests and tree life history. J. Trop. Ecol. 18: 151–159.

Lloyd J., Grace J., Miranda A.C., Meir P., Wong S.C., Miranda H.S., Wright I.R., Gash J.H.C. and McIntyre J. 1996. A simple calibrated model of Amazon rainforest productivity based on leaf biochemical properties. Plant, Cell Environ. 18: 1129–1145.

Luizão R.C.C., Luizão F.J., Paiva R.Q., Monteiro T.F., Sousa L.S. and Kruijt B. 2004. Variation of carbon and nitrogen cycling processes along a topographic gradient in a central Amazonian forest. Global Change Biol. 10: 592–600, doi: 10.1111/j.1529-8817.2003.00757.x.

Malhi Y., Baldocchi D.D. and Jarvis P.G. 1999. The carbon balance of tropical, temperate and boreal forests. Plant, Cell Environ. 22: 715–740.

Marengo J.A. and Hastenrath S. 1993. Case studies of climatic events in Amazon basin. J. Climate 6(4): 617–627.

Marengo J.A. 1992. Interannual variability of surface climate in the Amazon basin. J. Climatol. 12(8): 853–863.

Martinelli L.A., Almeida S., Brown I.F., Moreira M.Z., Victoria R.L., Sternberg L.S.L., Ferreira C.A.C. and Thomas W.W. 1998. Stable carbon isotope ratio of tree leaves, boles and fine litter in a tropical forest in Rondônia, Brazil. Oecologia 114: 170–179.

Martinelli L.A., Piccolo M.C., Townsend A.R., Vitousek P.M., Cuevas E., Mcdowell W., Robertson G.P., Santos O.C. and Treseder K. 1999. Nitrogen stable isotopic composition of leaves and soil: tropical versus temperate forests. Biogeochemistry 46(1–3): 45–65.

McKey D. 1994. Legumes and nitrogen: the evolutionary ecology of a nitrogen-demanding lifestyle. In: Sprent J.I. and McKey D. (eds), Advances in Legume Systematics 5: The Nitrogen Factor, Royal Botanic Gardens, Kew, UK, pp. 221–228.

Medina E. and Minchin P. 1980. Stratification of $\delta^{13}C$ values of leaves in Amazonian rainforests. Oecologia 45: 355–378.

Medina E., Sternberg L. and Cuevas E. 1991. Vertical stratification of $\delta^{13}C$ values in closed and natural plantation forests in the *Luquillo mountains*, Puerto Rico. Oecologia 87: 369–372.

Meints V.W., Boone L.V. and Kurtz L.T. 1975. Natural 15N abundance in soil, leaves, and grain as influenced by long term additions of fertilizer N at several rates. J. Environ. Qual. 4: 486–490.

van der Merwe N.J. and Medina E. 1989. Photosynthesis and $^{13}C/^{12}C$ ratios in Amazonian rain forests. Geochim. Cosmochim. Acta 53: 1091–1094.

Miller S.D., Goulden M.L., Menton M.C., Rocha H.R., Freitas H.C., Figueira A.M.S. and Sousa C.A.D. 2004. Biometric and Micrometeorological Measurements of Tropical Forest Carbon Balance. Ecol. Appl. 14(4): S114–S126 Supplement.

Moreira de Souza F.M., da Silva M.F. and de Faria S.M. 1992. Occurrence of nodulation in legume species in the Amazon region of Brazil. New Phytol. 121: 563–570.

Natelhoffer K.J. and Fry B. 1988. Controls on natural nitrogen-15 and carbon-13 abundances in forest soil organic matter. Soil Sci. Soc. Am. J. 52: 1633–1640.

Obregon G. and Nobre C.A. 1990. Principal component analysis of precipitation fields over Amazon river basin. Climanálise 5(7): 35–46.

Ometto J.P.H.B., Flanagan L., Martinelli L.A., Moreira M.Z., Higuchi N. and Ehleringer J.R. 2002. Carbon isotope discrimination in forest and pasture ecosystems of the Amazon Basin, Brazil. Global Biogeochem. Cycles 16: 1109.

Panek J.A. 1996. Correlations between stable carbon-isotope abundance and hydraulic conductivity in Douglas-fir across a climate gradient in Oregon, USA. Tree Physiol. 16(9): 747–755.

Robinson D. 2001. $\delta^{15}N$ as an integrator of the nitrogen cycle. Trends Ecol. Evol. 16: 153–162.

Roggy J.C., Preâvost M.F., Garbaye J. and Domenach A.M. 1999. Nitrogen cycling in the tropical rain forest of French Guiana: comparison of two sites with contrasting soil types using d15N. J. Trop. Ecol. 15: 1–22.

Roggy J.C., Prevost M.F., Gourbiere F., Casabianca H., Garbaye J. and Domenach A.M. 1999. Leaf natural "δN abundance and total Nconcentration as potential indicators of plant N nutrition in legumes and pioneer species in a rain forest of French Guiana. Oecologia 120: 171–182.

Saleska S.R., Miller S.D., Matross D.M., Goulden M.L., Wofsy S.C., da Roacha H.R., de Camargo P.B., Crill P., Daube B.C., de Freitas H.C., Hutyra L., Keller M., Kirchoff V., Menton M., Munger J.W., Pyle E.H., Rice A.H. and Silva H. 2003. Carbon in Amazon forests: unexpected seasonal fluxes and disturbance-induced losses. Science 302: 1554–1557.

Schimel D.S. 1995. Terrestrial ecosystems and the carbon-cycle. Global Change Biol. 1: 77–91.

Souza L.A.G., de Silva M.F. and da und Moreira F.W. 1994. Capacidade de nodulação de cem Leguminosas da Amazônia. Acta Amazônica 24(1/2): 9–19.

Sprent J.I. 1995. Legume trees and shrubs in the tropics: N_2 fixation in perspective. Soil Biol. Biochem. 7: 401–407.

Sternberg L.S.L., Mulkey S.S. and Wright S.J. 1989. Ecological interpretation of leaf isotope ratios: influence of respired carbon dioxide. Ecology 70: 1317–1324.

Stewart G.R., Joly A.C. and Smirnoff N. 1992. Partitioning of inorganic nitrogen assimilation between roots and shoots of cerrado and forest trees of contrasting plant communities of South East Brazil. Oecologia 91: 511–517.

Sylvester-Bradley R., Oliveira L.A., Podestá Filho J.A. and St. John T.V. 1980. Nodulation of legumes, nitrogenase activity of roots and occurrence of nitrogen-fixing *Azospirillum* spp. in representative soils of Central Amazonia. Agro-Ecosystems 6: 249–266.

Telles E.C.C., de Camargo P.B., Martinelli L.A., Trumbore S.E., da Costa E.S., Santos J., Higuchi N. and Oliveira Jr R.O. 2003. Influence of soil texture on carbon dynamics and storage potential in tropical forest soils of Amazonia. Global Biogeochem. Cycles 17-1040 doi: 10.1029/2002GB001953.

Vieira S., Camargo P.B., Selhorst D., Silva R., Hutyra L., Chambers J.Q., Brown I.F., Higuchi N., Santos J., Wofsy S.C., Trumbore S.E. and Martinelli L.A. 2004. Forest structure and carbon dynamics in Amazonian tropical rain forests. Oecologia 140: 468–479, doi: 10.1007/s00442-004-1598-z.

Vitousek P.M, Cassman K., Cleveland C., Crews T., Field C.B., Grimm N.B., Howarth R.W., Marino R., Martinelli L., Rastetter E.B., Sprent J.I. (2002) Towards an ecological understanding of biological nitrogen fixation. Biogeochemistry 57/58: 1–45.

Walcroft A.S., Silvester W.B., Grace J.C., Carson S.D. and Waring R.H. 1996. Effects of branch length on carbon isotope discrimination in *Pinus radiata*. Tree Physiol. 16: 281–286.

Walcroft A.S., Whitehead D., Silvester W.B. and Kelliher F.M. 1997. Determination of photosynthetic model parameters in response to temperature and nitrogen concentration in Pinus radiata D. Don. Plant, Cell Environ. 20: 1338–1348.

Yoneyama T., Muraoka T., Murakami T. and Boonkerd N. 1993. Natural abundance of ^{15}N in tropical plants with emphasis on tree legumes. Plant Soil 153: 295–304.